ALLE ZEIT WACH 1842

CAM

Developments in Computer-Integrated Manufacturing

Edited by
D. Kochan

With 157 Figures and 33 Tables

Springer-Verlag
Berlin Heidelberg NewYork Tokyo

Prof. Dr. D. Kochan
TU Dresden
Sektion Fertigungstechnik
und Werkzeugmaschinen
Mommsenstraße 13
DDR 8027 Dresden

ISBN-13: 978-3-642-70274-7 e-ISBN-13: 978-3-642-70272-3
DOI: 10.1007/978-3-642-70272-3

Library of Congress Cataloging-in-Publication Data
Main entry under title:
CAM: developments in computer-integrated manufacturing.
Bibliography: p.
Includes index.
1. Computer integrated manufacturing systems.
I. Kochan, Detlef.
TS155.6.C374 1985 670.42'7 85-22073

Softcover reprint of the hardcover 1st edition 1986

Printing: Beltz Offsetdruck, Hemsbach/Bergstraße
Bookbinding: J. Schäffer OHG, Grünstadt

2145/3140-543210

List of Coauthors

Coauthor	Section
Dr. Merchant, United States	3.4.2.2
Dr. Kozar, Czechoslovakia	3.4.2.5
Prof. Kochan, Dr. Schaller, German Democratic Republic	3.5
Prof. Hutchinson, United States	4.2
Prof. Kochan, Prof. Olling, United States	5.4.1
Dr. Semenkov, Dr. Klimov, USSR	6
Prof. Spur, Prof. Krause, Dr. Pistorius, Federal Republic of Germany	7
Prof. Crestin, France	8

Preface

"Developments in Computer-Integrated Manufacturing" arose from the joint work of members of the IFIP-Working Group 5.3 — Discrete Manufacturing, and other IFIP members.
Within the Technical Committee 5 of the International Federation of Information Processing (IFIP) the aim of this Working Group is the advancement of computers and their application to the field of discrete part manufacturing.
Capabilities will be expanded in the general areas of planning, selection, and control of manufacturing equipment and systems. Tools for problem solution include: mathematics, geometry, algorithms, computer techniques, and manufacturing technology. This technology will influence many industries — machine tool, automation, aircraft, appliance, and electronics, to name but a few.
The Working Group undertook the following specific tasks:

1. To maintain liaison with other national and international organizations working in the same field, cooperating with them whenever desirable to further the common goal
2. To be responsible for the IFIP's work in organizing and presenting the PROLAMAT Conferences
3. To conduct other working conferences and symposia as deemed appropriate in furthering its mission
4. To develop and sponsor research and industrial and social studies into the various aspects of its mission.

The book can be regarded as an attempt to underline the main aspects of technology from the point of view of its software and hardware realization. Because of limitations in size and the availability of literature, the problems of robotics and quality control are not described in detail.
Promoting the fundamental aim of sponsoring international cooperation, Working Group 5.3 essentially organized the three annual PROLAMAT conferences. Working conferences on special topics have recently been held, too. Past and planned activities are listed on p. 1.
The international scientific cooperation of the past few years has led to certain conclusions. In discussions of new results and trends in science it is sometimes difficult to reach a level of understanding conducive to agreement. This is true even with regard to apparently unambiguous terminology used in connection with new technical principles, methods, and solutions of advanced technological develop-

ment. Senior members of IFIP Working Groups saw a need to produce a volume on the "state of the art — CAM", to characterize the highest level in total computer-aided manufacturing. Some other aims were:

1. To outline the range of advanced computer-aided manufacturing and the key problems of technical development
2. To trace and describe the main trends leading to computer-integrated manufacturing; as it was not possible to describe every aspect of the state of the art in depth, selected examples were chosen from developed industrial countries
3. To point out the need for cooperation between specialists in various fields, such as manufacturing and mechanical engineers, production and industrial engineers, economists, designers, software and system engineers, mathematicians, social scientists. Teamwork is necessary to characterize and classify the advanced knowledge and essential terminology in this scientific and technological development.

The starting points of relevant technical development are illustrated and the fundamental terms explained at the beginning of each chapter, especially NC techniques, which were the basis of these developments. The international authorship includes contributions from the major industrial countries. The extreme distances made it difficult to coordinate and integrate the authors' contributions; I am very interested in receiving critical comments from readers. It is anticipated that this book will give a survey of the advances made in computer-aided manufacturing toward the goal of computer-integrated manufacturing. The problems and challenges of manufacturing development should stimulate young scientists to participate in finding solutions and in the efficient application of computer techniques and automation.

I would like to thank all of the IFIP coauthors, especially those of IFIP-WG 5.3, for their creative and effective teamwork. In addition, I wish to thank the publishers of Springer-Verlag, represented by Mr. Rossbach, for their cooperation and for the excellent quality of the editorial and printing work.

Dresden, October 1985 D. Kochan

Contents

1 Introduction

Computer-aided manufacturing (CAM) originated in the development of information processing, especially in connection with the control of machine tools. Parallel to this development, and to some extent independent of the automation of manufacturing processes in the 1950s, successful automation of intellectual routines by means of electronic data processing was also accomplished.

In addition to many single ideas, rules, and parameters, only a few methods of group technology were available for dealing with complex manufacturing processes in the metalworking industry.

The problems of production planning, supervision, and operational accounting were considered separately and independent of the direct manufacture.

The dynamic development of information processing in its different modifications for large computers, medium-sized computers, and minicomputers, and as programed logic or free programmable control, gave strong impetus to the entire field of manufacturing. In less than 20 years of development there have been four generations of numerical control systems for machine tools. Proceeding from the first programming languages and going on to NC programming and other technological programs for determining operations and stages of operation, the computer has become an indispensable aid to the engineer.

Information processing solves complex problems such as process elaboration, classification of single parts and group technology, recognizing of repetition parts, amendment service, and simulation of process run. The informational link to production planning, supervision, and operational accounting, and with different auxiliary manufacturing processes such as transport and storage — which are becoming increasingly automated — is becoming closer. Thus, the various items of informational feedback effectively contribute to the faster manufacture of industrial products which are better in quality.

The field of computer-aided manufacturing has reached a considerably high level; therefore, for international understanding, it is imperative to give a systematic survey of important definitions and to explain them in detail.

2 Computer-Aided Manufacturing (CAM)

2.1 Fundamentals, Definitions, and Philosophy

This field of work includes those principal problems and complexes of activities which are carried out with the assistance of computers, i.e., which are automated, in the preparation for manufacturing and in manufacturing itself. These are problems prevailing in the manufacture of discrete parts in the metalworking industry; they are transferable, to a certain extent, to similar processes in the manufacture of piece goods.

CAM — discrete manufacturing is defined as computer-aided preparation of manufacturing including (see Fig. 1) decision-making, process and operational planning, software design techniques and artificial intelligence, and manufacturing

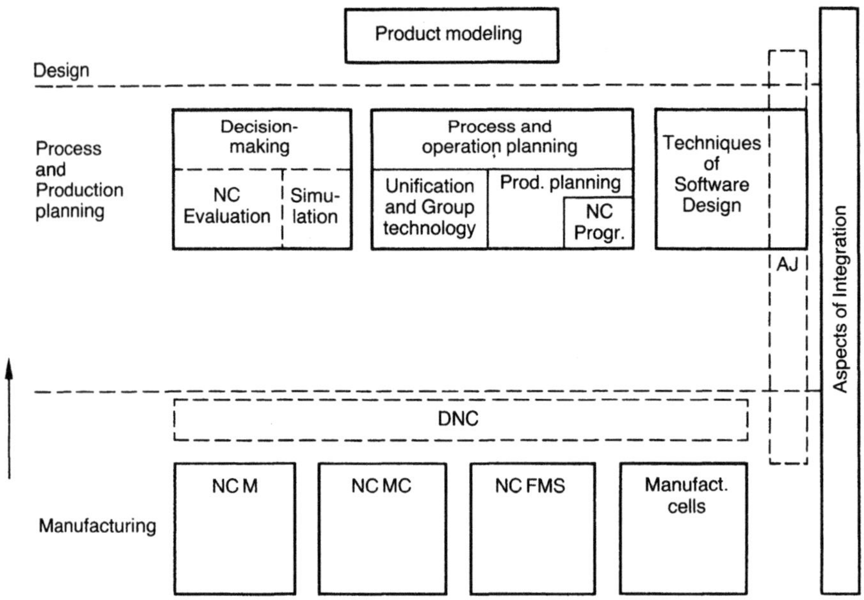

Fig. 1. The scope of computer-aided manufacturing

with different types of automation (NC-M, NC-MC, NC-FMS, NC machining cells), and different types of realization (CNC — single unit technology, DNC — group technology).
Proceeding from the fact that automation of manufacture decisively influenced computer-aided processes in the preceding stages of production, and that at present this influence prevails, this subject will be the center of interest. Very often, the concept of CAM is used only with respect to automated manufacturing but this is not the viewpoint of the IFIP-WG 5.3.

2.2 Main Historical Sources

In general, computer-aided manufacturing has three main sources:

1. The research in numerical control at the Massachusetts Institute of Technology in 1953 and the dynamic increase in the output of electronic data processing were fundamental prerequisites for this development. Thus, the necessity for computer-aided programming was caused by the progress in numerical control. The first programming language, APT, created at MIT, was the pattern for many further developments, especially in the western industrial countries.
2. The fundamental work carried out at the Soviet Academy of Sciences at the same time formed another substantial basis. The aim of these activities was to use electronic data processing to rationalize engineering projects and to plan the production of conventional machine tools for the first time. Outstanding results were achieved in rationalizing the delivery of working information for turret lathes with the aid of computers in the mid 1950s.
3. The work on unification and group technology originating in scientific activities in Leningrad in the 1930s were continued in the 1950s by means of modern technical aids. The activities and achievements of Professor Mitrofanov, and his worldwide influence on the progressive design of technological processes through his work on group technology, are recognized all over the world today.

3 Stages of Development in Flexible Manufacturing

3.1 General Remarks

For decades, in industrial-scale manufacture and mass manufacture the so-called rigid forms of automation prevailed, such as automated lines and transfer lines in the automobile and consumer goods industries, cam-controlled automatic machines in the mass production of standard components (screws, nuts, bolts), special machines or single-purpose machines for special manufacturing problems.
All these types of automation are characterized by the fact that, due to the high cost of setting up the manufacturing process the revision of production of a certain part or an entire product is profitable only after some weeks, or even after some years for special machines.
It is therefore imperative to develop more flexible forms of automation, adjustable to alternating requirements. The small and medium-sized pieces account for 75%-80% of the parts manufactured in all industrial countries for mechanical engineering. Following the Second World War the aeronautics, automobile, and consumer goods industries developed rapidly, increasing the demand for complicated geometric parts. And with increased demand for power and accuracy the manufacturing processes of turning, drilling, and milling have taken on added importance.

3.2 Historical Development and Modifications in Flexible Manufacturing

3.2.1 Technical Advances

In the eighteenth century the first musical clocks and music boxes were developed, and certain programs were carried out by pinned rolls which were traced mechanically.
In the nineteenth century M. J. Jacquard elaborated punched cardboard patterns to automatically control certain functions of looms.

In the twentieth century the tendency moved to more flexible types of automation by the principle of cam control.
The term "cam control" must be understood as a general term for a great number of program controls, where either mechanical trip cams, slots or punched patterns, or film work as dimension storage or impulse machines.
On cam-controlled machine tools the path dimensions are represented by a limited number of cams which are variably mounted on a cam shaft (cam plate, drum). During the operation signal pulses are emitted with microswitches by the cam, which can be utilized to select the functions and to coordinate them according to the prescribed program.
The principle of cam control may offer a way to solve an automation problem seen in industry, where — based on the same fundamental principle — electronic components are increasingly used in place of mechanical components.

3.2.2 Origin of Data Processing and Numerical Control

Automated information processing and electronic control have the same origin.

1945 – 1946

The first digital electronic computer with practical application was developed by Drs. John W. Mauchly and J. Presper Eckert. It was called ENIAC (Electrical Numerical Integrator and Calculator). It was large and massive, utilized energy-consuming vacuum tubes, and was difficult to program. However, it was the basis of electronic data processing used to carry out scientific computations.
This project was first promoted by the US Air Force, and later received considerable support from the government.

1942 – 1952

The genesis of numerical control, which can be termed one of the most significant developments of our century in manufacturing is illustrated by some examples [65]. Numerical control, the first answer to new aerospace machining requirements, was developed by a small aircraft industry subcontractor, the Parsons Corporation, from northern Michigan, USA. Early in 1942, the Parsons organization, then a stamping company located in Detroit, was awarded a government contract to manufacture bombs for the war then being waged. John Parsons knew that bombs were strictly a wartime emergency product, and that other, long-range plans would have to be made. So the Parsons Corporation was created, to act as a contract facility for major airframe builders.
One of the first contracts called for the production of helicopter rotor blades. Along with the contract came patterns for checking the contours of the finished part. The technique used a punchcard to check the helicopter blade airfoil patterns by generating coordinate tables of the contours. The final results showed that the furnished patterns deviated considerably from a true airfoil curve.
Later, tabulating equipment was used to generate a set of tables so that no point was more than 0.5 % of the chord length from the precoding point. Working with

a tabulating machine, multiplier, collator, and card sorter was somewhat cumbersome, but the equipment served the purpose until a fully digital computer became commercially available in the early 1950s.

With the generated tables of coordinate information Parsons and Stulen proceeded to make checking patterns and to generate some two-axis airfoil machining-path data. A factor was even entered for the cutter diameter. The machining coordinate tables were taken to a simple milling machine and the X and Y points were called out to one operator who ran the X-axis handwheel and to another who controlled the Y-axis. The resultant machined curves were far more accurate than anything that had been achieved up to that time. They generated 200 points accurate to plus or minus 0.0015 inch.

With the necessity of generating three-axis-curve data — which was impossible to do with the early-generation tabulating equipment — came the birth of numerical control.

At MIT, numerical control was realized in 1952 as a vertical-spindle Cincinnati Hydrotel, with a lab-constructed control unit; programmed with machine instructions on a straight binary perforated tape, it successfully executed simultaneous three-axis cutting-tool movements.

Following further developmental work, Numerical Control was announced to the public in November 1954. Based on Parsons' patent, the first industrial numerically controlled machine tool was manufactured by the Bendix-Cooperation (USA).

1957

In manufacture, the US Air Force utilized the first numerically controlled milling machines.

1960

Other industrial countries — the Federal Republic of Germany was a forerunner in Europe — began to develop, produce, and use numerically controlled machine tools.

The steady increase in production of NC machines during the past two decades emphasizes the general importance of this type of flexible automation (see Fig. 2).

In the metalworking industry there has been a shift in production from the original milling machines to turning machines; this is because turning machines have a larger assortment of parts.

Numerical control is one of the most important basic innovations of our century; in the meantime it has gone far beyond the original cutting-machine tools and has revolutionized manufacturing and other areas of human productive activity (see also Fig. 3.)

With respect to the main application of NC to the production of machine tools, after 20 years of development nearly four generations of control systems have been applied in practice. Every year, on an international scale, more than one hundred different types of control systems are put on the market, together with different types of machine tools and numerous modifications in equipment.

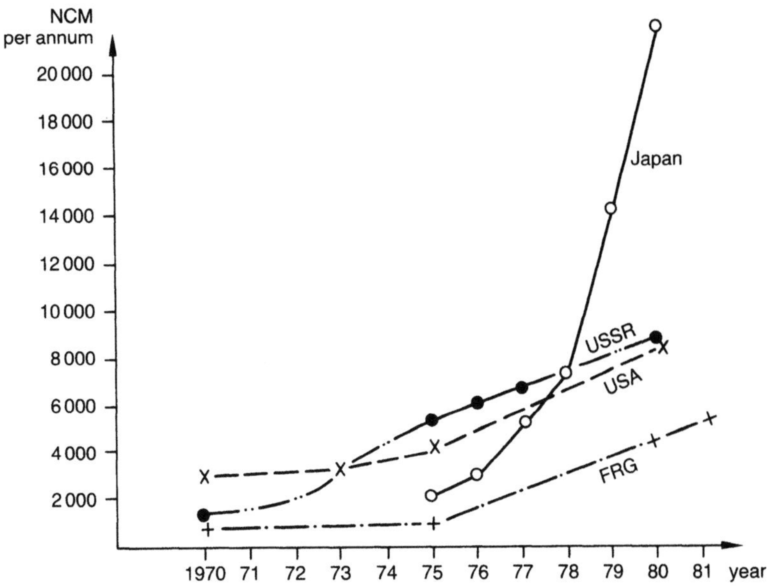

Fig. 2. Annual production of NCM

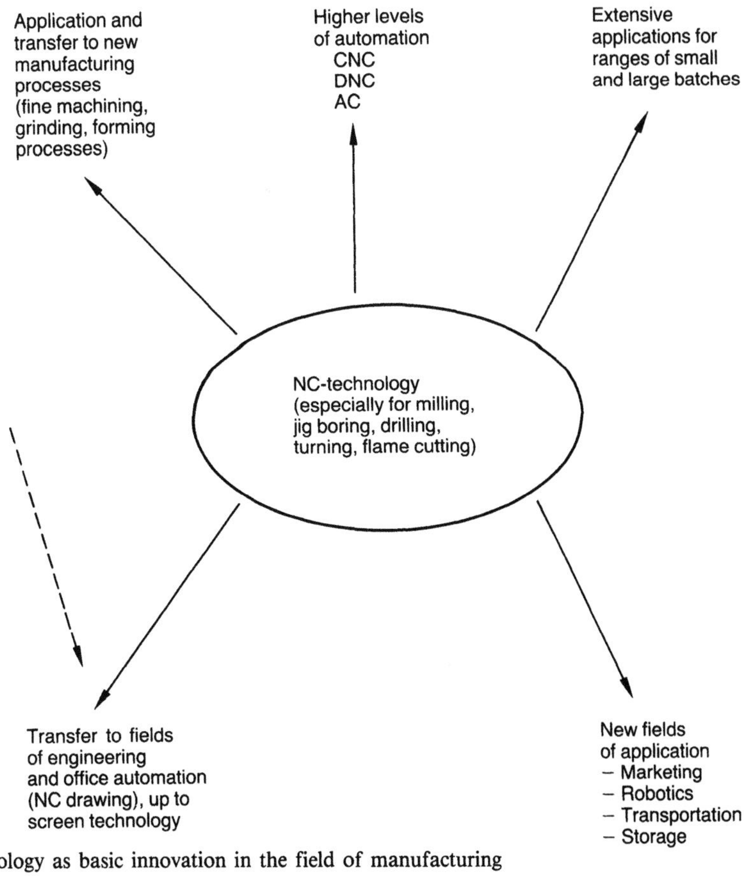

Fig. 3. NC technology as basic innovation in the field of manufacturing

3.2.3 Definitions, Characteristics, and Stages of Development

The first automated realization of a function with binary-coded information produced the following definition of numerical control, still valid today:
Numerically controlled machine tools (NCM) are machine tools equipped with control systems capable of logically processing prescribed programs (these are coded instructions using numbers or other symbols).
A more general definition comes from Temming and Seeliger [227]; it is based on the former one and considers the further technical development:

A numerical control system is a hierarchically arranged quantity of signal-interpreting, signal-storing, signal-processing, and signal-transferring elements, where the signals are offered in different forms: as letters, to describe geometric and technological circumstances and to define axles of machines on orders; as dimension figures, to describe the geometry; as characteristic values, to input coded information.

This general definition indicates that a nearly universal geometric and technological level of performance is attained by the numerical control systems available at present.

The succession of generations in numerical control is characterized by two principal stages of development:

— Wired programming controls
— Storage programming controls

1. *Hard-wired.* All control functions were realized by means of the circuitry (i.e., the hardware). This meant that all functions such as interpolation, tape format recognition, absolute or incremental positioning, character code recognition, and so on, were determined by the physical electronic elements built into the control unit. From the progressive point of view this stage of development can already be defined as "classical" numerical control.
2. *Soft-wired.* What was formerly accomplished by hardware design was replaced with complete computer logic that had infinitely more capability. With programming and data input done by the user, the 4th generation of control systems is prepared for specific problems. In this context the term "free programmable" is very often applied. This means that the system has a considerably greater range of functions.

The succession of four generations of control systems with their components is illustrated in Fig. 4. The cost account for the components and for the control systems in general is shown in Fig. 5.

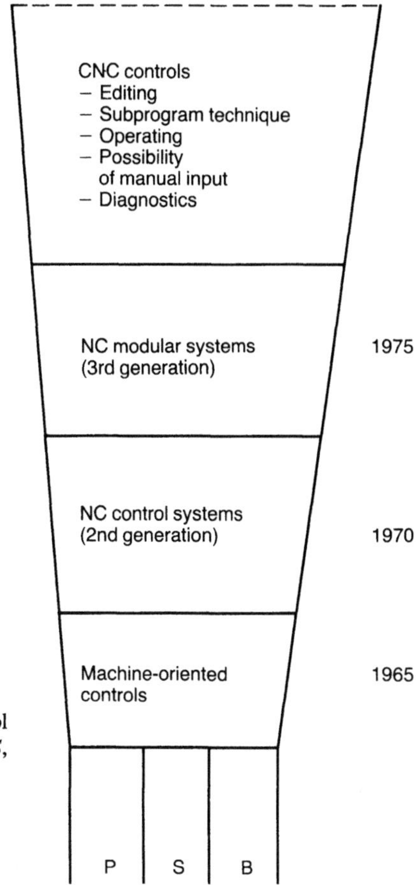

Fig. 4. Succession of generations in numerical control — important characteristics. *P*, point-to-point control; *S*, straight-line control; *B*, continuous-path control

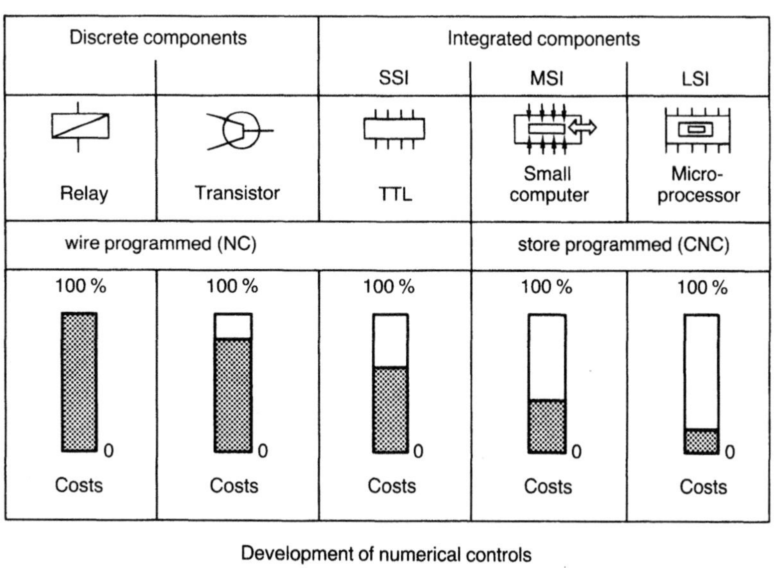

Fig. 5. Cost relation for discrete and integrated circuits

3.3 General Principles and Definitions of Numerical Control

In conformity with the concern of the "state of the art" report, the most important principles are presented here in condensed form.

3.3.1 General

There are five elements to the principle of numerical control:
1. Input of work information
2. Control and matching parts
3. Drives
4. Machine and controlled components
5. Measuring systems for each controlled axis

The combination of these basic elements is illustrated in Fig. 6.

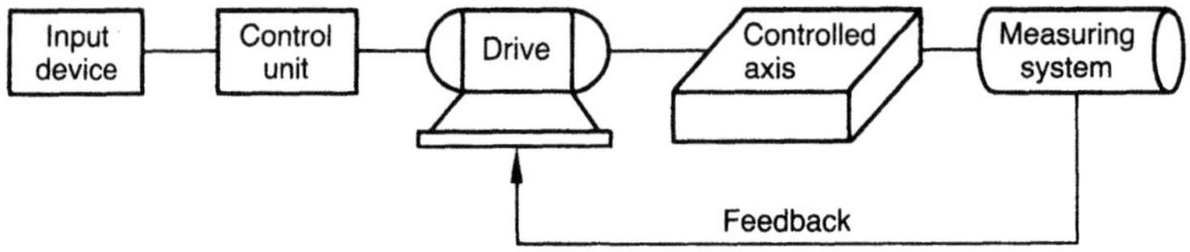

Fig. 6. Elements of numerical control

It must be taken into account that all these system elements are continuously being further developed.
Preparation and inputting of work information, which is classified as path information or switching information, is an important task for the manufacturing engineer; it will be described in Sect. 5.4. Machines and variants of equipment for controlled systems are the subject of Sect. 3.4.

3.3.2 Principles of Control

3.3.2.1 Fundamental Definitions

The optioned path-information W (commands) are compared with the actual value measured (controlled variable x) in the comparator to form a specific component of the control unit.
Basically, two types of circuits are used with numerically controlled machines. These are:

1. The cut-off circuit —
   The comparator emits a signal in the breakdown circuit for the case of coincidence
   $X - W = 0$
2. The regulating circuit —
   Signals are sent to the drive from the regulating circuit when
   $X - W \neq 0$

The principle of the cut-off circuit is applied in *point-to-point control* and in *straight-line control,* while the regulating circuit is used in continuous-path control.

In point-to-point control (see Fig. 7 a) there is no functional relation between the coordinate axes; the setting aimed at must be positioned before the real processing is started. The adjustment of the machine elements to the required positions is not detailed here; it can be realized simultaneously or subsequently (Fig. 7 a). This type of control is often used for drilling.

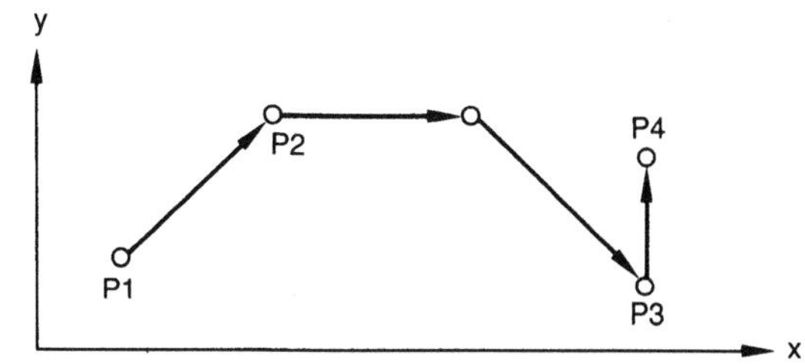

Fig. 7a. Sequence of motions during point-to-point control

Straight-line control (see Fig. 7 b) is characterized by the fact that after positioning the setting of a coordinate axis, the positioning of the 2nd axis is carried out by simultaneous processing. Cycles of rectangles, and in special cases certain levels, are handled. Straight-line control is especially applicable for cutting and turning.

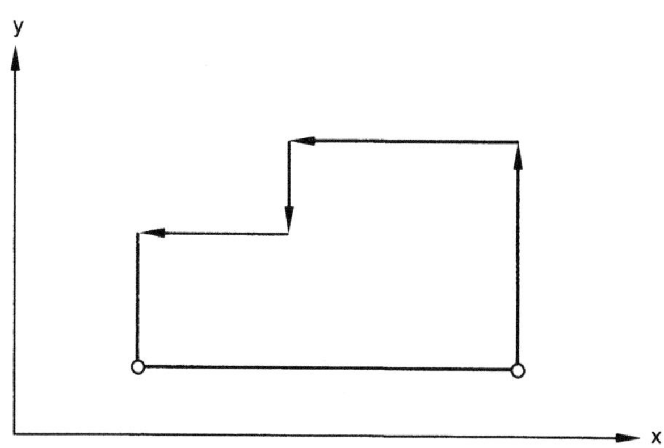

Fig. 7b. Sequence of motions during straight-line control

During continuous-path control (see. Fig. 7 c) there is a functional relation between the individual coordinate axes; handling and positioning are simultaneous (Fig. 7 c). To calculate the dots of the flat or the three-dimensional graph an interpolator (i.e., a digital or analogous minicomputer) is required.
Machines to process workpieces with problems of continuous-path control are available for cutting, turning, profile grinding, and flame cutting.

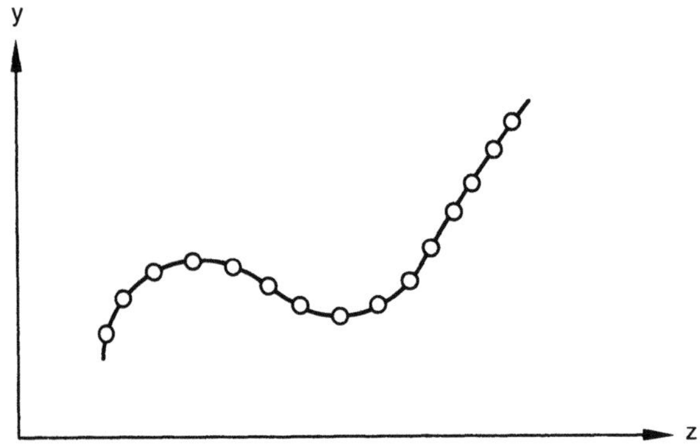

Fig. 7c. Sequence of motions of the continuous-path control in the plane

Continuous-path is the most universal principle of control, by which the highest accuracy is realized when excellent feed drives with regulating circuits for positioning and speed are used.
Depending on the number of axes controllable simultaneously and independent of each other, there can be:
2-axis controls in the plane
2.5-axis controls in the plane
3-axis controls in the plane
4-axis controls in the plane
5-axis controls in the plane
In substituting a 2-D control within the three main axes (x, y, z of the Cartesian coordinate system) the principle of the 2.5-D control (2 axes are continuous-path controlled, the 3rd is point-to-point controlled or straight-line controlled) is realized. With a 3-axis control the simultaneous interpolation is carried out in all coordinate axes included (Fig. 7 d).

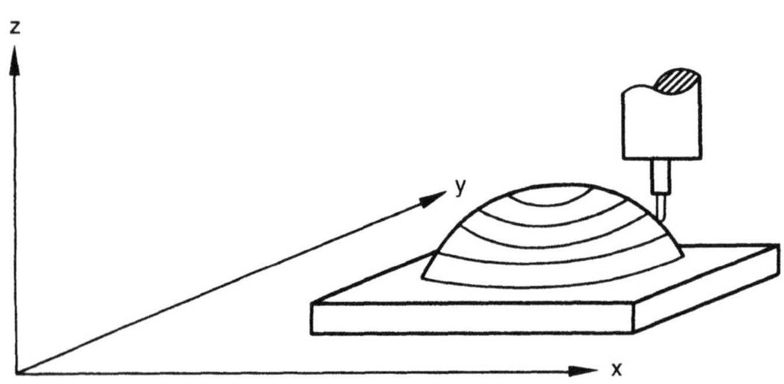

Fig. 7d. 3-D continuous-path control

Thus, the frequent spatial machining problems are kept under control.
In addition, the 5-axis continuous-path control is of special importance where the simultaneous movement in three coordinate axes is completed by the simultaneous movement of a turntable and swiveling of the tool mounting (Fig. 7 e).
The advantage of 5-D processing as the most developed form of NC manufacture at present is that in the cutting operation, the tool tip can be guided mathematically in the perpendicular direction to any double-curved plane. Thus, 5-D processing is especially suitable for the accurate and economical manufacture of double-curved planes (e.g., turbines, vanes, blades).

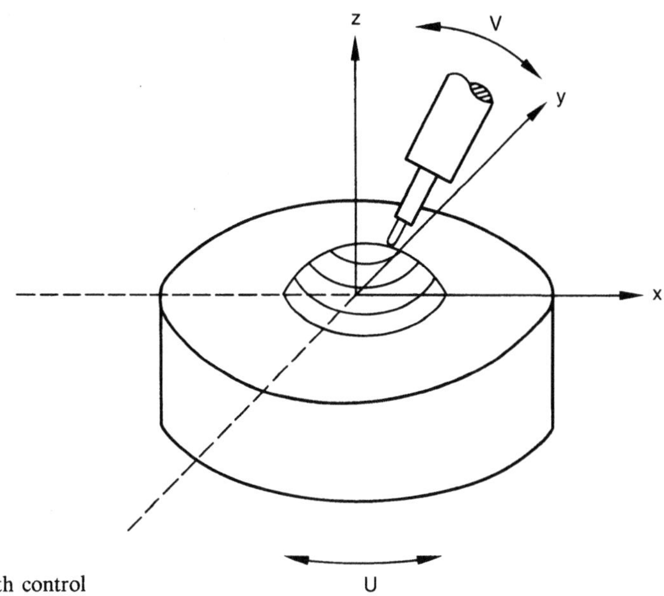

Fig. 7e. 5-D continous-path control

3.3.2.2 Types of Interpolation

To carry out continuous-path controls and definite cycles of movement in the plane or in the space, an interpolator to approach the desired ideal path curve and a speed controller for feed drives are required to realize the necessary relation of velocities in the individual directions of the axes. To coordinate the axis movements in the NC controls various interpolations are taken as a basis:

1. *Linear interpolation* is the prevailing principle. In linear interpolation the approximation to a prescribed path of movement is achieved by straight-line segments (see Fig. 8).
   The approximation to a desired path curve or space curve is achieved by a sufficient number of points of support. In accordance with the resolution power of the control (1/100 mm or 1/1000 mm), proceeding from the prescribed points of support in each case the dots are calculated by the interpolator. To achieve a very accurate approximation to the path course, the number of points of support to be prescribed must logically be increased.

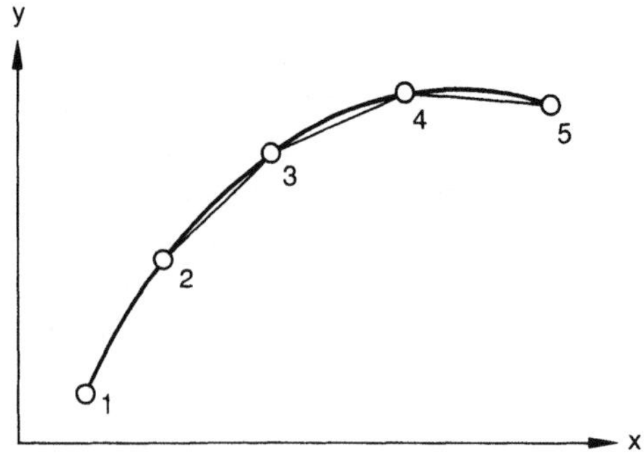

Fig. 8. Principle of linear interpolation

Linear interpolation is carried out for the coordinate axes x, y, z and also for the rotary and swivel movements with a justifiable technical expenditure; therefore, this interpolation has stood the test and is applied very often.

2. *Circular interpolation* is used to reduce the amount of data required. Proceeding from the equation of the circle the approximation to the path curve is achieved either by circular segments for continuous-path control problems or as a special modification.
3. *Helical interpolation* is circular interpolation with assignment of an additional constant in the 3rd axis to control space curves (for example, for the circular cutting of large bore holes on cutting machines.
4. *Parabola interpolation* gives an approximation to space curves.

3.3.3 Feed Drives

According to the allowances of the program (reference values) the required positioning into specified values or path points is realized by feed drives. For each controlled axis a particular drive is required.

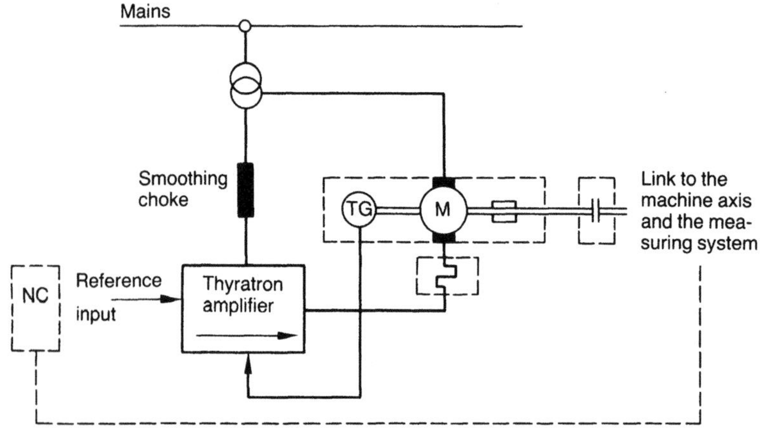

Fig. 9. General wiring diagram of a thyratron-feed drive [4]

When numerical control was first developed, hydraulic feed drives prevailed, but when power thyratrons and power transistors were developed, low-priced and fast dc drives were produced, which almost completely replaced the hydraulic drives [4].

Consequently, a feed drive of NC machines in general consists of the following components:

— Motor and tachometer (possibly with brake)
— Thyratron amplifier
— Power transformer with smoothing inductances
— Coupling between motor and machine
— Protection of the motor (overload protection)

In addition, the combinations of measuring systems and control system must be mentioned (allowed reference value). Figure 9 illustrates the coordination of these components.

Table 1 delineates the development of feed-drive performance over the past 20 years.

Table 1. Development of feed-drive performance [3]

Year	1960	1965	1968	1970	1975	1977	1980
Process speed in m/min	2.5	4	6	10	20	40	60
Switch-selected gear stages	1	4	1	0	0	0	0
Process time (a)	25	10	6	2	1	0.9	0.6
+ positioning time (s) for 200 mm	+ 20	+ 6	+ 3	+ 0.3	+ 0.2	+ 0.2	+ 0.15

3.3.4 Principles of Coordinate Positioning

Numerically controlled machines consist of three characteristic elements:

1. Machine tool
2. Control unit
3. Measuring system

There are three main features for the combination of these elements.

1. *The signal* can be digital or analogue (see Fig. 10).
   With digital controls the measured values are registered within determined segments, or "increments". The essential advantages of the digital signals are that a very high accuracy can be reached and direct evaluation and further processing in the control computer is possible. For these reasons digital controls predominate.

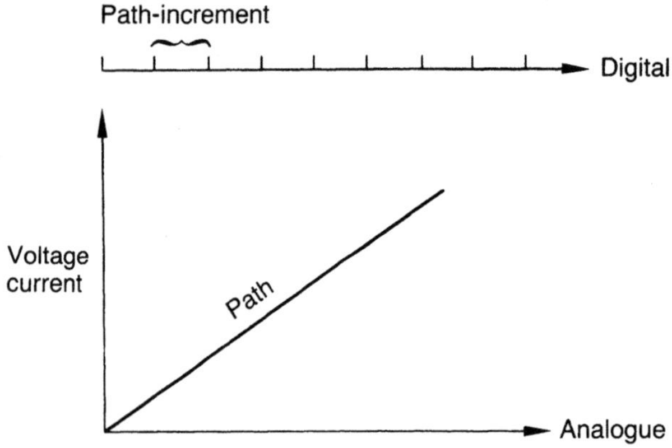

Fig. 10

Analogue signals are characterized by a continuous transformation of measured values. Generally, these signals are used less for path measuring than for other measuring sizes in NC.

2. *Size processing* has two basic modes: absolute and incremental (see Fig. 11). Early numerically controlled machine tools were exclusively one or the other. Most control systems today accept program information in either an incremental or an absolute mode.
   "Absolute" means the value is independent of the time of measuring and the driven path. The distance from the origin of the coordinate system can be indicated unambiguously.
   In the *incremental* mode the momentary growth is registered by appreciation counters. The value of the counter depends on the preceding course of motion, and the distance to the origin of the coordinate system can be indicated only conditionally.

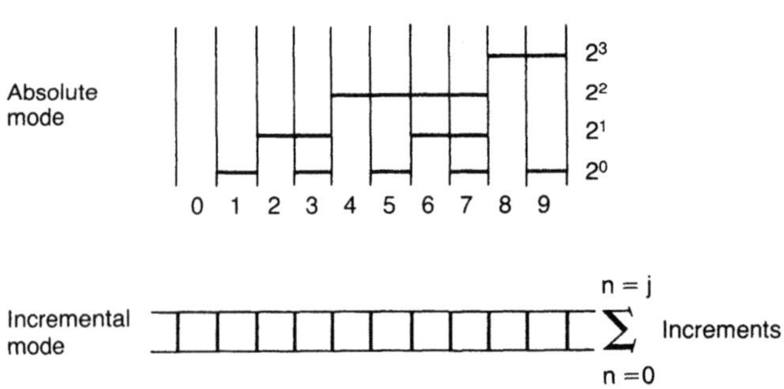

Fig. 11. Size processing in NCM

3. *The reading of sizes* can be done *directly,* e.g., using a measuring head or a rule (see Fig. 12), or *indirectly,* reading the length as an angle from a disk. The determined value of an angle is interpreted as a defined path element. For instance, one rotation of the input shaft of a measuring system can be divided into 1000 sections by the measuring disk. Presuming a spindle taper of 10 mm, one angle of 0.36° corresponds to a path of 10 μm.

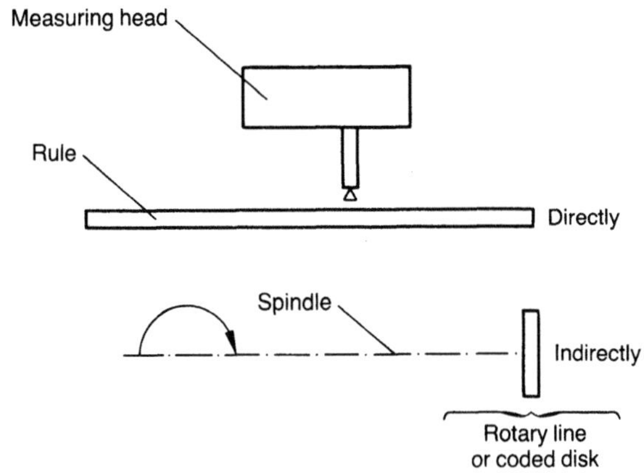

Fig. 12. Reading of sizes for NCM

3.4 Modifications of Numerically Controlled Variants of Equipment

With respect to the complexity of controlled functions, the development of the numerical control technique has proceeded from (a) numerically controlled individual machines to (b) numerically controlled machining centers to (c) numerically controlled flexible manufacturing systems to (d) numerically controlled machining cells. The development of the degree of integration and the level of automation are illustrated in Fig. 13. In this connection, the term "degree of integration" characterizes the extent of automated machining.

Degree of Integration

Automated machining	Automated machining sequence	Automated manufacturing sequence
Includes **Machining** with – positioning – measuring	Includes **Machining** with – positioning – measuring **tool exchange**	Includes **Machining** with – positioning – measuring **tool exchange** **Workpiece change** **Storage and transfer** – workpiece (-tools) **Total control** – flow of information – flow of material
NC M		
NC MC (machining centers)		
NC MS (FMS) (flexible manufacturing systems)		

Fig. 13. Level of automation of different NC manufacturing systems

Machining. In a single workpiece setup the machining process, including setting, positioning, and measuring, is automated with the aid of numerical control.
Machining sequence. All the operations required for machining a workpiece are executed automatically. Therefore, the machining stations have to be equipped with automated tool-changing mechanisms. In this way, several (cutting) processes are executed in a single workpiece setup.
Manufacturing sequence. An automated manufacturing sequence for the machining of different workpieces in any sequence also requires that they are linked with storage and transfer facilities, in addition to the automated sequence of operations at the machining stations.

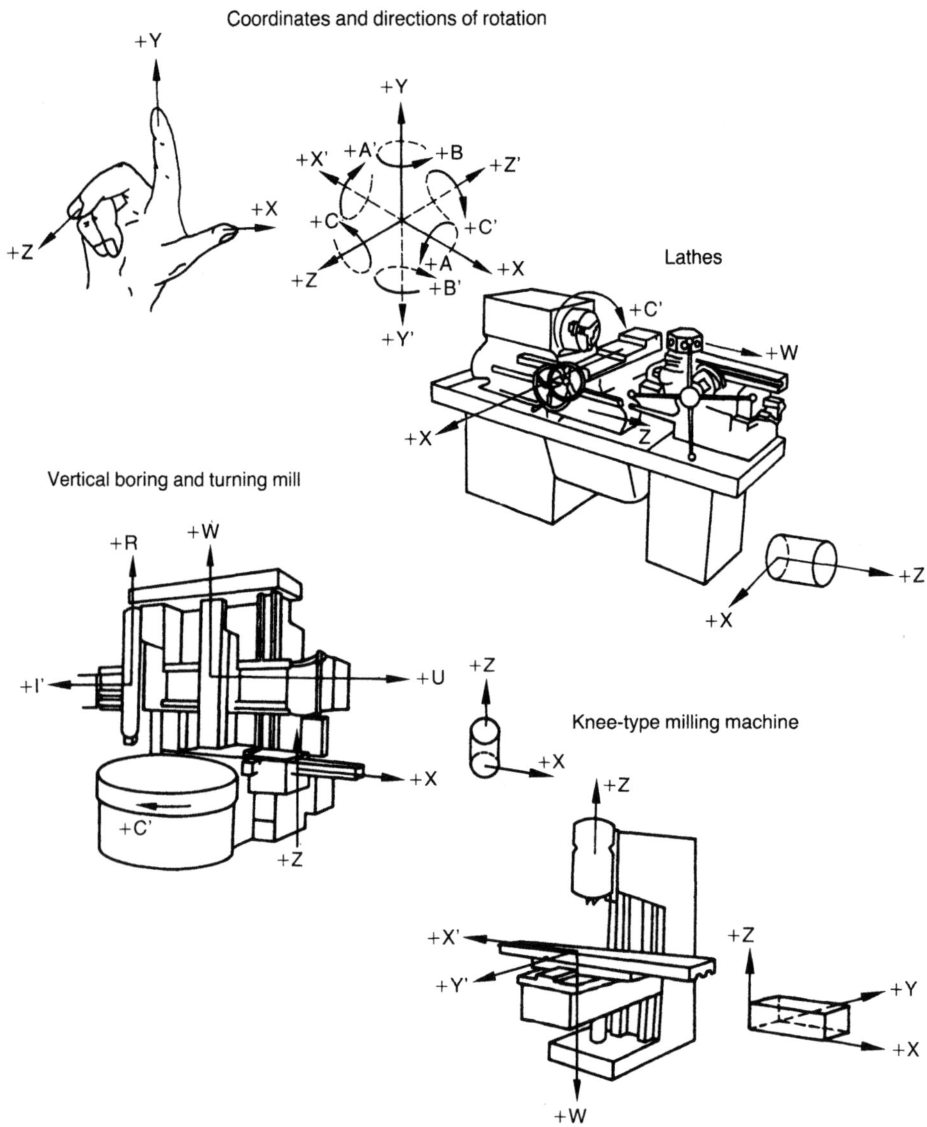

Fig. 14. Axes and directions of motion for NCM

In connection with the trends of development towards manufacture with a minimum of operators, the term "machining cell" was coined for machining stations with a high level of automation, making possible process monitoring and automatic workpiece change as specific additional features.
To a high degree, present-day development activities focus upon improving the machining cells.

3.4.1 Numerically Controlled Machine Tools

As illustrated in Fig. 13, the immediate machining process, i.e., the guidance of the tool movement, is automated by NC individual machines. This principle of numerical control was first applied to milling, then later to turning, flame cutting, drilling, and grinding. It is used more and more for other manufacturing processes, such as forming (fine forging, ringing, rolling etc.), engraving, and laser lettering.
During the first few years of development each machine-building enterprise used its own terms and definitions to designate the possible axial motions, but uniform designations have since been accepted on the basis of international recommendations (ISO RS 267-A). These conventions are illustrated in Fig. 14.

3.4.1.1 Specific Features and Advantages of CNC Controls

With the generation of computer-integrated control (CNC) efficiency has been considerably increased. An extended and more varied spectrum of tasks can be controlled than is possible with the wired programming controls. This is precisely the purpose of control engineering — to further increase flexibility and unattended manufacture.
The hardware and software solutions now available for storage programming controls provide a considerably more comprehensive capacity with respect to both quality and quantity than do the generation of wired programming controls still in use. Fig. 15 is a schematic comparison of capacity.
The increase of capacity manifests itself in both the considerable extension of external assistance for the operator by display screen technique and the comprehensive internal control functions. The differences shown in Fig. 15 are characterized especially by the following features:

1. Facilitation of control in connection with the display of different operational states or functions on the screen.
2. The possibility of correction and program improvement directly at the machine (program editing), programming in situ being possible through the same manual input. (This advantage of manual input should not be overemphasized, however, since for the highly productive manufacturing plants unproductive delays have to be reduced to a minimum.)
3. The availability of workpiece program storage obviates the necessity of intermittent and repeated reading-in of one and the same workpiece program.

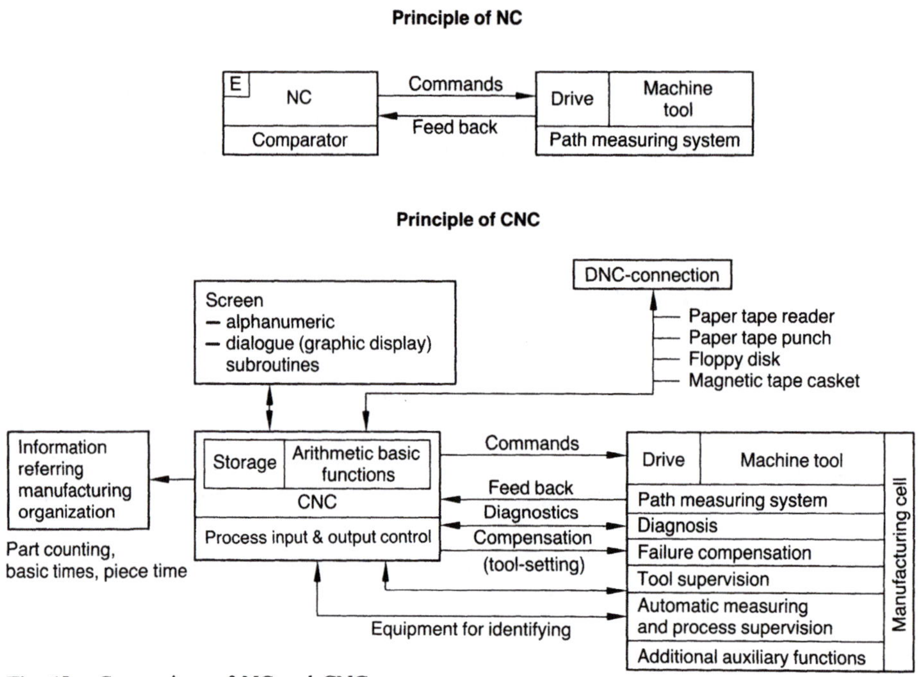

Fig. 15. Comparison of NC and CNC

The following components or functional groups of a CNC control are essential:

1. *Display screen units* for alphanumeric and/or graphic indication. The trend toward the employment of multicolored graphic display screens is growing (see Fig. 16).

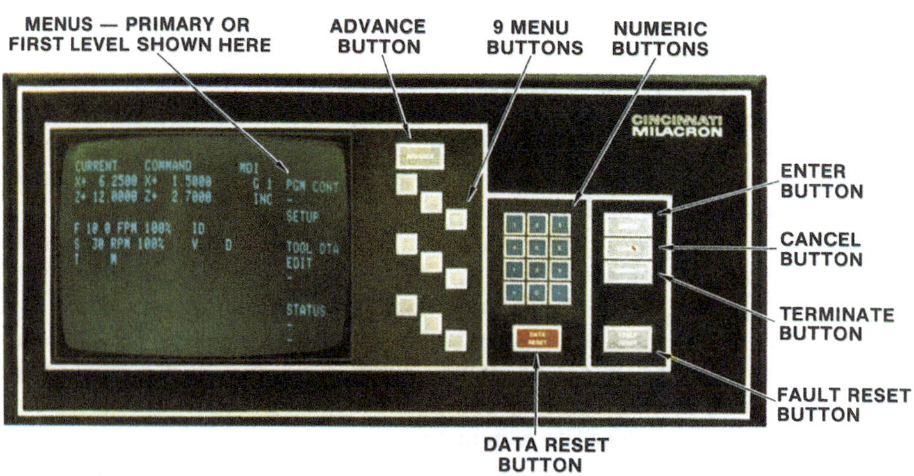

Fig. 16. Display screen unit

2. *Information input* via:
   a) A universal alphanumeric keyboard or special operational keyboards for manual input (see Fig. 17).
   b) Data carriers; supplementary to the punched tape data carrier, which was common at the beginning of numerical control engineering, more and more information carriers with high storage capacity are being used, e.g., magnetic tapes, floppy disks, or bubble memories.

Fig. 17. Universal alphanumeric keyboard

c) Direct data input in the DNC mode; the internationally standardized interface V24 is the most common.
3. *Control units* with different structures. Though all modern CNC controls are based on microprocessors and are similar with respect to their functional contents, one can characterize different structures (cf. Fig. 18). Each of the structures has its advantages and drawbacks, which are represented in Table 2.

The considerably simpler direct program generation is supported by the use of production-technological (standard) subprogram, parameter programming, the use

Table 2. Comparison of advantages and drawbacks of different CNC structures

CNC structures	Advantages	Drawbacks
Single-processor system	• Simple programming • Clearly arranged program structure	• Insufficient capacity for more comprehensive control tasks
Multiprocessor system with a fixed range of functions Examples: — FANUC and Siemens series — CNC 600 (GDR)	• Sufficient capacity for the respective control tasks • Developed service for operation and diagnosis	• In case of great performance volume — unused functions at the user with MACRO, unfavorable price/capacity ratio • Small volume of performance — small rage of application
Multiprocessor system with random assignment of functions	• High flexibility • Adaptability to user requirements	• Communication between microprocessors constitutes critical interfaces • High expense of operating system • In cases of low demand, a large portion of unused hardware and software complexes
MPST	• Optimum solution for different user requirements by means of a standardized modular system • Upward compatibility for higher forms of manufacture (MC, FMS) — low cost of modification	• Free microprocessors cannot intervene in other microprocessors if queues are formed
Examples: — Procontic — IBH-control — Micromodus — EPM/Gildemeister	• Uniform operating system • For users: equal software structure, equal control panel, operator prompting and programming • Always well-balanced price/performance ratios	• User control possible only when unit is being manufactured; initial expense during production of control unit very high (thereafter, only expense is connected with organization of production)

Kind of structure	Model structure (exemplary)	Charact. number of μ pr	Param. numb. of function	Explanation
Single processor systems		1	const.	
Multiple processor systems	Determined functional volume 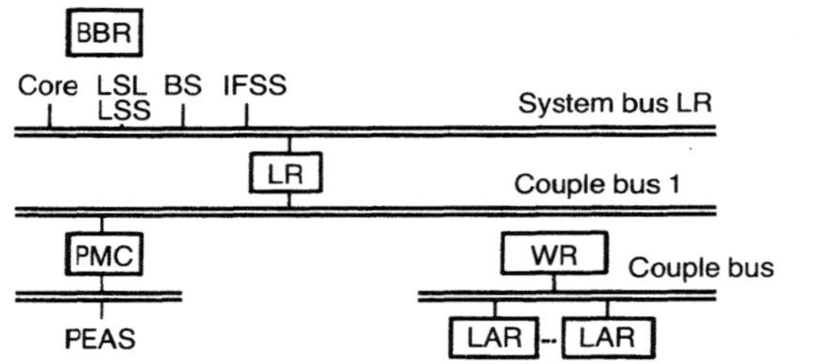	const.	const.	A maximum of functions is to be garanteed by a minimum of μ pr.
Modular multiple processor controlling	Optional function assignment 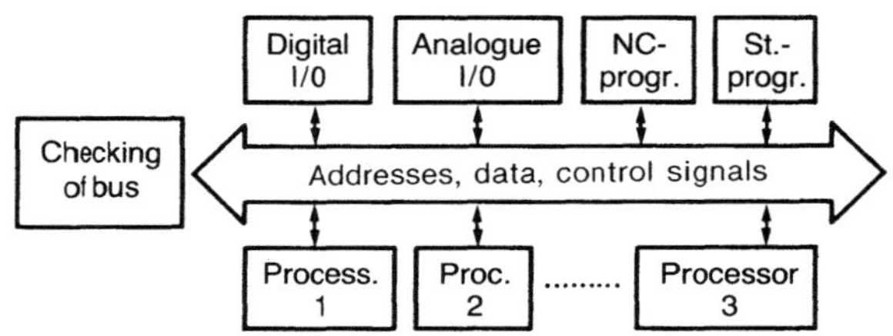	const.	varying	The present function is assigned to the unoccupied μ pr.
Modular multiple processor controlling system (MPST)	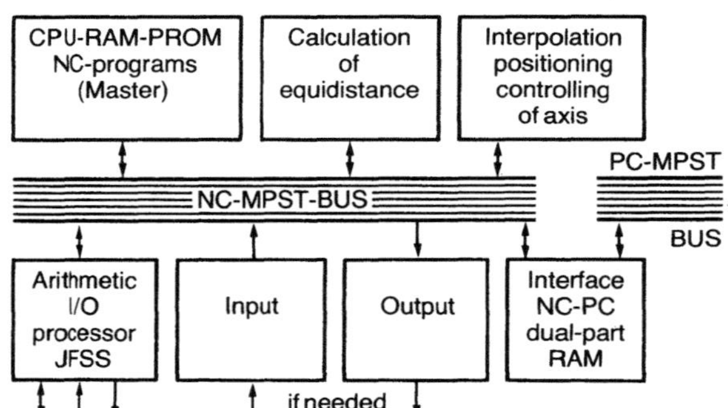	varying	varying	Functions which are unambiguously assigned to the μ pr.

Fig. 18. Overview of some CNC structures

of arithmetical/geometrical definitions for contour description, and automatic determination of equidistance. The availability of the control is improved by internal test and diagnostic routines. Direct programming allows compensation for machine errors, tool modifications (adaptive control), and thermal and other influences. It can supply the desired values and process the actual values at high speed, along paths which are three to four times higher and at path resolution of 1 μm. It is possible to control a continuously increasing number of NC axes with direct programming. Connections can be established to superior computers (hierarchies), and programs can be further developed to include other automated functions, such as measuring operations for workpieces and tools.

A survey of the range of functions of presently available CNC controls is given in Table 3. With these characteristic features and advantages offered by control based on microcomputers the conditions have been created for the control-technological mastering of other functional complexes which must be automatically controlled in manufacturing cells (cf. Sect. 3.4.3).

Table 3. Range of functions of CNC controls (as of March, 1984)

Characteristic feature	State of realization
— Programming facilitation	○
Diameter program	○
Radius program	○
Progr. zero shift	○
Definition of instructions for computation	○
Definitions of geometry	○
Correction of equidistance	○
Cutting subdivision	□ (autoprogramming)
Constant cutting speed	○
Subroutines	○
Symbol keyboard	□
— Determination of machining data	□ (Mazatrol)
— Interpolation for several NC axes	○
— Integrated PMC	○
— Diagnostic routines	○
— Interactive guiding of operators	○
— Control of tool change	○
— DNC connection	□
— Machine error correction	○
— Graphical assistance	□
— Modularity with respect to functional extensions	□
• Automatic zeroing of tools	(by offering a
• Integrated measuring control (workpiece, tool)	large number of
• Automatic return of reference points	controls in control
• Adaptive control	families with
• Software limit switch	different
• Parallel programming	efficiencies)
— Editing	○
— Storage medium with long-term archivating	□

○ Realized
□ Partly realized

From this can be seen the trend toward multiprocessor systems and modular concepts (MPC: modular program controls).

4. *Essential Operator Functions*
- Feed and speed override (stepless adjustability by selector switches); almost all CNC controls are equipped with this, and in general it amounts to
  0%-120% for feed
  75%-120% for speed
- Zero points which are put into the control directly through the data carriers or through the keyboard are an expression of the convenience of programming, especially for drilling and milling. With a programmable zero shift or new zero declarations the external computation of a large number of values, and thus the danger of computing errors is avoided.
- Tool programming; under present conditions it is possible to code about 120 tools, with a fixed or variable coding.
- Tool readjustment values are essential for simplifying program generation. About 200 tool-correction switches are available.

3.4.1.2 Machining Centers: NC/MC Characteristics and Features

Machining centers are typical of NC machining today. The machining center executes a number of different operations, such as milling, drilling, boring, facing, spotting, counterboring, threading, and tapping in a single setup on several surfaces of a workpiece. Machining centers are mostly capable of machining on five sides of a prismatic workpiece.

The typical additional part is the automatic tool changer, by means of which a cutting tool is automatically selected from a storage magazine and inserted into the machine spindle. The number of tool storage locations is 24, 48, 60, 120, or sometimes higher. There is no absolute theoretical limit. Several kinds of tool magazines are explained in Table 4. Moreover, for specific cases of application complete toll heads (e.g., drill heads) can be automatically exchanged.

Table 4. Examples of magazine types for automatic tool change

Type	Feature
Disk magazine	A ring-shaped disk at the side of the machine or above the machining spindle
Ring magazine	Tools are arranged in two concentric rings
Chain magazine	In general at the side of the machine; can be enlarged if necessary, partly also in a double or triple arrangement, and thus has the greatest capacity of all magazine types

The automatic tool change within machining centers requires *tool coding* for fast search, supply, and correct classification in the magazines. At present, several methods are used for this:

1. *The switching magazine* (turret for drilling machines or lathes), with a fixed arrangement inside the turret, is the simplest form, but in general limited to 8, 12, or 16 tools. No coding is required.
2. *With location coding* the individual magazine locations are coded from 1 to n. Each tool is assigned to one location, which is defined in such a way that the tool returns to the same location after use.
3. *With tool coding* each tool is unambiguously identified, e.g., at its shank with the aid of coding rings. The advantage of the arbitrary classification of the tools in the magazine is achieved at a relatively great expenditure of time for coding and finding in the magazine [3].

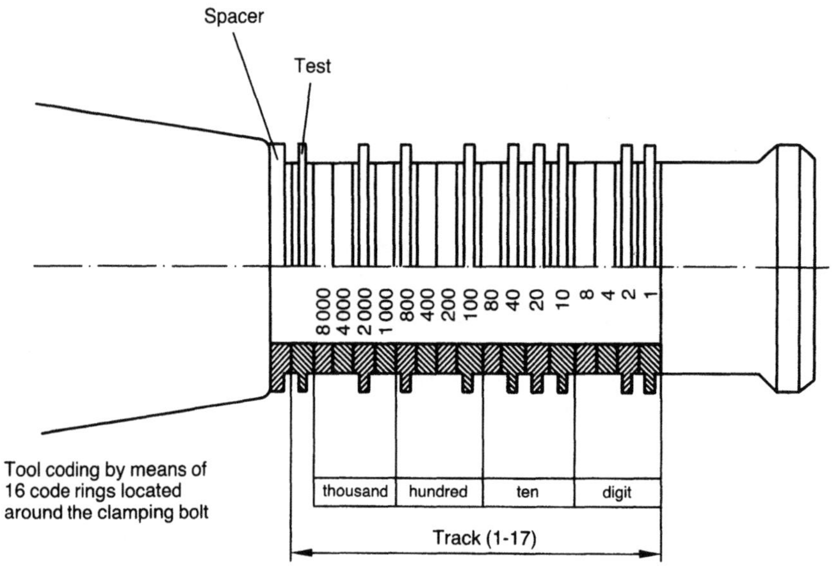

Fig. 19. Principle of tool coding by means of code rings

4. *Variable Location Coding:* If a special NC logic is available, the advantages of the above-mentioned methods (location coding and tool coding) are utilized. The application of a specific form of coding is dependent on several things. Therefore, practically all the principles mentioned are applied.

In their typical fields of employment, especially in milling and drilling, the machining centers offer a number of essential advantages, such as savings in fewer jig and fixture requirements and shorter lease time. Product designers are finding ways of redesigning many parts into a single component machined from a forging, casting, plate or bar stock. Design configurations that were previously impossible are now possible if enough axes of motion can be simultaneously controlled to develop the configuration.

Because of these important advantages machining centers are widely used. Hundreds of different types of machining centers are produced throughout the world today for use in industry. Two examples are given here, flexible manufacturing systems (Kearney & Trecker, Milwaukee, USA) and manufacturing cells.

3.4.2 Flexible Manufacturing Systems

3.4.2.1 General Explanations

As illustrated in Fig. 20, flexible manufacturing systems (FMS) achieve a higher scope of functions, and therefore a higher level of automation than do single NC machines and machining centers.

In all leading industrial countries the concurring aims of these developments are:

1. A steady increase in labor productivity
2. Better utilization of fixed assets with respect to time
3. Reliable, flexible, automated manufacture for smaller and smaller lot sizes when the part assortments are always changing
4. Abolition of monotonous work that is physically hard or dangerous
5. A high level of integration of manufacturing sequences
6. Consideration for environment and operators with respect to noise, dirt, and maintenance of the automated manufacturing processes.

These aims are especially reached by the automated tool change, the automated intermediate storage and transfer of workpieces and partly of tools, and the computer-aided overall control of information flow and material flow within the system. The principle of the flexible manufacturing systems, therefore, consists in the combined action of four flows:

1. The parts flow, carried out by means of transport paths and pallets; chutes, raceways, air-cushion transport paths, rail-guided transfer robots, and robocarriers (induction-loop-guided transfer cars) are used as transfer facilities, and pallets may be used for work accommodation in lots (e.g., rotationally symmetrical workpieces) or for work accommodation in pieces (larger prismatic workpieces).
2. The tool flow, mostly within the machining stations (machining centers), automated, and strictly organized within the overall system
3. The information flow, which proceeds in general in the DNC operation (Sect. 3.4). One process computer is used to supply the working information, a second for the optimization of occupation (per week or per shift) and the control of the material flow.
4. The energy flow to coordinate the function of all main and accessory drives in the system

The functional structure of flexible manufacturing systems is illustrated in Fig. 21. The concept of flexibility as the central characteristic of efficient manufacturing systems, with respect to their adaptation to different requirements, has been defined more closely by Freist et al. [75] and Kegg [123]. According to them, the following types of flexibility are distinguished:

1. Mix flexibility

   This is the ability to process various members of a well defined family of parts without the loss of time for setups.
2. Parts flexibility

   This represents the ability to add new parts to the family. It broadens the restrictions which must be placed on the parts family.

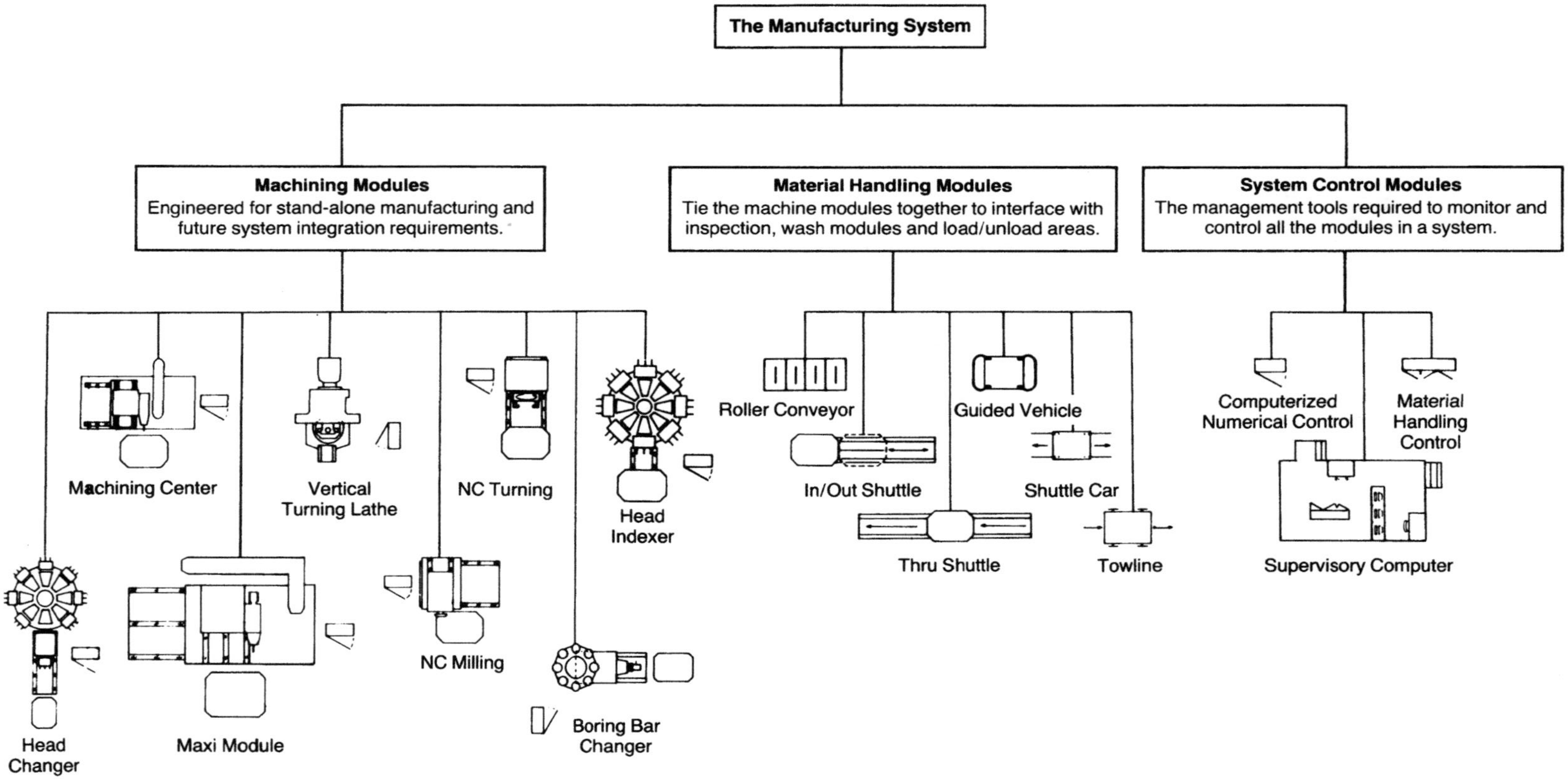

Fig. 20. The fundamental building blocks of FMS

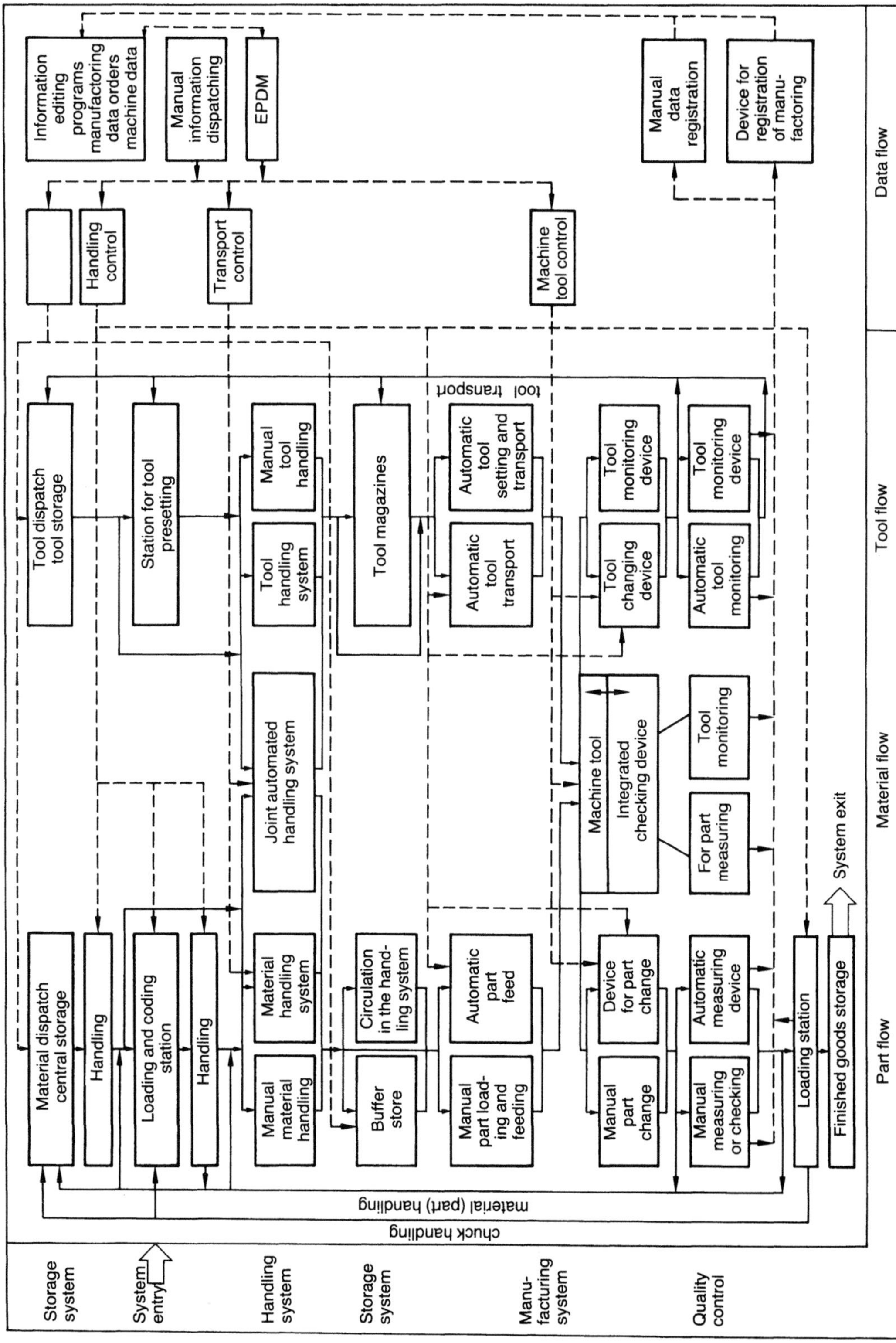

Fig. 21. Functional structure of FMS

3. Routing flexibility
   There are two slightly different types of routing flexibility. The first is the ability to move different parts to the individual work stations in different sequences. The second is the ability to move a given part number through different routes depending on machine availability.
4. Design change flexibility
   This represents the ease of implementation of engineering design changes to parts which are already being manufactured.
5. Volume flexibility
   This measures the efficiency with which changes in the master production schedule can be accommodated. It is the ability to respond to these volume changes without any increase in unit workpiece cost, loss of productivity, or reduction in equipment utilization.
6. Factory systems flexibility
   This represents the ease with which the FMS hardware, software, and information systems can accommodate changes in the factory hardware or information systems which will almost certainly come in the future.

All of these influence factors add up to a requirement for extreme flexibility in the factory data base and software systems. Today, FMS constitute the top technical level of manufacturing engineering. FMS have a broad effect on engineering, since they influence or advance the future development of machine tools, NC engineering, transfer, technical preparation of production, production control and organization, and material flow. FMS are important for the entire field of parts manufacture of the metalworking industry, but mainly for small- and medium-batch production.

3.4.2.2 FMS — Characteristics and Features

The International Institution for Production Engineering Research (CIRP) recently completed a set of definitions of automated manufacturing production systems, of which flexible manufacturing systems (FMS), or, more correctly, flexible manufacturing production systems (FMPS), are a part. An automated manufacturing production system is a unit within a manufacturing system consisting of an integrated assembly of machines and/or associated equipment necessary to carry out production with a minimum of manual attention, together with the means for transferring components automatically through the system, all operating under fully programmable control. In engineering manufacture, an automated production system usually includes machine tools for machining or forming materials but may also contain other machines, such as those for washing or for assembly of the product, and/or other equipment (e.g., for painting), and may also include computer-aided scheduling, planning, and other capabilities.
Two main types of automated manufacturing production systems can be distinguished: dedicated systems and flexible systems.

1. *Dedicated manufacturing production system*
   A dedicated manufacturing production system (DMPS) is an automated system designed for the production of one product only, which cannot readily be adapted for the production of other products.
2. *Flexible manufacturing production system*
   A flexible manufacturing system (FMPS) is an automated system which is capable, with a minimum of manual intervention, of producing any of a range or family of products. The flexibility is usually restricted to the family of products for which the system was designed.

Fig. 22 is an artist's sketch of a typical modern FMS, this one being installed for machining aircraft parts.

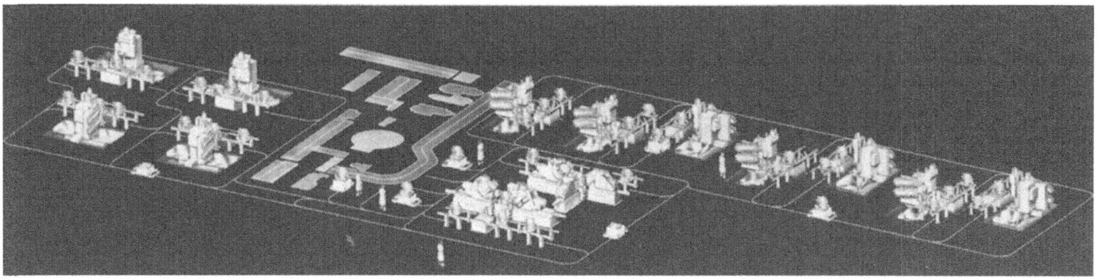

Fig. 22. Artist's rendering of a typical modern flexible manufacturing system (Courtesy Cincinnati Milacron)

Flexible manufacturing systems first began to appear about 1967 in the United States; since then they have proliferated worldwide. Hatvany [94] recently edited a world survey of computer-aided manufacturing in which it is estimated that by 1980 there were about 25 FMS installed in industry in eastern Europe, about 25 in western Europe, about 40 in Japan, and about 25 in the United States. There were a few such systems in use in lesser developed countries as well. Thus, there must have been on the order of 125 FMS in use in world manufacturing industry at that time.

Concerning the rate of proliferation of such systems, Hutchinson [103] has presented a very good description of the seven FMS systems in actual operation in United States industry up to 1977. (Other experimental and research systems had been built by then, but had not been operated in an actual production environment.) Tables 5 and 6 are Hutchinson's summary of the characteristics and features of these seven systems. Plotting data from Table 5 results in the trend line shown in Fig. 23. The rate of increase in the number of installed systems is exponential, and the number of systems was doubling approximately every 3 years. Considering that by 1980 there were an estimated 25 systems operating in United States industry, that rate of growth even appears to have been increasing somewhat in recent years. This is consistent with a forecast made recently by the Society of Manufacturing Engineers [37] using the Delphi method of technological forecasting. It predicts that some 15% of total machine tool production will be going into FMS by 1987.

Table 5. Characteristics of FMS in use in the USA up to 1977

System	Approximate year	Vendor	Mission	Class	Workstations	Approximate cost	Material handling
Ingersol-Rand Omniline	1972	White-Sundstrand	180 parts, generally in batches, but service parts as well; 3-ft cube; 70000 per year, up to 16 parts simultaneously	FMS	2 4-axis mills 2 5-axis mills 2 4-axis mills	Unknown	Roller conveyor of palletized parts. Buffer position at each machine
Caterpillar	1974	White-Sundstrand	Cast-iron case and cover for tractor transmission; 6 parts; 1200 per year; approximately 3-ft cube	FMS	5 Omnimills 2 G&L turret lathes 3 Omnimills 1 DEA inspection	Unknown	Palletized parts, 2 carts move parts between machines and load area, which also serves as buffer. Carts perform shuttling operations, use straight-line track
Sundstrand Aviation	1967	White-Sundstrand	Aluminium pump parts and magnesium castings for aircraft speed drive housings, approximately 16-in cube. 2000 per month in batches of 25—300	FMS	8 OM-2 Omnimil machining centers 2 Multiple spindle drilling machines	Unknown	Palletized workpieces have code to direct to next station on roller conveyor (power and free)
Allis-Chalmers	1971	Kearney-Trecker	Produces cast-iron tractor parts for direct assigning; 23600 per year; approximately 3-ft cube	FMS	5 Machining centers 1 Mill 4 Duplex multispindle head indexers	$5 M	Palletized workpieces moved under computer control to workstation on towed carts. Local, multi-position queue at workstations and carts serves as buffer pool. Complex network
North American Rockwell	1973	Kearney-Trecker	Cast-iron differential carriers for trucks in lots of 10—50; 33 parts; 1.5-ft cube; 8 parts simultaneously; 24000 per year	FMS	8 Machining centers	Unknown	Same as Allis-Chalmers, except a simple loop with spur rather than a network
Avco Lycoming	1976	Kearney-Trecker	Aluminium aircraft engine (4 and 6 cylinder) crankcase halves	FMS	2 Simplex multispindle head indexers 1 Duplex multispindle head indexer 9 Machining centers	$10 M	Same as Allis-Chalmers, except some machines have a single position queue
Caterpillar	1977	Kearney-Trecker	Family of large-axle housings for loaders up to 4000 Ib and 106 in	Batch transfer	2 Simplex head indexers 1 Vertical mill 1 Duplex machining station 1 Simplex machining station	Unknown	Palletized workpieces move on conveyor. Empty pallets returned to load area

Table 6. Features of FMS in use in the USA up to 1977

System	Tooling	System management and control	Comments
Ingersol-Rand Omniline	Local to machines, approximately 200 in system, but capacity is 360	Fixed sequence within system. Foreman balances workload by dispatching; has multiple alternative routings	50% reduction in direct labor. Up to 70% reduction of part costs. 16—18 machines (estimated) would be required if conventional
Caterpillar	Local to machines, 350 in system	Fixed sequence due to tooling limitations. Production to monthly master plan. System performance improved due to investigations using simulation	Tooling constraints. Carts are weak link in system, replacement planned. Utilization of machines doubled. Labor reduced from 13 (estimated) to 7
Sundstrand Aviation	Local to machines, maximum of 39	Limited versatility, fixed sequence at introduction to system	Replaces estimated 100 conventional machines
Allis-Chalmers	Local to machine. Head indexers have total of 864 tools on 73 heads	Work stations dynamically assigned by computer to balance load. Computer assists supervisor in scheduling	Incrementally installed. Work mix changed drastically between design and use. 3 weeks from release by production engineers to production. Engineering changes within 20 h elapsed time
North American Rockwell	Local to machines	Same as Allis-Chalmers	Original installation included inspection station. Greatly reduced set-up time and batch sizes, reducing work in progress. Shares with a DNC system having 5 lathes
Avco Lycoming	Local to machines	Same as Allis-Chalmers. Every operation has alternative work station	Reduced 30 setups per part to 5. Direct labor and floor space reduced. Largest, in terms of workstations, FMS. Incremental installation
Caterpillar	Local to machines	Batch production within family. Changeover time to next part is 2 h	Output to increase 37 times over old system

In view of all this, it appears that the period when large numbers of these systems are installed worldwide has now arrived.

Concerning the relative attributes of FMS systems in the United States, Hutchinson's 1979 assessment was as follows:

Although it is doubtful that the U. S. has the most sophisticated hardware systems, they have the greatest variety of systems and probably the best in an economic sense. They are operating in a variety of production environments and a large number of physical logical configurations. There is no lack of technology but the design and management skills and techniques must be developed if those systems are to be more fully utilized in the U. S. [103].

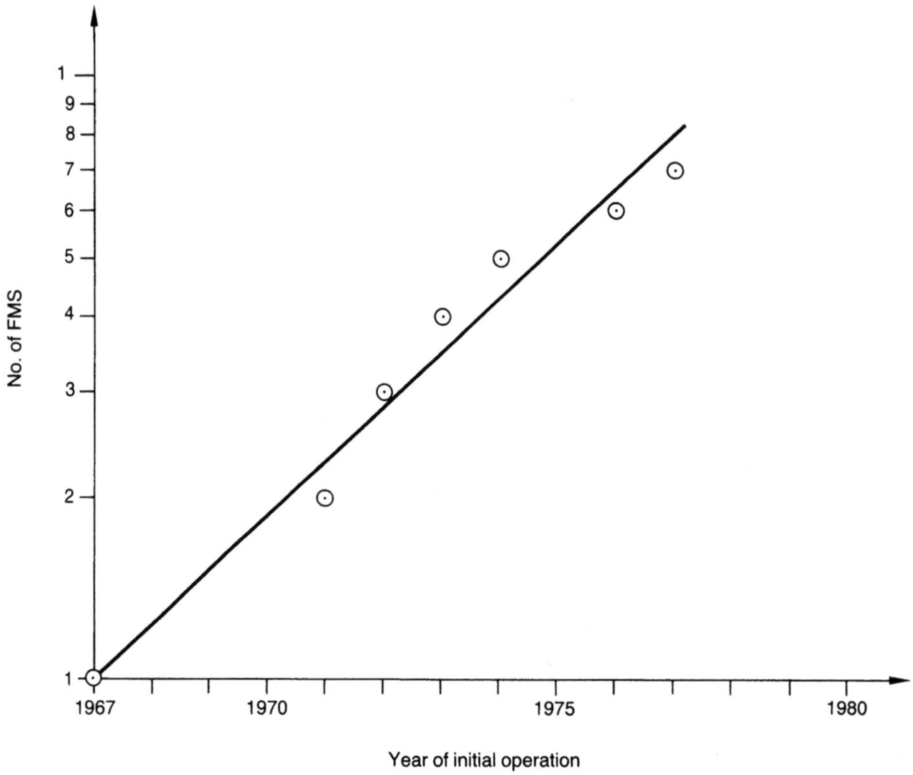

Fig. 23. Early proliferation of FMS in the United States

What are the technological characteristics of FMS that have evolved over the years since they first appeared? The changes in these systems in recent years have been rapid — and fascinating — and we can expect even more exciting changes in the next few years. One of the most noticeable has been the shift from systems which are mainly dedictated to producing a rather small family of quite similar parts (often processing those parts in fixed order or even in batches) to systems which can process a large family of parts having a wide range of characteristics, and do so in random order. Thus, these newer systems are beginning to take full advantage of the potential of the computer to provide flexible programming which can be varied on-line moment by moment.

One of the major trends accompanying this shift has been the move from design and construction of essentially monolithic systems to modular systems — modular with respect to both hardware and software. This makes possible the use in such systems of primarily standard machine tools, standard modular workpiece and tool transport and handling equipment, and standard computer and control systems. These can then be readily configured into systems whose features are optimized to the short- and long-term needs of the user. This also makes the systems readily reconfigurable as long-term needs change. The modular approach provides the added advantage of allowing the user the freedom to move toward the implementation of one or more FMS in his plant in a phased manner, starting with stand-alone machines and gradually integrating them by adding additional modules of other types as appropriate.

A very important trend observed as a corollary to the foregoing is that toward full computer simulation of FMS. In the move toward modular systems it was quickly found that configuring various modules into systems optimized to the needs of the user was a task not at all suited to manual solution by human brainpower. The possible permutations and combinations of workpieces, tools, pallets, transport vehicles, transport routes, operations, etc., and their resulting performance, are almost endless. Computer simulation has become an absolute necessity in the design of practical systems, and the trend toward broadening its capabilities is continuing as systems move to encompass more and more of the factory. This capability of simulating a system for design purposes has an added advantage: The software simulating the operation of a system, once designed, can become a part of the operating software of that system, providing on-line the scheduling and production control needed to ensure optimized performance moment by moment.

The trends toward modular systems and computer simulation of such have at this stage resulted in a preferred set of modules observed in most of the more recent and more advanced FMS in the world. These are a combination of standard versatile machining centers as workstations, fixtured pallets for handling the workpieces, computer-guided carts for transporting the workpiece pallets (and sometimes tools as well), and a hierarchical DNC/CNC control system. This is the type of system illustrated in Fig. 22. However, research and development presently underway suggest trends which may considerably broaden that spectrum. Work aimed at incorporating turning centers, grinding machines, sheet metal-forming machinery, plastic-processing machinery, assembly machinery, and many other types of workstations into FMS is already well advanced and they have in fact been applied in some instances. Industrial robots have also already been applied in some systems to do tool and workpiece handling, automatic palletizing of parts, and even limited transport, and this capability will certainly increase. Work is being done in a number of places to develop general purpose, flexible, adaptable pallets, or adaptable fixtures on the machine tools, to eliminate the need for special-purpose fixtures or even pallets themselves.

Another major worldwide trend is the aim of making FMS more intelligent. At this stage most of this effort is centered around making the workstations more intelligent. This includes development and application of such features as on-line diagnostics combined with feedback to accomplish self-correction of actual or incipient failures, in-process sensing and feedback control (adaptive control) of machining and cutting tool conditions and of the location of workpiece surfaces, and in-process sensing and control of workpiece accuracy, surface quality, etc. (in-process inspection and control). While some of this capability is already in existence and being applied, much of it is still being actively developed.

A corollary to the foregoing is the trend toward unattended systems. Already, with some of the intelligence capability in existence to a limited degree, a few systems in several countries are running unmanned on the night shift and with limited manning on the day shifts. However, as most of the intelligence capabilities enumerated above become a reality, it is likely that many FMS will develop the capability to run essentially unmanned around the clock.

3.4.2.3 FMS — Examples of Applications in the United States and Japan

As indicated in the preceding section, FMS installations are now proliferating rapidly throughout the world. Thus, there are myriad examples which could be described. An excellent survey of the various types of such systems in use throughout the world, replete with examples, has been published by Dr. Dupont-Gatelmand [54]. To date, however, not a great deal of information has been published about the *benefits* being experienced from application of such systems. Yet, at this stage in the development of CAM, it is very important to have such information. The reason is that, although complete integration of the element of CAM into a total system of computer-integrated manufacturing (CIM) has not yet been accomplished anywhere in the world, nevertheless, looking at an FMS in relation to CIM, it can be recognized in its most advanced form to be almost a complete microcosm of the shop-floor portion of the future total computer-integrated manufacturing system — but available now. Thus, it offers the opportunity both to evaluate some of the potential future economic benefits of total CIM, today, and to lay groundwork and gain experience for future integration of all the elements of full computer-integrated manufacturing. In this section we present examples of a variety of FMS installations for which information on the economic and performance benefits obtained is available.

The first of these, described by Dronsek [53], is the FMS now operating at Messerschmitt-Bölkow-Blohm in Augsburg, Federal Republic of Germany — probably one of the most sophisticated in the world today. At Augsburg, MBB builds the center section ("wing box" and related structure) of the swing-wing Tornado fighter plane. The Tornado is being produced at a rate of about ten per month against a total initial order quantity of 800.

In 1975, when the plans were made for going into production on the Tornado at Augsburg, it was decided to go the computer-integrated manufacturing route, starting with as fully automated machining of the titanium parts as possible. This was motived partly by the fact that, contrary to the situation with more conventional aircraft, 40% of the cost of manufacturing the plane would be in the machined parts, even using the most advanced computer automation. MBB then developed their long-range master plan for computer-integrated manufacturing called Computer Integrated and Automated Manufacturing (CIAM). The overall goal, long-term, is reduction of manufacturing costs by integration of design, production planning and control, and production operations into a single computer-automated system.

The major initial development in the CIAM program was the FMS, fully operational for the past several years, for the machining of titanium parts for the Tornado center section. The basis elements of the system are

a) 28 numerically controlled (NC) machining centers and multi-spindle gantry and traveling-column machines,
b) a fully automated tool flow and changing system,
c) an automated workpiece transport system, and
d) coordinated computer control of all these elements.

Figure 24 shows the main components of the system. One of the 28 numerically controlled machine tools is seen in the center of the figure. The automated workpiece transfer system brings workpieces to and from each machine tool by means of computer-controlled carts, such as shown in the lower right-hand corner of the figure. The fully automated tool transport and tool changing system brings tools to each machine via an overhead transport system, seen in the upper part of the figure. It then automatically transfers these to a continuous elevator tool storage, seen at the left of the figure, which in turn interfaces with the automatic tool changing mechanism of the machine tool. All three of these subsystems — the machine tools, the work transfer system, and the tool transfer system — are coordinated, controlled and automated by a hierarchical distributed-computer system (DNC/CNC).

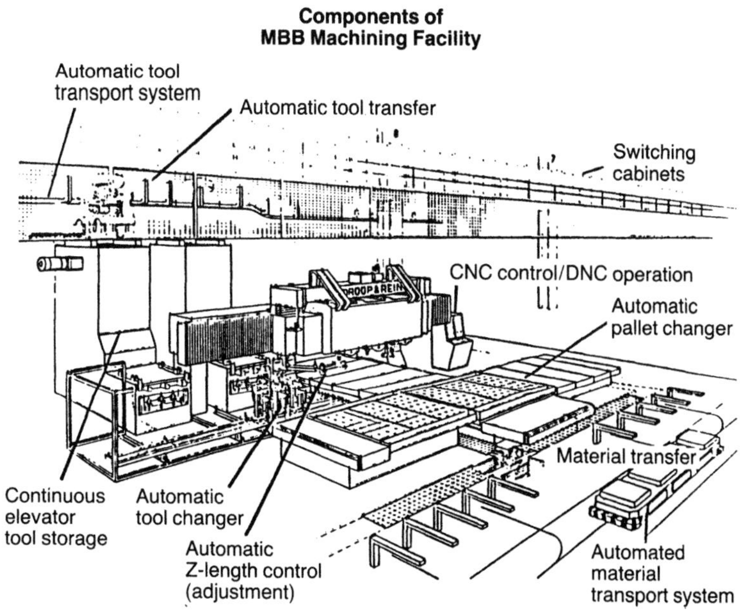

Fig. 24. Five-axis controlled multiple-spindle CNC milling machine in bridge design with automated peripheral equipment

The system has demonstrated remarkable performance benefits. MBB find that the machines in the system are cutting metal, on average, about 75% or more of the time, i.e., machine utilization is 75% or better. Lead time for production of a Tornado is only 18 months, compared with about 30 months for planes produced by more conventional means. The system reduced the number of NC machines required (compared with doing the same job with unintegrated stand-alone NC machines) by 52.6%, required personnel by 52.6%, required floor space by 42%, part through-put time by 25%, total production time by 52.6%, tooling cost by 30%, total annual costs by 24%, and capital investment costs by 10%. This last fact alone illustrates that the idle capital associated with normally underutilized machines, freed up by application of computer-integrated manufacturing, is more than sufficient to provide the additional sophisticated capital equipment needed to implement FMS.

The increased productivity represented by all these figures is most impressive. In addition, there are nonquantified benefits experienced with this system. Quality has been increased, manifesting itself in the form of higher accuracy and reproducibility, lower rework costs, and lower scrap rates. This in turn has resulted in lower quality-assurance costs. In addition, adherence to production schedules is much improved and the usual flood of paper work has been considerably decreased (no order-chasing). Furthermore, working conditions are improved due to the decreased risk of accidents, the relief from heavy physical labor, and the more challenging nature of the work. The increased flexibility has made the operation more independent of batch size, of the types of parts, and of production quantities.

Turning now to Japan, with the implementation of FMS already advancing quite rapidly there, efforts are now being directed toward developing capability for such

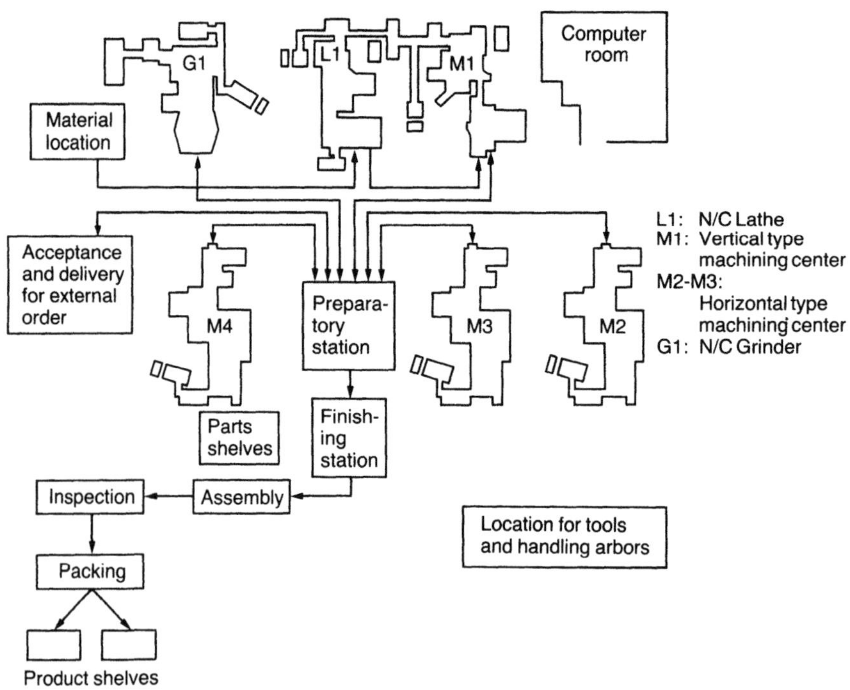

Fig. 25. Layout of Toshiba Tungaloy FMS

systems to run unattended at night. One of the first systems to accomplish this at least partially, described in a paper by Takeyama et al. [225], was developed by Toshiba Tungaloy Company, Ltd. and is located at their Kawasaki plant. It was put into service in August, 1980, and produces milling cutter and single-point cutting tool bodies. It consists of one NC lathe, one vertical machining center, three five-axes machining centers, and one multi-spindle NC grinder and operates under hierarchical computer control. The layout of the system is shown in Fig. 25. It can handle some 4000 different types of parts weighing up to about 200 kg, in lengths from 30 to 300 mm and diameters from 50 to 300 mm, and can produce parts to an accuracy of 0.025 mm. Its special feature is that the three five-axes machining centers can run unattended at night, from about 9:00 p.m. to 8:00 a.m., machining a variety of parts automatically throughout that period. The parameters monitored for obtaining this unattended operation are spindle motor current and touch-type length sensing of small drills. This shuts down a machine whenever a tool breaks or becomes excessively worn. Normally, however, tools are changed automatically after their cutting time reaches a given preprogrammed value, avoiding excessive wear.

The performance results being obtained with the system are striking. These were compared with the results obtained in another Tungaloy factory producing a similar range of cutting tools with a mix of unintegrated conventional and NC stand-alone machines: The required number of machine tools has been reduced from 50 to six (88% reduction); the personnel has been reduced from 70 to 16 persons (77% reduction); the product yield has been increased from 95% to 98%; the average percent utilization of the machines has been increased from 20% to 73%; the required floor space has been reduced from 1500 square meters to 350 square meters (76% reduction); the average number of operations per part has been reduced from 15 to eight (47% reduction); and the total average processing time from raw stock to finished product has been reduced from 18.6 days to 4.2 days (77% reduction). Thus, very large benefits of increased productivity, decreased costs, increased utilization of capital equipment, and reduction of work in process are being realized.

Another example from Japan is that of a system operating at Niigata Engineering Company's Internal Combustion Engine Plant in Niigata, Japan [176]. Niigata experimented over a period of years with running individual computer-controlled machine tools unattended at night, and achieved excellent capability. The success of this effort led them to develop what is probably the world's first full FMS capable of running unattended for a long period of time. It was put into service in 1981. The layout of the system is shown in Fig. 26. It machines 30 different types of cylinder heads in lot sizes ranging from six to 30 parts. The system runs 21 h a day, including unattended operation at night, and the unattended operation extends even to automatic mounting (and demounting) of many of the simpler types of workpieces on pallets by a robot. The results obtained are impressive: The number of machines required to produce these parts has been reduced from 31 (including six NC machines) to six (an 81% reduction); the number of operators has been reduced from 31 to four (an 87% reduction); the time during which parts are being produced has been increased from 9 h per day to 21 and the lead time for these parts has been reduced from 16 days to 4. This again illustrates the tre-

Layout of cylinder head machining system

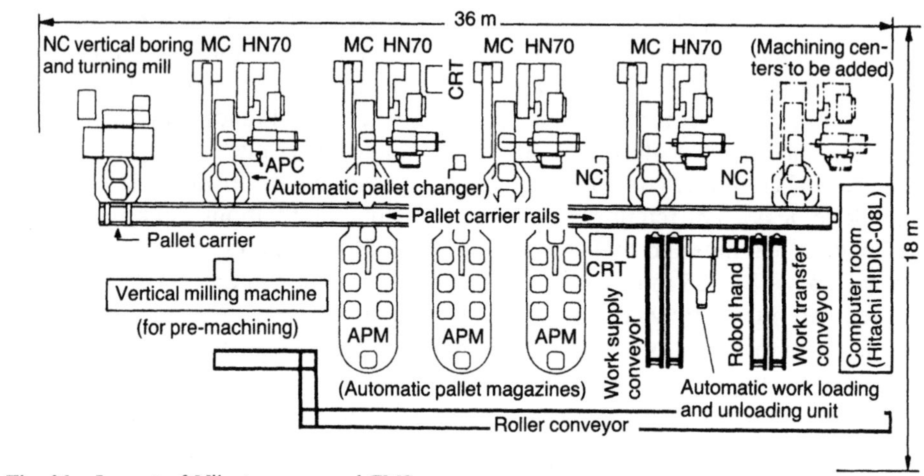

Fig. 26. Layout of Niigata unmanned FMS

mendous capability of computer-integrated manufacturing to increase the utilization of machines and thus of the capital investment which they represent, as well as to increase manufacturing productivity in general.

A major factor in Niigata's success with running individual machine tools and this system unattended at night is the *total* responsibility and authority of the "operator" (actually supervisor) of each for controlling the variables which determine its success or failure. He or she does all the NC programming for it: does all the NC tape try-outs on first-run parts; specifies the tooling, including required uniformity; carries out final inspection of all tools and rejects any that are found not acceptable; has final say on acceptance of the workpiece metallurgical properties; controls and supervises all maintenance on the system, etc. Thus, if a machine or the system shuts down automatically at night before it has finished its supply of parts, it is the operator who is held responsible. As a result, that almost never happens.

The same FMS technology evidenced in the operation of these unattended systems is now being applied by a number of other Japanese companies to accomplish even more startling improvements in manufacturing productivity. For example, as described by Usui [235], Fanuc Ltd. has recently put into operation an entire FMS factory for producing robots, small machining centers, and wire-cut electrodischarge (EDM) machines, the machining section of which operates almost wholly unattended on the night shift. There are 29 machining cells in this section (22 machining centers served by automatic pallet changers and seven robot-served cells), each consisting of one or more NC machine tools and a robot or pallet changer to keep the machine or machines loaded with parts during the night. Workpieces are transported on pallets to and from the cells and computer-controlled automatic stacker cranes by computer-controlled wire-guided carts.

During the day there are a total of 19 workers on the machining floor. On the night shift there is no one on the machining floor and only one man in the control room, with the machines producing parts unattended during the night. During the

24-h period machine availability runs close to 100% and machine utilization averages 65%-70%. As a result, this 20000 square meter factory is producing 100 robots, 75 machining centers, and 75 EDM wire-cut machines per month with a crew of only 100 workers! Quite a number of other Japanese companies are in various stages of implementing full FMS factories capable of computer-controlled, unattended night-shift operations.

Having that capability, many of these companies are now beginning to take advantage not only of the 24-h operation but also of the great versatility of these computer-controlled FMS to accomplish just-in-time (*kanban*) production. Here, they let the requirements of the assembly floor of the manufacturing facility "pull" the needed parts through the factory, rather than depending solely on scheduling to "push" them through. They accomplish this by loading these highly flexible systems in one 24-h period with just the mix of parts which will be needed on the assembly floor in the next 24-h period. Thus, work in progress is reduced virtually to zero, freeing up tremendous amounts of idle capital.

The remarkable performance benefits being obtained with FMS throughout the world, illustrated by the foregoing examples, emphasize a tremendously significant fact about the economics of such systems. Simply stated, this is that the acquisition of such equipment can be accomplished at essentially *no net cost* if the systems are properly utilized! First, the reduced capital investment in workstations (a much smaller number is required because of their greatly increased utilization compared with stand-alone, non-integrated workstations) more than pays for the required additional supporting, integrating facilities. These latter include not only such entities as the control computer and the work and tool transport equipment, but also the software system. Second, the drastic reduction of inventory of work in process and that in stock waiting to be assembled to virtually zero (due to the capability of these flexible systems to produce just whatever mix of parts is required for immediate assembly) frees up sufficient capital previously lying idle in inventory to more than pay for the total FMS. Another way of looking at this is that the just-in-time production made possible by truly flexible automation allows a plant to turn over its total inventory many more times per year than is possible with a conventional plant.

This fact can be well illustrated by the final example chosen, one from the United States. This is an FMS for Vought Aerospace, a subsidiary of the LTV Corporation in Dallas, Texas [240]. Significantly, the system is so conceived as to be able to machine "parts" in purely random order, handle both similar and dissimilar parts, and accommodate a variety of sizes and shapes. A single one-of-a-type part can be processed without disrupting the production flow. Because of this versatility, a complete set of parts for a given application can be made in sequence and delivered promptly, thus reducing materials inventory and storage requirements to a minimum. Further, the some 60 aircraft parts which are initial candidates for machining on the system would require some 200000 machining hours by conventional methods. On the automated system this will be reduced to 70000, due primarily to the increased machine utilization. As a result, although the cost of the system is some $10 million, Vought projects a saving of some $25 million through its use. The system, the layout of which is shown in Fig. 27, includes eight machining centers, each with automatic tool changing and a 90-tool magazine; an automated

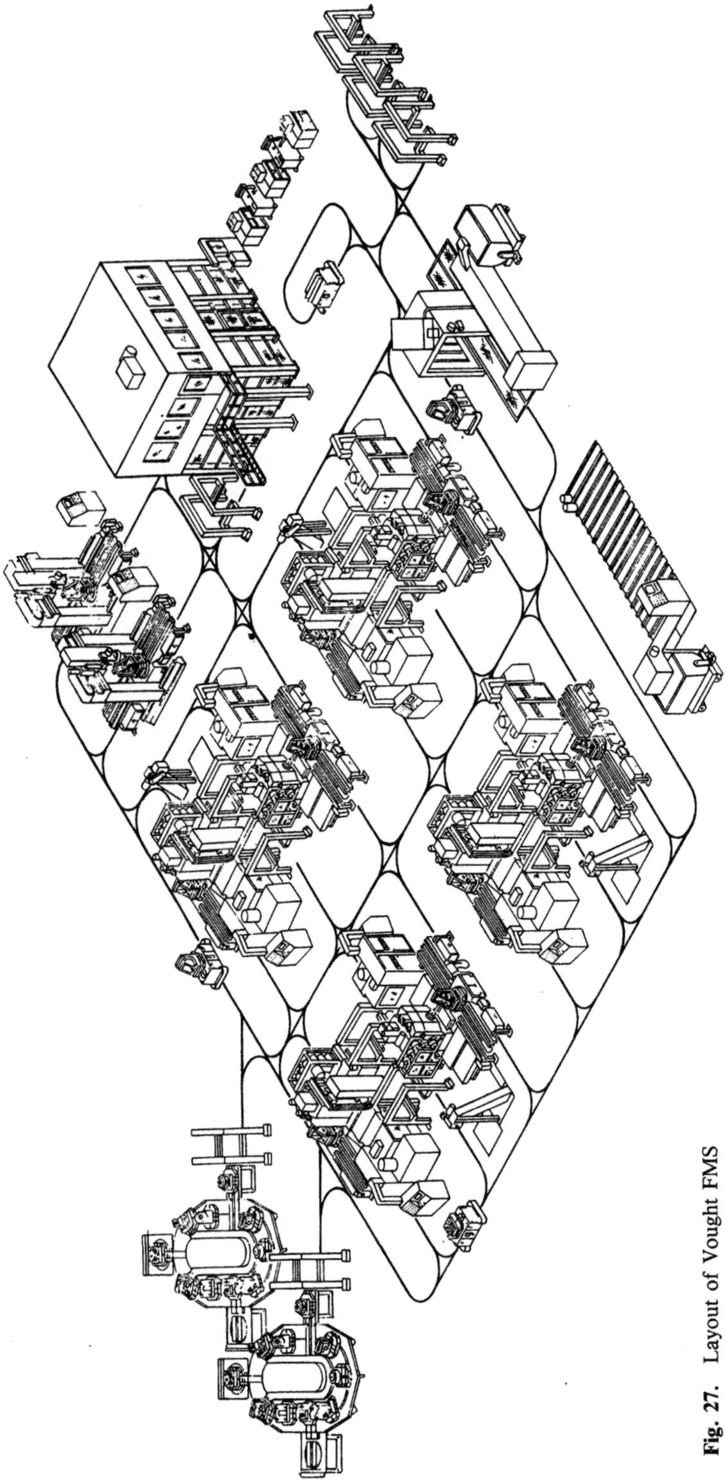

Fig. 27. Layout of Vought FMS

cleaning station; two automated inspection stations capable of checking parts to engineering design tolerances; two 10-pallet carousels for parts pickup and delivery; and a central chip-collecting system which separates ferrous and nonferrous materials for reclamation. Four small robot carriers, each capable of carrying 2300 kg, shuttle parts-laden pallets, tools, and other equipment throughout the system, following a network of wires embedded in the floor.
All elements of the system are directed by a master computer station capable of monitoring more than 200 machine functions, performing fault-analysis and simulating work processing cycles to permit accurate determination of parts-scheduling requirements.
The operation of the system begins with manual loading of work pieces on pallets on the carousels. From that point, no manual intervention is required, and automatic operation proceeds through a three-shift workday.

3.4.2.4 FMS in Japan[1]

The development of FMS in Japan has been very active, and they are widely used in various fields of manufacturing at present. The number of FMS in Japan since 1981 is estimated to be more than 70, classified into FMS for machining and FMS for assembly. About 35% of all systems for machining are for diesel engines and machine tools in Japan, and application covers a wide range, including food-manufacturing machines and sewing machines. About 38% of the systems for assembly are for mechanical units of home electric goods such as fans, video cassettes, and cassette decks.
Although there is no quantitative evaluation method for flexibility at present, at least, the following two capabilities have to be considered: capacity for manufacturing various parts and products within the same category such as box-type parts, and various parts and products of different types, such as box, plate, cylinder, and disk. These two capacities are termed here versatility and flexibility respectively. Most of the systems developed have versatility, but only 5% of all machining systems in the world have real flexibility.
The most frequently used material (about 50%) is cast iron, followed by steel. Chip disposal does not present a problem with the former, while it is a major concern with the latter.

3.4.2.4.1 Basic Constitution of FMS Hardware

In this section, FMS hardware is classified according to the manufacturing and the materials-flow functions, which are the most important of the five fundamental functions in FMS.
In general, the manufacturing system can be described by the directed graph, the node of the graph representing the manufacturing function and the branch, the

[1] This section is based mainly on Iwata [116]

materials-flow function. The FMS hardware is classified according to the type of graph. The classification includes (a) a simple series system, in which the parts can be transferred in one direction through a group of manufacturing functions arranged in a line; (b) a circular system, in which the manufacturing functions are arranged in a loop; (c) a free transportation system, in which the parts can be transferred freely within the whole group of the manufacturing functions, and so on. The classification is simplified by referring to the existing systems, summarized as

a) linear-flow type,
b) loop-flow type,
c) random-access type, and
d) combined robot-and-FMC type.

Most of the systems developed belong to the linear-flow type, while there are a relatively small number of random-access type systems. For most of the FMS, the manufacturing functions are accomplished by the combination of the machining centers and the general machine tools in machining, and by the combination of the assembly centers and the robots in assembly.

Typical FMS for machining and assembly in Japan are:

1. Linear-flow type for machining
   Fig. 28 shows a typical linear-flow type system, which was manufactured by the Toyoda Machine Tool Co. for the Maekawa Manufacturing Co. The system was developed in 1979 to manufacture box-type parts for refrigerators of medium and large size. There are three machining centers and a cart for transportation. Four pallets can be accommodated at the loading station, and eight pallets at the buffer station.
2. Random-access type for machining
   Fig. 29 shows a random-access type system developed by Okuma Machining Works Co. Seven machining centers are installed in the system. The work materials set on the pallets are transported directly to the machining centers by the

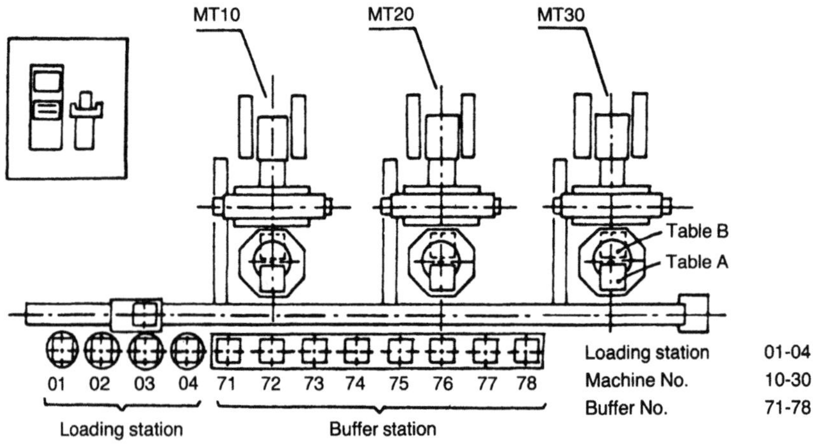

Fig. 28. Example of linear-flow type FMS (TIPROS)

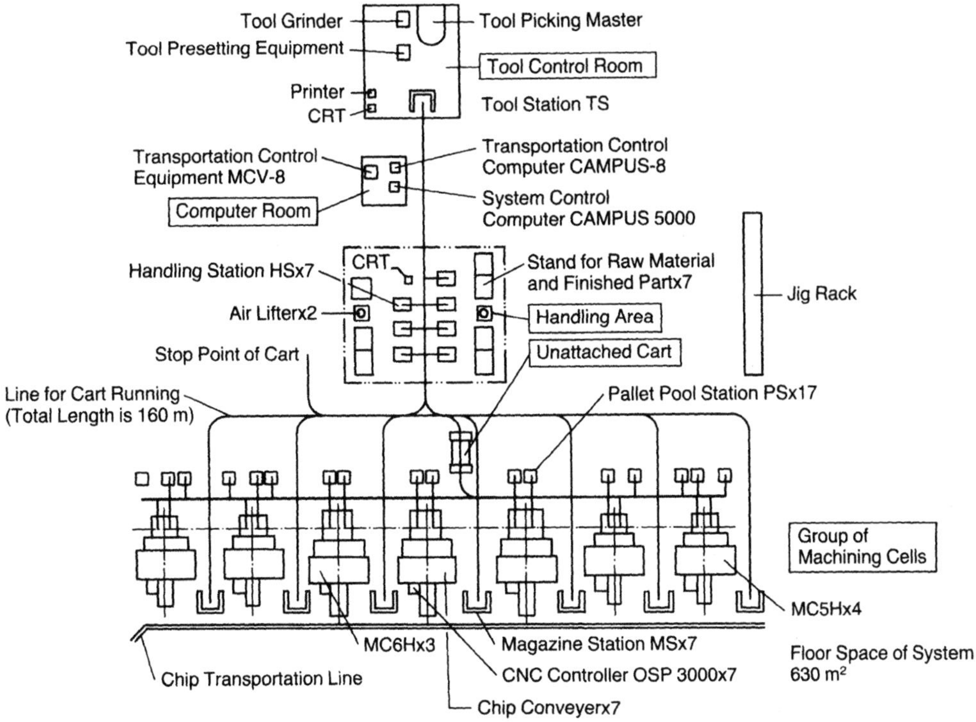

Fig. 29. Random-access type FMS (Okuma Machinery Works Co.)

unattached cart called "Robotrailer", and also to the pallet pool station for unattached machining during the night. The finished materials are automatically collected at the handling station by the unattached cart after the automatic machining. The preparation of the tools for each machine tool is executed according to indications by the tool control room via the terminal, which is synchronized with the schedule. The tools set on the pallet are transported to the magazine station beside the machining center. The entire system is controlled by the system management computer.

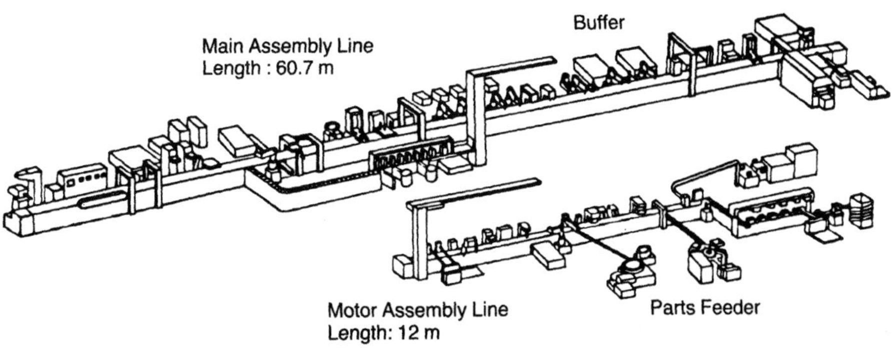

Fig. 30. Linear-flow type FMS (Toshiba Electric Co.)

3. Linear-flow type for assembly
   Most of the FMS for assembly in Japan are linear flow. An example is shown in Fig. 30. The system was developed in 1982 to assemble 24 kinds of electric fans by the Toshiba Electric Co. Seventeen assembly robots are installed, some of which are used for wiring and soldering of electric circuits. Sub-assemblies are transported by the conveyor on the pallet.
4. Random-access type for assembly
   Fig. 31 shows a random-access system which was developed by the Tohoku Oki Electric Co. in 1982 to assemble three kinds of dot-matrix printers. The assembly function is executed by 12 NC robots and 47 sequential robots, while the transportation function is realized by four carts. The system also includes an automatic warehouse function.

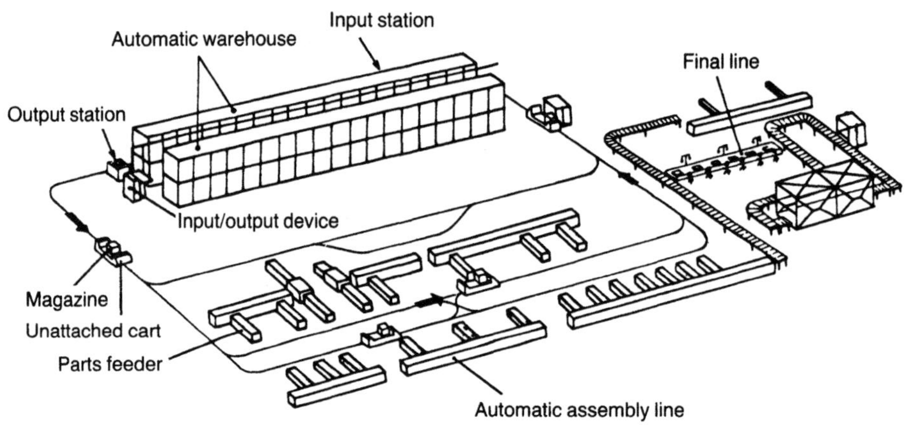

Fig. 31. Random-access type FMS (Tohoku Oki Electric Co.)

3.4.2.4.2 Economic Effects of FMS

The FMS has now been widely introduced to many industries as a manufacturing system which is suited to small- and middle-volume, large and medium variety products. The examination of practical economic effects obtained with the system has just begun. The main published results are summarized in Table 7.
The figures in the table indicate a drastic decrease in the number of machines and employees (70% - 80% on average) as well as a reduction in the overall process (by about 1/4). Although the investment is large for construction of facilities, there is a tendency toward reduction of machining costs.
Yamazaki Machinery Works is one of the world's leading machine tool producers. The sales growth in the past few years has been extremly fast.
Assuming 1978 = 100% the growth rates are 1982, 360% (realized), 1984, 450% (planned).

Table 7. Comparison of previous process with FMS

	Previous Process	FMS	Effect
[Kawasaki Plant of Toshiba Tungaloy] (FMS for Machining)			
Number of machine tools	50	6	Reduced by about 88%
Employees	70	16	Reduced by about 77%
Yield of product	95%	99%	Increased by about 4%
Rate of operation	20%	70%	Increased about 3.5 times
Floor space	1480 m^2	350 m^2	Reduced by about 76%
Number of processes	15	8	Reduced to about 1/2
Total process time	18.6 days	4.2 days	Reduced to about 1/4
Cost of system	¥ 130 ~ 140 million	about ¥ 500 million	Increased about 4 times
Manufacturing cost	—	—	Reduced to about 1/2 in 2—3 years
[Brother Industry Co.] (FMS for Machining)			
Number of machine tools	42	25	Reduced by about 40%
Employees	24	2	Reduced by about 92%
Floor space	600 m^2	650 m^2	Increased by about 8%
Total process time	—	2 h	—
Cost of system	—	about ¥ 300 million	—
Manufacturing cost	—	—	Reduced by about 30%
[Niigata Internal Combustion Plant of Niigata Iron Work Co.] (FMS for Machining)			
Number of machine tools	31	6	Reduced by about 81%
Employees	31	4 during the day	Reduced by about 87%
Operting hours	9 h/day	21 h/day	Increased 2.3 times
Number of processes	8	4	Reduced to 1/2
Total process time	about 16 days	about 4 days	Reduced to about 1/4
Cost of system	about ¥ 600 million	about ¥ 600 million	No change
Manufacturing cost	—	—	No change
[Yamazaki Machinery Works Ltd.] (FMS for Machining)			
Number of machine tools	68 Conventional machine tools	18 Machining centers	Reduced by about 74%
Employees	215	12	Reduced by about 94%
Total process time A-line	35 days	1.5 days	Reduced by about 96%
B-line	90 days	3 days	Reduced by about 97%
Floor space	6500 m^2	2790 m^2	Reduced by about 57%
[Tohoku Oki Electric Corp.] (FMS for assembly)			
Number of robots	—	59	
Employees	60	15	Reduced by about 75%
Tact time	240	40	Reduced by about 89%
Production rate		about 2 times	Increased 2 times
Production load time	160 h	2 h	Reduced by about 13%
Setting time	4 h	10 min	Reduced by about 96%

One explanation for this progress is the nearly total orientation toward FMS in parts manufacturing. Since 1981 the "Flexible Manufacturing Factory — FMF" has been in operation with two FMS. The economic results are included in Table 7. Since 1983 the "Factory of the 21st century" with eight FMS has been operating. The manufacturing und control information for the entire factory is provided by the central CAD/CAM center in the main building, over 20 km away, via fiber-optic lines.

3.4.2.4.3 Development Trends of FMS in Japan [115]

There are five directions in the development of FMS:
1. Toward large FMS
2. Toward low-cost FMaC (flexible machining cells — see Sect. 3.3.4)
3. Application of FMaC to the production of military supplies
4. Diversion of the concept of FMS to other metal processing procedures
5. Diversion of the concept of flexible manufacturing to other industries, e.g., baking

At present the development of large-size FMS and low-cost FMaC is being actively pursued in Japan. Knowledge about the kernel functions of the FMS is necessary to design and manufacture FMS suited to the production requirements. At the very least, these are machining, transportation and integrated software control. A lot of research has been and is still being done on these kernel functions. Research in the following areas is unecessary in order to forecast the future evolution of FMS:
1. Design and evaluation methods
2. Design of simulation software, including for system evaluation
3. Development of in-process recognition devices, such as the in-process sensor of cutting conditions, which is already available

In future it will be necessary to design system functions to be compatible with FMS, rather than the other way around.

3.4.2.5 The Development of CAM-Type Automated Manufacturing Systems in the Czechoslovak Engineering Industry

During the 1970s, there was considerable development of flexible automation of manufacturing processes based on the use of NC technology in the Czechoslovak engineering industry. In the framework of the State Technology Policy Program, research and development of NC machine tools, their components, and control systems was undertaken. Considerable efforts were also made in research on process engineering for NC machine tools (including tooling), in the preparation of control programs (AUTOPROG, ČKD-APT, KOVOPROG systems), and in the training of specialists in NC technology.
These efforts were followed by research aimed at the basic problems of manufacturing systems and their integrated solution, i.e., the CAM basics. Three basic stages in the development and application of CAM systems can be distinguished in the process of implementation of these automated manufacturing systems.
During the first stage (1974-1977), integrated manufacturing sections under the designation "IVU" were designed, developed, tested in operation, and built for the machining of nonrotational parts (four IVU sizes), for the machining of rotational parts (one IVU size), for the forming of sheet-metal parts (four IVU sizes), and for the manufacture of electronic components (one IVU size).

Integrated manufacturing sections are systems characterized by the level of automation appropriate for small-batch, irregularly repeated manufacture of parts distinguished by the similarity of technological processes. Appropriate IVU sizes were determined on the basis of an extensive analysis of machine parts manufactured in most engineering plants. Thanks to this approach to the problem, repeated applications in other engineering plants characterized by a suitable part mix were possible, once the first types of integrated manufacturing sections had been tested.

The basic feature of IVU consists in the concentrated introduction of NC manufacturing technology interlinked by means of an automated conveying and handling system and by automated control of the production process by a computer. As a rule, the system also includes an automated store of semi-products, work-in-progress, and finished parts, inspection and measuring devices and, if required, complementary equipment. Automated tool handling is a feature typical for only a small number of IVU systems. All these systems are characterized by the inevitable presence of operators attending to groups of working sites.

These manufacturing systems substantially increase productivity and output per unit of shop-floor area, reduce the flow process time, and allow for speedy changes to the manufacture of other parts. They set man free from physical work and reduce the extent of routine mental activities. Man's immediate participation in the process of production is shifted to the sphere of process preparation and checking, and to the prevention of failures. Characteristics of two of these IVU (IVU 320 and IVU 800) are presented later in this section.

The second stage of the development of CAM-type automated manufacturing systems is represented by "flexible machining systems". These are flexible manufacturing systems of a higher level of automation which display the following basic features:

1. Technological working sites are represented by computer-controlled automated machine tools characterized by flexible setting-up and automated work and tool handling during operations; they are capable of taking over work and production aids from the handling and conveying system and of removing process waste without man's interference.
2. Intersystem handling is automatic and computer controlled.
3. Direct computer control of the process of production in real time is applied within the manufacturing system; it is based on short-term scheduling, and reacts to the current state of the production process.
4. Man's activity is restricted to the setting-up of technological working sites, production process preparation, supervision, and elimination of failures.

Within the framework of this program, the building of manufacturing systems for the following workpieces will be completed in the period 1983-1985:

1. Box-type parts up to 400x400x400 mm
2. Box-type parts up to 800x800x800 mm
3. Bar-stock rotational parts up to 80 mm in diameter
4. Flange-type parts up to 250 mm in diameter
5. Shafts up to 120 mm in diameter
6. Large rotational parts up to 4000 mm in diameter

According to an investigation performed by our institute in engineering and electrical plants in Czechoslovakia, 34 CAM systems of the two types mentioned are presently in operation with a total of 322 NC machine tools; another 40 manufacturing systems are in the design stage or in partial implementation. Of this total, 34 % are designed for the manufacture of rotational parts, 28 % for the manufacture of nonrotational parts, and the rest for the manufacture of both rotational and nonrotational parts or for other processes (welding, forming, surface treatment).

The investigation confirmed that the overwhelming majority of automated manufacturing systems operates in small-batch, repeated engineering production. In the case of the systems for the manufacture of rotational parts, average lot sizes range from 20 to 200 parts, while in the case of the systems for the manufacture of nonrotational parts, they range from 10 to 120 (maximum 180) parts.

The third stage is represented by the creation of plant-level CAM systems. The most extensive FMS at the plant level is at the stage of advanced implementation in the TOS Hostivař model plant. The entire main-line production is designed as a system of mutually interlinked integrated manufacturing sections. The description of one such IVU (IVU 500) appears later in this section.

In the main Hostivař plant the system includes 13 technological, auxiliary, and storage units — from starting stock stores, parts machining and surface treatment, and in-process stores, up to sub-assemblies and final assembly. Two main principles are applied here:

1. Based on decisions by the control center, workpieces and production aids are supplied to the input point of every working site; working sites do not send requests for jobs, nor do they provide for the transportation to their input point and from their output point.
2. In-process inventories of parts are kept in the in-process store and the ready-to-use tooling sets in the tooling store; only the plant's control center can dispose of them, and the in-plant handling system is the executing agent.

In the four basic production units, which are mainly for NC machine tools, the handling of workpieces and production aids is automated by means of rack-loading machines and automatic control. The handling between units is performed by induction-controlled carts. They link input and output points of the basic production units and are computer controlled from the plant's control center. Decision-making concerning the tasks to be assigned to the handling between production units is automatic.

Direct control of the process of production at the level of a production unit consists of:

1. Simulation of the material flow, of handling batches of workpieces, and of the sets of production aids in the computer's memory, by the flow of their information blocks with the simultaneous recording of the state of transport batches. Information is scanned automatically, in real time, so that the control computer knows at any given moment what is being stored or transported and where, and the status of any batch or set.
2. Decisions on the tasks to be assigned to technological working sites are made in real time, i.e., according to the real state of manufacture at the moment of

decision-making. This is complemented by prediction of the developments for the near future, which serves as a basis for direct control of the preparation of production aids. Based on the decision as to which task is to be assigned to a technological working site, instructions for the handling and conveying system are formulated. Valuable knowledge for the further development of flexible automation was derived from the design, testing, and current operation of FMS. The following are the most substantive conclusions:

1. Economical operation of such manufacturing systems calls both for the selection of suitable parts (of high complexity) and for the analysis of their adaptability to manufacture with NC technology. This applies in particular to the use of standard design elements, to the reduction of the number of the types of holes, threads, etc. At the same time, it should be guaranteed that the part will be repeatedly manufactured in lot sizes required for the smooth flow of assembly.
2. Within the system, only those methods should be used for the manufacture of parts which have been found optimal in advance, and which make full use of the tool's parameters.
3. At least one pair of fungible technological working sites should be included in the manufacturing system. This is necessary not only to compensate for possible failure of one of the machine tools, but also from the point of view of direct control of the process of production ("the lineup of tasks").
4. To increase economic efficiency, it is advisable to replace the inspection of the machined surface performed *after* the completion of the operation concerned by active inspection performed *in the course of* the technological process; in a number of cases, it is advantageous to make use of the clamping of the workpiece on the work-holding pallet, which carries the workpiece through the prescribed technological working sites.
5. The rack store for work-in-progress is of great importance for multi-shift operation; it should facilitate operation during the afternoon and night shifts (and, if required, on Saturdays and Sundays) in the absence of the operator at the technological working site.
6. The number of storage cells is inversely proportional to operation times; in the case of relatively long operational times, the handling system can fulfill the compensating function of the rack store. The handling system is controlled by the NC system with direct linkage to control computers.
7. High operational reliability of individual elements of such manufacturing systems is an absolute necessity. Failures should be prevented by a system of preventive inspections; in case of failure, trained personnel equipped with the necessary technical means and knowledge of diagnostic methods should be available for quick correction.
8. Manufacturing systems consisting of NC machine tools are highly capital intensive; maximum utilization is necessary to achieve desirable efficiency. They should work 24 h a day, a minimum of 5 days a week; the flow of parts through such systems should be optimized according to the criterion of maximal utilization of the most capital-intensive working sites.

3.4.2.5.1 Examples of Operational CAM-type FMS

1. IVU 320 integrated manufacturing section for the machining of rotational parts up to 320 mm in diameter

With the government's financial support, an automated flexible CAM system for the machining of rotational parts up to 320 mm in diameter (IVU 320) was developed, tested in operation, and made operational in the Kovosvit Sezimovo Ústí plant. It is among the most extensive FMS in the Czechoslovak engineering industry, both in terms of the number of NC machine tools installed [25] and the range of manufactured parts (more than 5000).
The system's manufacturing program includes rotational parts of both the flange type (50-320 mm in diameter) and the shaft type (up to 120 mm in diameter and 500 mm long). The operations performed are: turning, drilling, boring, milling, and squaring up and hole centering of shaft-type parts.
The parts manufactured are made mostly of steel and cast iron, less frequently of light or nonferrous metals. They are manufactured in lots of 10-100 pieces (average lot size is 38 pieces) which are repeated several times a year. The parts are designed mostly for the manufacture of machine tools.
The "IVU 320" FMS system (Fig. 32) is built in a hall 78x18 m. It consists of 25 NC machine tools, 22 of which are designed for turning work; others are an NC drilling machine, an NC milling machine, and an NC machining center. Eighteen NC machine tools in this section are DNC controlled.
The machine tools are arranged on both sides along the transport system so that an operator handles two machine tools (Fig. 33). Each of these working sites (i.e., pairs of NC machine tools) has four handling points to and from which workpieces and production aids are supplied and removed by the transport system.
The control center is located in an elevated cabin at the head of the manufacturing system; the control minicomputer is equipped with working storage of 32 K words. Apart from this, the control center includes other peripheral devices for the control of technological and auxiliary working sites and the transport system. A signaling panel, which provides for emergency control of the section as a whole in case of computer failure, is also located here. The following functions are performed by the computer:

1. Simulation of the flow of material through the manufacturing system
2. Job assignment to technological working sites in accordance with the production plan which has been prepared in advance
3. Control of the transport system in conjunction with storage
4. Supply of information to the technician in charge on nonstandard situations, and participation in their solution
5. Supply of information on the status of the orders in progress, on storage, etc., upon the request of the technician in charge
6. Monitoring of the load of individual machine tools
7. Under the DNC mode of operation, full data transfer

The integrated manufacturing system has made possible the reduction of labor intensity in standard hours by 43.7%, the increase of productivity per worker by

Fig. 32. IVU 320 flexible automated integrated manufacturing system in the Kovosvit plant, Sezimovo Ústí

Fig. 33. Multi-machine attendance in the IVU 320 system

92%, the reduction of the flow process time by 60%, a savings of 17% of the shop floor area and of 50 conventional machine tools, and a savings of Kčs 2.6 million annually.

2. IVU 800 integrated manufacturing section for the machining of box-type parts up to 800 mm

With the financial support of the government, two CAM-type integrated manufacturing sections for the machining of nonrotational box-type parts have been built in the TOS Kuřim plant, the largest manufacturer of special-purpose machine tools, unit-type machine tools, and transfer lines in Czechoslovakia. "IVU 800" is designed for parts of 800 mm maximum dimension (cube edge of the machined part) while "IVU 1250" is designed for parts of 1250 mm maximum dimension. The two automated manufacturing sections form a complex FMS at the shop level which manufactures basic parts for the machine tools produced by the plant.
Within the system, parts made of gray cast iron, steel, or light alloys of maximum dimensions 800x800x800 mm are machined in lots of 5-20 pieces; the lots are repeated several times a year. These are parts requiring high labor intensity and high machining accuracy. Parts are clamped on work-holding pallets of two sizes: 800x800 mm and 800x360 mm.
The IVU 800 system (Fig. 34) consists of four machining centers, two of which are general purpose; the other two are special-purpose machining centers for milling and finishing of boring operations. Each of the machining centers features four tool magazines of 20 tools each. The machine tools are arranged along one side of the track for the handling and conveying cart.
The system for handling work-holding pallets consists of carts and rotary manipulators near the machine tools. For the control of the manufacturing section, an RPP 16 S Czechoslovak control minicomputer and two RPP 16 M-type minicomputers are used.
Castings (or other semi-products) are clamped on work-holding pallets at a specialized working site (Fig. 35). A coordinated measuring machine is used for accurate and quick workpiece location. Certain pieces of equipment which are required for the operation of both manufacturing systems are shared by IVU 800 and IVU 1250. These are, for instance, fixed-bed milling machines for the machining of base surfaces, cleaning machines, the measuring machine, and the clamping working site.
The system manufactures 70 different parts annually, reduces labor intensity by 50%-55% and the flow process time by 40%, and leads to annual savings of Kčs 1.4 million.

3. IVU 500 integrated manufacturing section for the machining of rotational parts up to 500 mm in diameter

Within the framework of the State Technology Policy Program an extensive innovation in the production base of the TOS Hostivař plant is being developed and

Fig. 34. IVU 800 flexible automated manufacturing system in the TOS Kuřim plant

tested in operation. The firm is the largest manufacturer of grinding machines in Czechoslovakia, with a long manufacturing tradition.

The "IVU 500" FMS is designed for the machining of shaft-type rotational parts (maximum length 800 mm, weight up to 40 kg) and flange-type (160-500 mm, weight up to 60 kg). Parts are made of steel, cast steel, gray cast iron, or aluminum alloys. Altogether, 300 different parts are manufactured here with a combined total of some 65000 pieces annually. Annual output of different parts ranges from 20 to 900 pieces, and part lots are repeated 2-17 times a year.

This automated manufacturing system (Fig. 36) consists of 16 technological working sites, seven of which are for NC machine tools. Technological working sites are arranged on both sides along the multi-story, double-row, in-process store, which is serviced by an automated loading machine. The loading machine trans-

Fig. 35. Detailed view of the work-holding pallet in IVU 800

ports pallets of 600x800 mm. Identical pallets are used for the transportation of production aids which are brought from the room designed for their preparation by means of a two-way, extra-level roller conveyor. Work-holding pallets and pallets carrying production aids are brought from the lower rack cells to individual working sites by means of roller handling devices.

Chip removal is ensured by handling devices from technological working sites to an under-floor channel which is connected with the central chip system.

The control of the production process in the IVU 500 system is automated and centralized (Fig. 37). The control computer, with extensive on-line peripheral devices, handles the overwhelming majority of control activities. The technician in charge of control performs mainly supervisory functions and intervenes in nonstandard situations. Apart from devices for communicating with the computer, the control center technician has a signaling model at his or her disposal.

Fig. 36. IVU 500 flexible automated integrated system in the TOS Hostivař plant, Prague

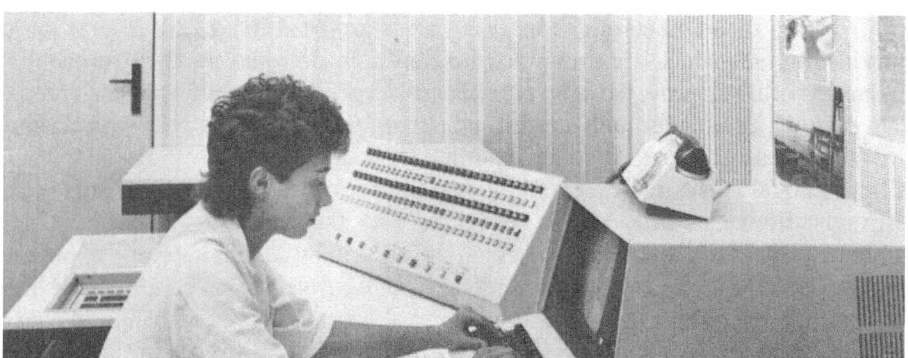

Fig. 37. Control center of the IVU 500 system

Based on the decision of the control center, material is brought into the system in production lot sizes, divided into transport lot sizes. Sets of production aids, including the NC programs and documentation required for the performance of the operation, are prepared in a special room.
With respect to ergonomics and working conditions, the arrangement of IVU 500 can be considered substantial progress compared with the conventional method of production and operation; practical experience has shown its optimal characteristics in terms of working environment, as well as hygienic standards and safety of operation.
The introduction of this system resulted in 170% increase in productivity, and a 32% reduction of labor intensity. With the shop floor area unchanged, the output has doubled.

3.4.2.6 FMS in the German Democratic Republic[1]

The German Democratic Republic ranks among the industrial countries which first designed FMS and employed them in production in the late 1960s and early 1970s. These include the selected machine systems which are represented in Table 8. Several publications have appeared concerning experience with the Prisma II system, which has been successfully used since 1972 for prismatic workpieces of 800 mmx1600 mmx1600 mm [74].
As an example of one of the largest machine systems with respect to the integrated machining stations for rotationally symmetrical parts the machine system FZ 200 is described, in Sect. 3.4.2.6.2.

3.4.2.6.1 The FMS 1000

Since 1985 the FMS 1000 has been operating in the parent plant of the Fritz Heckert Combine, i.e., in the same plant which has been using the Prisma II system successfully since 1972. The FMS 1000 is used for the machining of the housing-shaped workpieces, measuring 500-1000 mm, of a machine tool producer.
The production-technological aim of the system design consisted in a high degree of utilization of FMS by the largely automatic machining within the system, from the feeding-in of the blank to the ready-to-assemble supply. In addition, the special condition was given that the plant where it is used was already automated to a high degree in machining, in process control, and in intra-enterprise transport. The task was to integrate the new manufacturing system with the one in operation and to adapt it to the existing solutions of automation without any cuts being made at the technical and economic levels.
An automatic coloring system had to be integrated to secure the production-technological basic aim of complete machining with almost 100% self-contained tech-

---

[1] The editor is grateful to the management of the Research Center of Machine Tool Building and to enterprises of the WMW Combine.

Table 8. FMS in Eastern Europe

Line		Current number	1	2
1		User		
2		Producer	VEB-WZMK 7. Oktober	VEB-WZMK 7. Oktober
3		Name of system	ROTA F-125-NC	ROTA W-Q-63/1
4		Year put into operation	1971	1971
5		Products	Machine tools	Cars
6	Part assortment	Kind of part	Chucked parts	Starters
7		Geometry	Ro1	Ro2
8		Type variety	9	
9		Lot sizes	10—50	
10		Max. size (mm)	125	200
11		Max. weight (N)		
12	Manufacturing system	Manufacturing	suppl., subst.	suppl.
13		Lathes	4	
14		Drilling machines		1
15		Milling machines	2	
16		Drilling/milling machines		
17		Grinding machines	1	2
18		Others		3
19		Total no. of machines	7	6
20		Machine tool control	NC	A, NC
21		Tool number		
22		Quality control	Manual/automated	
23	Material flow system	Conveyor	SE	UF
24		Part carrier	Magazine	Magazine
25		Number of part carriers		
26		Transport loading	un	un
27		Transport control	PR	RE
28		Storage positioning	ze, de	de
29		Handling	SE	
30	Info.-system	DNC	x	(x)
31		On-line registration of manufacturing data	x	
32		Optimization of dispatch		

Abbreviations: x = in application, (x) = in preparation, SE = special equipment, UF = under-floor chain conveyor, PR = process computer, RE = relay control, ge = tight, de = decentralized, ze = centralized, un = loose, K = crane, Roll = roller conveyor

3	4	5	6	7
VEB-WZMK 7. Oktober, Berlin	VEB-WZMK F. Heckert Karl-Marx-Stadt	VEB-WZMK Auerbach	VEB KuT Erfurt	Stolowa Huta, People's Republic of Poland
VEB-WZMK 7. Oktober	VEB-WZMK F. Heckert	VEB-WZMK F. Heckert	Skoda x Pilsen, CSSR	VEB WZMK 7. Oktober Berlin
ROTA FZ-200	Prisma II	Auerbach M 250/01 DNC		ROTA FZ-200
1972	1972	1972	1978	1982
Machine tools	Machine tools	Machine tools	Machine tools	Agricultural machines
Gearbox, wheels	Housings	Housings	Frames, casings	Gears
Ro1	p1	p1	Prismatic	Ro1
2000	15			70
	1			
200	1600	315	4000	200
	16000		2100	
suppl., subst.	suppl., subst.	suppl.		
3				11
		1		22
	5	1	3	
	2			16
5	2			14
8	9	2		63
NC	NC	NC	NC	NC
	450	50		
Manual	Automated	Manual	Manual/automated	Manual/automated
SE	SE	SE	SW	SE
Magazine, pallet	Pallet	Pallet	Pallet	Pallet, magazine
25	16			25
un	ge	ge		
PR	PR	M	PRO	PR
ze, de	ze	ze		
SE	SE	SE	SE	SE
x	x	x	x	x
x	x	x	x	x
A	A		A	A

nology. Such a flow of parts was aimed at that coloring was to be carried out prior to machining. The layout of the machining range of the system, inclusive of the store, is represented in Fig. 38. The feeding-in station and the automatic coloring system are not shown.
The workpieces are transferred into the system in the store of the enterprise. The transport between the store, the system-internal high-shelf unit, and the coloring system is carried out by a guide-line-controlled transport robot. The principle is illustrated by Fig 39.
After the automatic storage of workpieces in the system-internal high-shelf unit they are made available for machining or for coloring in the clamping range of the system. Workpieces are clamped manually, the clamping fixture being actuated in a largely mechanized way by means of hydraulic orienting and clamping devices. In the machining center the workpiece clamping is checked before each operation.

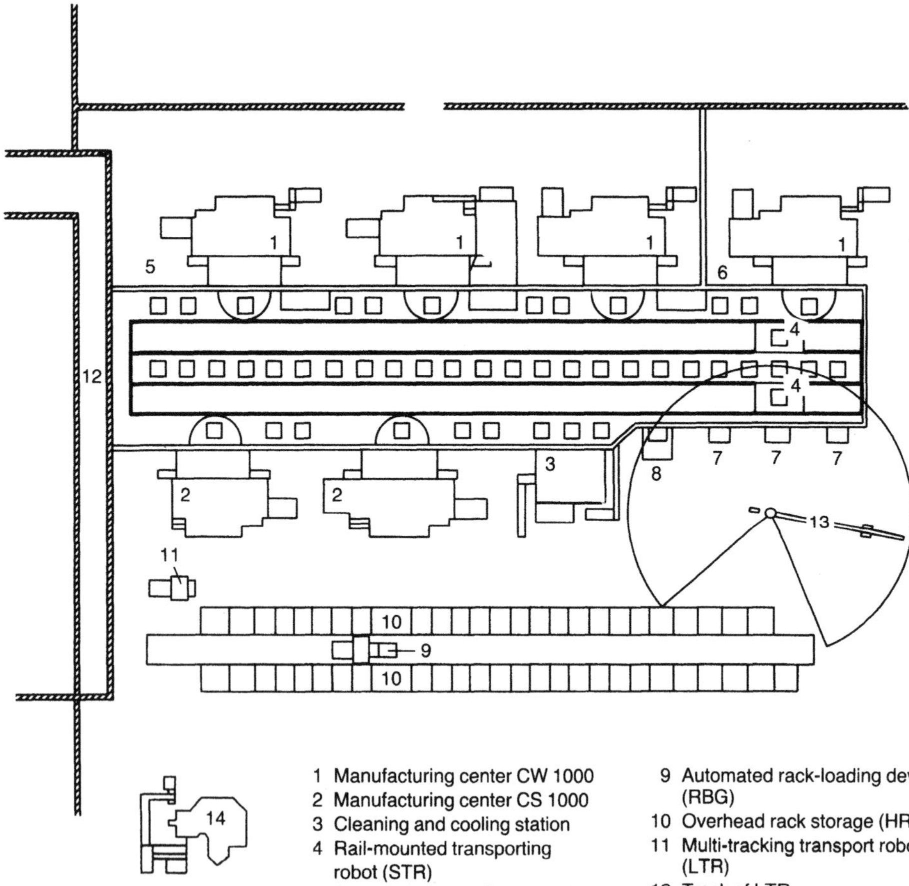

1 Manufacturing center CW 1000
2 Manufacturing center CS 1000
3 Cleaning and cooling station
4 Rail-mounted transporting robot (STR)
5 Storage locations for part pallets
6 Storage locations for tool pallets
7 Loading station
8 Service station
9 Automated rack-loading device (RBG)
10 Overhead rack storage (HRL)
11 Multi-tracking transport robot (LTR)
12 Track of LTR
13 Post-mounted swing jib crane (SDK)
14 Vertical boring and turning mill

Fig. 38. Flexible manufacturing system 1000

By redesign and standardization the originally more than 200 different workpieces were reduced to a total of 28 parts.
After two or three passages through the system the ready-to-assemble parts are transported by the guide-line-controlled transport robot to the store and there made available for assembly.
The subsystem for machining consists of two horizontal machining centers, each with 60 tool storage locations for rough machining, two horizontal machining centers, each with 160 tool storage locations for finish machining, and two vertical machining centers. Because of the extreme workpiece accuracy required, machining had to be done in two setups; centers for both horizontal and vertical machining were necessary, as some workpieces could not be reset.
The following system components are integrated:

1 washing and cooling station
2 rail-mounted transport robots
36 storage locations
3 clamping stations
1 servicing station

The rail-mounted transport robots execute automatic transport of the workpiece carriers between the machining centers, the washing and cooling station, the storage locations, the servicing station, and the clamping stations. To provide good accessibility to the workpieces during setting-up and removing, the clamping stations have to be designed in such a way that they can be lowered and swivelled, and they must be equipped with a special safety device.
Tool transport from the tool presetting to the individual machining centers is carried out by a guide-line-controlled transport robot. It makes the work pallet available close to the tool store of the machining centers. The exchange between the tool pallet and the tool store of the machining centers is executed manually. Direct automatic tool change into the spindle of the machining centers in possible for very heavy and oversized tools; the tools are accessible through special stores that are arranged on workpiece pallets and are made available for the machining centers, as is done with workpieces.

The control of the FMS 1000 is subdivided into three levels:

1. Intra-enterprise planning and preparation
2. Production control
3. Control of devices

On the first level the manufacture is planned, along with determination of demands, determination of production tasks, filling out of work documents, and balancing and accounting of production. This level of control is performed by the superior plant computer.

On the second level a microcomputer system is used as the process-control computer, which has to perform among others, the following tasks:

1. Workplace occupation
2. Control of the sequence of manufacturing operations
3. Control of the sequence of transport operations

4. Store disposition
5. Disposition of fixtures, tools, and measuring devices
6. Supply with DNC-programs (2 K 1630) — process computer
   — control computer

On the third level mainly CNC and PC controls are used to control the machining centers, the washing and cooling station, the transport system, the systems of supply and removal, and the clamping stations. The controls of the system stations and the transport control are coupled with each other in the DNC complex. The whole system is linked with a large computer ES 1055.
Due to the extreme flexibility of the total system it is possible to manufacture the entire workpiece assortment without being forced to produce series. Thus, there is always an adequate supply of workpieces for assembly. In addition, the time required for a workpiece, from the feeding-in of a blank to the delivery of a ready-to-assemble part, is now only about 4 days, which means a considerable reduction in working capital.
In Fig. 38 the system structure characteristic of FMS 1000 is schematically represented. Due to the large area covered it was necessary to build underground passages for the attendants and operators. Industrial television is used for better visual observation.
In order to improve utilization of space, storage columns are arranged between the individual stations. In addition to workpiece pallets, tools pallets are stored here at selected locations; the operator can then pull them out to reset.

The machine-technical equipment consists of:
11 NC lathes with industrial robots
16 grinding machines (hole and face grinding), ten of them with industrial robots
19 hobbing machines with industrial robots
3 gear-tooth chamfering and deburring machines with industrial robots
5 gear-shaving machines with industrial robots
2 vertical broaching machines with industrial robots
2 electrochemical deburring machines with industrial robots
3 automatic metal-cleaning machines
2 test stations
All of these stations are integrated with the automatic flow of workpieces; the same applies to the test cabin used for the final checking.

3.4.2.6.2 Machine System FZ 200

This system consists of 63 machining stations (see Fig. 39) which are arranged on both sides of the transport paths of the rack-operating devices and the storage racks. The two rows of machines are connected by a cross store with 220 storage locations. Fifty-two machining stations for the process stages of turning, hole and face grinding, hobbing, deburring, electrochemical deburring, shaving, and broaching are equipped for tool changes and with pallet timing devices, so that the automatic operation of workpiece pallets with 24 or 48 parts per pallet is ensured.

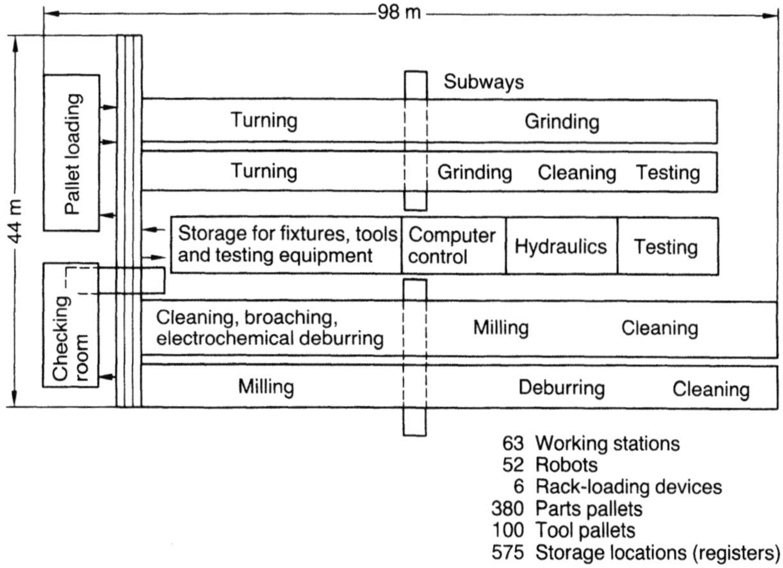

Fig. 39. Structure of the machining system FZ 200 of the Stolowa Huta enterprise, Poland

The system includes three additional automatically loaded washing machines as well as two testing stations, two surface-grinding machines, and four hole-grinding machines with manual loading. The project treatment was based on a parts assortment of about 270000 spur gears in about 70 positions, partly with internal profile and internal toothing.

To secure the required manufacturing accuracy with a minimum of operators, more rigorous demands are made than for conventional manufacture. As early as during turning, the basis for the final quality of the spur gears must be created. The following measures serve this purpose:

— Securing a defined spindle position in order to avoid the clamping of workpieces with rough surfaces at only two points in the three-jaw chuck
— Ensuring that the chuck is free of chips by intensive blowing cut
— Ensuring the facing setup of workpieces in the chuck using pressing-disks with an adjustable force
— Minimizing the deviations from truth of rotation and plan movement (25 μm) for high clamping forces (60-70 kN) by internal turning of the chuck jaw pads
— Matching of the cutting values to the nominal tool lives and to the different strength values of workpieces with the aim of machining parts of two to three pallets without a tool change

To ensure the required quality in the system the workpieces are checked on the machine. Calls for testing are given by the computer to the machine. After turning, the parts are checked again at a testing station. A final checking of gears is performed after unloading in a test cabin. The toothing of the spur gear assortment is manufactured at the customer's location. This places increased demands on the

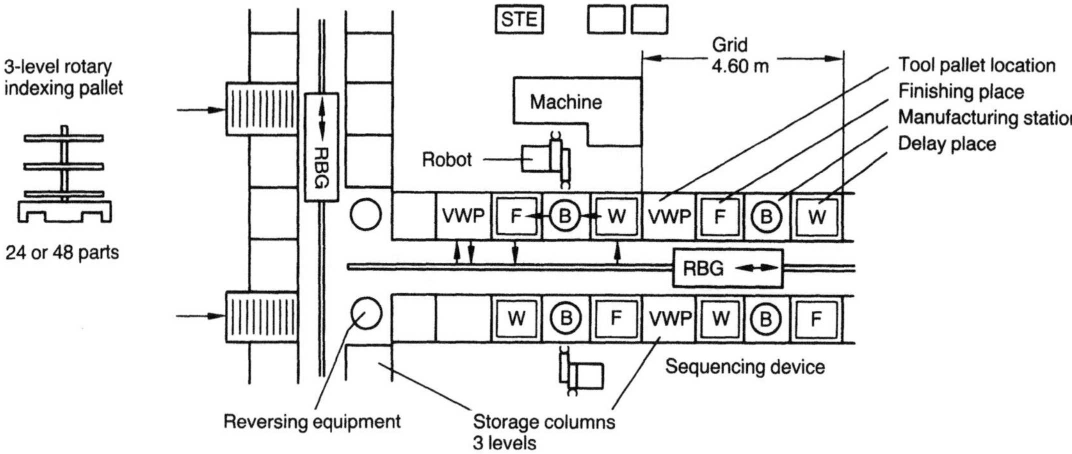

Fig. 40. Partial structure of the machining system at Stolowa Huta, Poland

techniques because of tenacious materials, relatively highly developed modules, and special reqiurements for profile modification.

For the tooth profile shaving with more accurate teeth, a profile deviation of 5 μm, and a profile modification with a strictly specified shape the positional and shape deviations of bore and faces in IT 7 (ISO Tolerance-class 7) must not be exceeded.

Machining control is of great importance for the efficiency of the system. It must guarantee that the machines are utilized to capacity and that disturbances in the processes of machining and transfer are detected and signaled within the framework of process monitoring and diagnosis. The process of manufacturing in the system is controlled by two coupled process computers of the KRS 4201 type with a drum store and with an extensive process periphery for the recording and output of binary signals (see Fig. 40). This is based on an extensive program system (see Figs. 41, 42), which includes 78 service programs for the system operation, 35 control programs, and 92 information and error printouts. These user programs occupy a storage capacity of less than 180 k bytes.

The computer for transport supervises the occupation of machines and stores and controls the rack-serving devices according to data from the machine occupation computer. It occupies the machines according to the stock of work and prepares orders of transfer by the computer. Available to it are the actual system data on the one hand, and the production-technological desired data in the form of sequences of operations for the machine groups on the other. Besides restricting required production-technological data, the changeover time between two lots or orders and the time of flow are criteria of machine occupation. The computation of the setting-up time proportions is based on the respective part- and machine-related classifiers of the setting-up time. In this way, the part which is favorable from the user's viewpoint will always be worked on the machine in question.

The tools are prepared after calls by the computer and are transferred to the machines by the rack-serving devices in the same way as are the workpiece pallets.

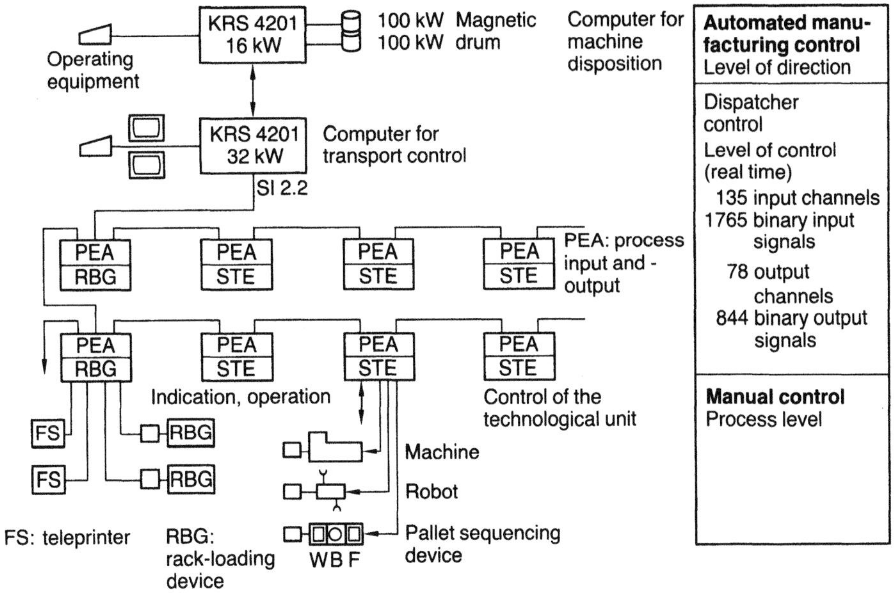

Fig. 41. Data flow system of the machining system at Stolowa Huta, Poland

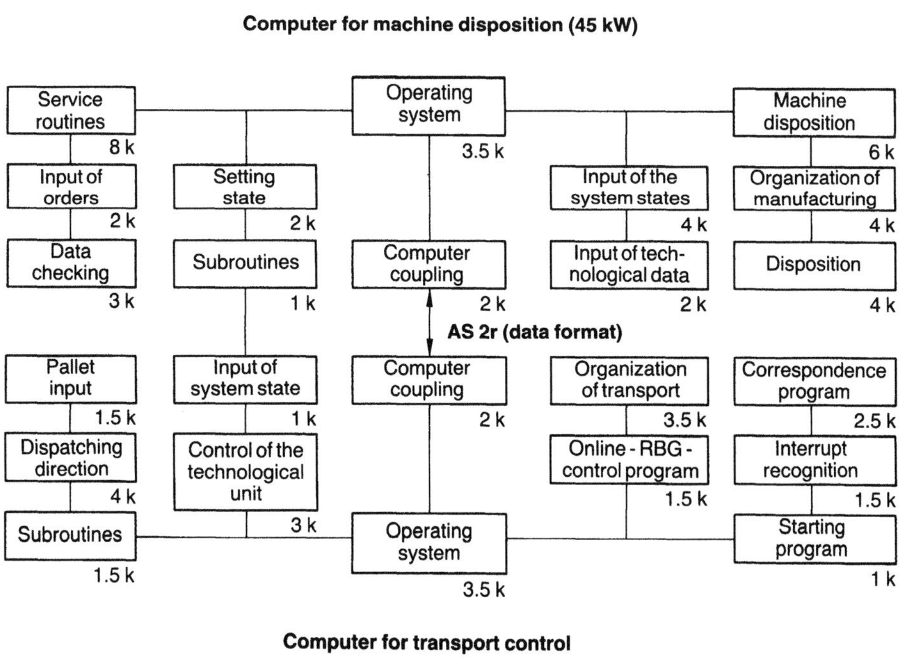

Fig. 42. Program system of the machining system at Stolowa Huta, Poland

Table 9. Flexible manufacturing systems in the Federal Republic of Germany [200]

Line		Current number	1	2	3
1		User	Heidelberger Druckm.	Heller, Nürtingen	Bauer, Esslingen
2		Producer	Burr/ Heidelberger Druckm.	Heller	Burr
3		Name of system	Heidel-berger/ Burr	Heller NC-FS 1	Bauer/ Burr
4		Year put into operation	1971	1974	1976
5		Products	Printing machines	Machine tools	Gearbox motors
6		Kind of part	Frames, housings	Housings	Housing covers
7	Part assortment	Geometry	p1, p2	p1	p1, p2
8		Type variety	50	150	250
9		Lot sizes	30—150	12	10—100
10		Max. size (mm)	1200	1000	700
11		Max. weight (N)	1000		800
12	Manufacturing system	Manufacturing	subst.	subst.	subst.
13		Lathes			
14		Drilling machines			
15		Milling machines			
16		Drilling/milling machines	10	2	9
17		Grinding machines			
18		Others			
19		Total number of machines	10	2	9
20		Machine tool control	NC	NC	NC
21		Tool number	1100	100	350
22		Quality control	Manual	Manual	Manual
23	Material flow system	Conveyor	Roll	SW	Roll
24		Part carrier	Pallet	Pallet	Pallet
25		Number of part carriers	30	24	30
26		Transport loading	ge	ge	ge
27		Transport control	RE	RE/M	RE
28		Storage positioning	de	de, ze	de
29		Handling	SE	SE	SE
30	Info.-system	DNC	x	x	
31		On-line registration of manufacturing data		x	
32		Optimization of dispatch	M	M	M

Abbreviations: x = in application, (x) = in preparation, SE = special equipment, UF = under-floor chain conveyor, PR = process computer, RE = relay control, ge = tight, de = decentralized, ze = centralized, un = loose, K = crane, Roll = roller conveyor

4	5	6	7	8	9	10
Zahnradfab. Friedrichs.	VFW, Varel	MBB, Augsburg	Daimler Benz, Untertürkh.	Zeiss, Oberkochen	VW, Kassel	Maschinenbau
ZF	Heller/ Heyligen- staedt/VFW	Droop & Rein MBB	Burkhard & Weber DIAG/DB	Steinel	Burkhard & Weber	Deckel/ Anwender
ZF 2	VFW	CIAM Zerspannung	Flexible Universal- fertigung	Flexibles Festigungs- system	Burkhard	Deckel
	1979	1979	1979		1979	1981
NKW gearbox	Aircraft	Aircraft	Cars	Optical instruments	Cars	Machine tools
Gears	Aircraft components	Aircraft components	Motor and gearbox components	Precisioning components	Housings	Machine tools components
Ro1	p2	p2	p1	p1, 2; Ro1, 2	p1	p1, p2
250	200	200	110	105	3	80
50—500	24	8—150	50	(10—800/a)	(60/d)	20—500
270	1200	4000		250	600	400
190	200	27000			500	
suppl., subst.	suppl., subst.	suppl., subst.	suppl., subst.	suppl., subst.	suppl., subst.	suppl., subst
4						
1		x				
5		x				
	6		7	3	5	2
4					2	
14	6	28	7	3	7	2
PC, NC	NC	NC	NC	NC	PC, NC	NC
312	576		1200	120	300	100
	Autom./man.		Man./autom.	Manual	Manual	Manual
K	SW	IW	IW	SW	Roll	SW
Magazine	Pallet	Pallet	Pallet	Pallet/mag.	Pallet	Pallet
200	26				23	
un	ge	un	ge	ge, un	ge	ge
PC/RE	PR	PR	M/PC	PC	RE	PC
ze, de	ze	ze, de	ze, de	ze	de	ze
IR	SE	M	SE	SE	SE	SE
x	x	x	x	x		
x	x		x	x		
A	A	M				M

Only when the tools in the system are available are the workpieces transferred to the machine.
In the operating system at Stolowa Huta, for example, 53 machines with workpiece changers (process-flexible robots) are used. Three times the labor productivity has been achieved as compared with a well-organized conventional series manufacture. An analogous machine system, FZ 200, was put into operation in Trencin, Czechoslovakia. There, 24 machining stations are equipped with workpiece changers.

3.4.2.7 FMS in the Federal Republic of Germany

The Federal Republic of Germany ranks among the leading producers and users of FMS in Europe; in the past few years especially, efforts to accelerate their development and to employ them efficiently have increased [200]. Special importance is attached to a high degree of flexibility, for a normal range of 50-250 different workpieces. Table 9 presents a survey with essential characteristic features. A characterization of the significant components of FMS in the Federal Republic is given in Table 10 [200].
The levels of automation of the information systems differ for the FMS listed. Four systems use no integrated process computers; the transport system is controlled via circuits with permanent wiring.
The punched tapes with the NC programs are manually fed into the machine control units. For the FMS with computer control that were put into operation in the past few years or are still being designed, the functions of administration and distribution of NC programs, transport and storage control, and the guidance of production are automated.

Table 10. State of development of FMS components in the Federal Republic of Germany [200]

System				
Machining	Transport	Storage	Manipulation	Information
Mainly machining centers	Roller tracks; induction-controlled cars, rail-mounted cars	Combination of centralized and decentralized stores	FMS for prismatic parts: pallet changers	Very different degree of automation
Mostly 5-10 machining systems	Workpiece carriers: mainly pallets with clamping fixtures for one workpiece	Trend toward computer-monitored chaotic systems	FMS for rotationally symmetrical parts: industrial robots and handling automated	Trend toward the integrated process computer system for administration and distribution of NC programs, transport and storage control, and guiding of manufacture
Mostly 20-40 tools per machine tool				
Mostly 300-400 tools per FMS	Prismatic parts: 20 – 30 workpiece carriers	Central capacity of 20 pallet locations, up to 800 magazine locations		

One system of international importance has been characterized in section 3.4.2.3. Supplementary to it, the following applications are presented.

3.4.2.7.1 FMS for Medium-sized Prismatic Workpieces

One of the newest manufacturing systems is the so-called duplex cell DFZ 630 of the firm Fritz Werner, West Berlin. This system is designed for pallet sizes of 630x630 mm, 800x800 mm, and 800x1000 mm [91, 164].
Two horizontal machining centers in standard design are linked by a rail-mounted pallet car in such a way that a common automatic workpiece supply is possible. The example shown has 20 pallet-depositing locations which are fed by a central setting-up and removing station, as well as by a separate resetting station (Fig. 43). This system may be extended, however, with respect to both the number of interlinked machines and the pallet-depositing locations.

Fig. 43. Duplex cell DFZ 630 for pallet sizes of 630x630 mm, can be extended for four and more machines and for pallet sizes of 800x800 or 800x1000 mm (Fritz Werner, West Berlin)

A control tool store at the back of the machines can accomodate up to 500 tools according to the stage of its development. By means of an NC-controlled handling device, the tools are automatically exchanged between the tool rack and the disk magazine of the machine. The disk magazine of the machine can accommodate 40 tools.

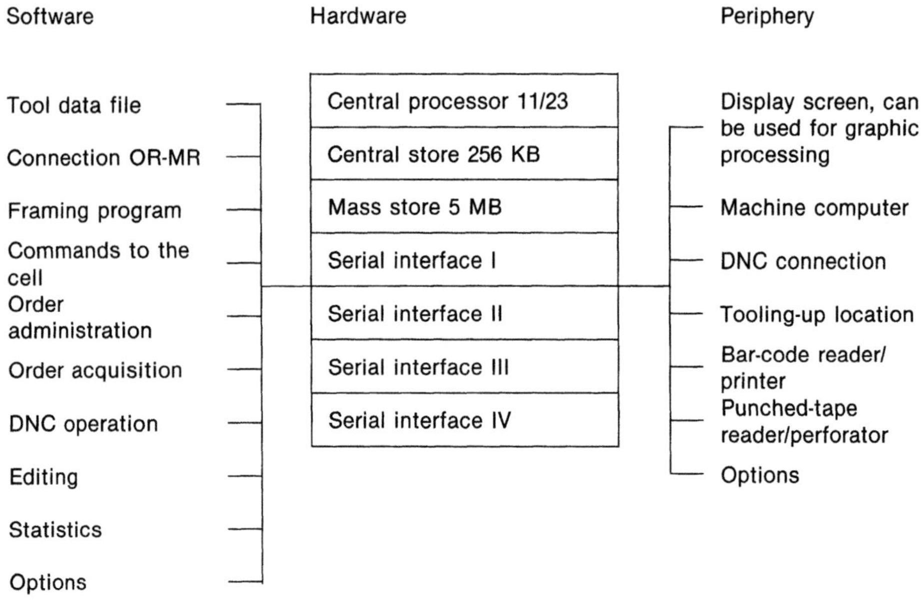

Fig. 44. Configuration of a cell computer for controlling a flexible manufacturing system (a Fritz Werner design)

Each of the machines and handling devices is equipped with an independent NC or PC control. Arranged in front of this is a so-called cell computer for storing all data and programs, controlling the transport motions, and operating the system. This computer performs the tasks listed as follows (Fig. 44).

1. *Storage of programs:* Up to 2000 NC programs can be kept in the data store
2. *Transfer of NC programs* to the NC controls and vice versa (DNC operation)
3. *Administration of tools* inclusive of the data assigned to them, such as identification number, correction values, tool lives, etc.; up to 1000 tool sets can be stored
4. *Keeping the list of orders:* Parallel administration of up to 100 orders is possible. Every order number is coupled with the pertinent NC program and the tool list. Moreover, information on the type and number of workpieces and on the work carriers is assigned. During the input of an order the availability of programs and tools is first checked by the computer.
5. *Administration of workpieces in the pallet pool:* The storage location and state of the workpieces are continuously recorded and a "timetable" is generated for controlling the transport vehicle.
6. *Exchange of tools* between the tool magazine of the machining center and the central tool store is carried out simultaneously during the machining of the preceding workpiece: the sequence of exchange as well as the process of exchange itself are controlled in such a time-efficient way that the machines are never idle.

7. *Automatic tool monitoring* is a prerequisite for the unattended operation in the second or third shift. The machines are equipped with adequate built-in sensors for detecting wear or failure. In case of tool failure machining is stopped at once, and the disturbance is indicated to the operator acoustically and on a display screen. During the unattended operation the tools and workpieces are automatically exchanged in this case. For the purpose of assessing the state of tool wear a continuously updated tool-life value is also kept in the tool data file. All worn tools are placed in a portable tool car by the handling device. The spare tools are manually substituted, a mechanical identification by a bar-code label being provided.
8. *Computer-guided presetting of tools:* For this purpose the cell computer is equipped with a direct connection to a tool presetting device. The new tools clamped in the holder are measured in the tool presetting device; the measured data are immediately transferred to the computer and to the machine control unit without any manual input. After the loading of the tool car and the system transfer, the putting-in to the tool store is automatically executed by the gripper.
9. *System operation and error diagnosis:* The machining cell has to be operated in a convenient, central, computer-guided way. The required software is available in the cell computer. The hierarchical structure of commands and the operator prompting in menu technique guarantee a plausible, process-related, and therefore almost error-free operation of the cell. The following peripheral devices are directly connected with the cell computer:
   - Operating console with display screen
   - Bar-code reader/printer
   - Tool presetting device
   - DNC connection
   - Punched-tape reader/perforator

   Besides the automatic tool and workpiece supply, making available the clamping accessories and NC programs must be integrated. Points 10 and 11 below are related to this.
10. *Making the clamping accessories available:* Mostly workpiece-specific jigs and fixtures are used to setup a workpiece. The design and construction of such clamping mechanisms are very expensive, however, and there are long waits for availability. For the conventional assortment of parts this drawback can be avoided by the use of a modular system of jig and fixtures.
11. *Mechanical generation of programs:* The economy of flexible manufacturing systems is considerably influenced by the amount spent on the NC programming. Many NC programs are required, especially in the manufacture of smaller quantities of parts to be machined. In the given case the programming language TC-APT (Terminology Comprehensive for Automatically Programmed Tools) is used. Optimum conditions for program generation are provided by the graphic representation of the tool movements and the dialogue input. The programming system TC-APT is implemented in a separate small computer, with which four display screen devices can be connected. The program generated and administrated at the programming location can be directly transferred to the cell computer.

3.4.2.8 Example of an FMS in the United Kingdom

3.4.2.8.1 The SCAMP FMS (1983)[1]

The FMS SCAMP is doubtless regarded as one of the most relevant innovations (development) in flexible manufacturing systems internationally. It is the largest FMS in the United Kingdom and all parts are British made except the Japanese made robots. It is installed at the Colchester Lathe factory in Essex, and was designed by the 600 Group (owners of Colchester Lathe).

The FMS SCAMP ("*S*ix-hundred group *C*omputer-*A*ided *M*anufacturing *P*roject") was developed for the production of typical components in general engineering manufacture. The SCAMP cell machines a relatively wide range of turned components consisting of a mix of shafts, gears, and disks in steel, cast-iron, and aluminum. The machining processes involved include turning, milling, drilling, tapping, cylindrical grinding, spline hobbing, and broaching (see Fig. 45).

1. *Work load:* There are 46 different shaft and gear components (rotational components) in batches of 80 for lathes.
2. The *aim* is to reduce work in progress. The maximum throughput time for any batch is 3 days, compared with 3 months for a conventional machine shop.
3. *Manning level:* There are three operators per shift, with 24-h/day operation. Operators are responsible for complete operations, i.e., NC programming, loading and unloading pallets, clamping worn tools, setting up non-NC machines, and inspection.
4. *Machines, conveyors, and robots:* Work stations 1 to 4 are first-operation machines, i.e., rough turning, facing, etc. Two are 2-axis CNC lathes, fitted with Block tooling, and two are 5-axis CNC lathes, which are basically the 2-axis machines adapated for milling and drilling and fitted with a robot vision system for separating nearly identical parts which differ in minor detail.

   The fixed-axis machines have an additional lower saddle with two axes of movement. They have a special 12-station turret which can carry six axial and six radial rotating tools such as milling cutters and drills, with a 1.5. H-P drive. The fifth axis is the lathe spindle itself, which can be positioned and rotated to allow drilling of cross holes, drilling of axial holes on a pitch circle, milling of slots and cams, and so on. The machines have even been used for drilling at an angle, using an angle head and combining the U and V axis feed.

   Work station 5 is a machine for chamfering and deburring all gears in the cell. It is basically a 2-axis CNC lathe fitted with a tooth-rounding and wire-brushing head.

   Work station 6 is a semi-NC gear shaper.

   Work station 7 is a manually set gear hobber for spline milling.

   Work station 8 is a manually set horizontal broaching machine.

   Work station 9 is a CNC grinding machine.

---

[1] This section is based on various published papers of Prof. Davies, UMIST, Manchester, England

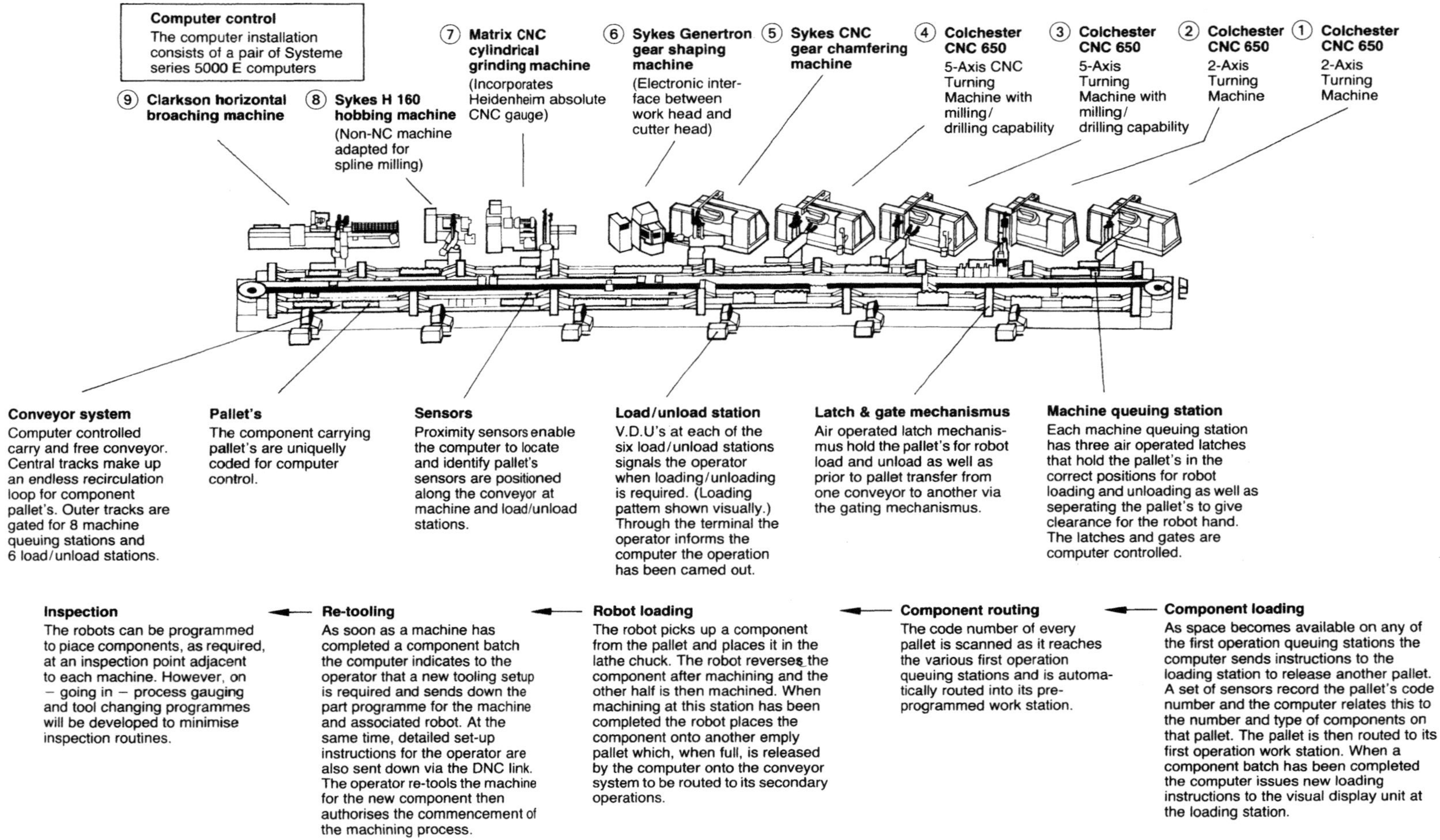

Fig. 45. SCAMP flexible manufacturing system

The machines are placed in a straight line. Each machine is served by a cylindrical axis, a 3-axis robot with a 2-axis wrist, and a "universal" 3-finger gripper.

Four parallel tracks carry and free conveyors (i.e., components are conveyed on pallets on a continuously moving conveyor unless the computer sets stop or gates to stop or diverts the pallet). The ends of the inner two tracks are connected with turntables, so that the inner two tracks operate as one continuously circulating track, carrying pallets.

The outer track nearest to the machines is divided into eight machine queue sidings for pallets. The other outer track is divided into six manual loading/unloading stations, each with its own control station and VDU. Pallets are transferred between inner and outer tracks by 14 shuttle mechanisms, one for each queue and for each loading/unloading station.

5. *Pallets:* There are 150 pallets, each with a binary-coded tag read by proximity switches at all queue loading/unloading stations. All pallet movements are controlled by one of two cell supervisory computers. Disk-type components are stacked on pallets without special fixtures. Shafts are carried in vertical magazines placed on pallets.

   The VDUs at loading/unloading stations instruct operators on part loading and machine setup; details given are the component type, part number, part quantity, tooling required, and diagram of part stacking on pallet. Once a pallet is loaded, the system reads the pallet code and identifies the components with the pallet.

   All machines are DNC linked to a second computer which downloads the preprepared NC program for the appropriate parts to the appropriate machines.
6. *Inspection:* Components are chosen for inspection on either a preprogrammed or a random basis. Robots are preprogrammed to remove a part ready for inspection and indicate this to the operator with a red flashing light. The machine is allowed to continue machining components on the current pallet. Until the inspection is complete, however, the system will not allow any more components in the batch on other pallets to be machined. If the inspected part is scrap or needs rectification the operator can signal to the computer to route all previously machined parts in the batch accordingly.
7. *Scheduling of work:* is carried out by the track supervisory computer; priorities are determined by the operator. Basically, there is a computerized Gantt chart system with two extra color VDUs in the computer room. Operators key in work load data and use the display to see the effects of scheduling decisions at each work station and on the whole cell. The computer also analyzes and displays production data, e.g., the total number of pallets required. Operators optimize scheduling by trying out different schedules.
8. *Cell status:* The track supervisory computer also monitors cell status on the same VDU. A colored MIMIC diagram of the system is used to monitor satisfactory operation, faults, and batches held up awaiting inspection.
9. *The vision system* is used to identify and separate components which have the same plan shape but differ in fine detail. Parts are placed in front of the camera in any angular orientation to ± 2 mm positional accuracy. The system takes a quick look at $3°$ intervals around the component. A second look at

preprogrammed positions showing distinctive details enables the component to position to ±0.5°.

10. *In-process measurement of grinding diameter:* The NC grinder is fitted with an NC control system containing canned cycles for plunge, traverse, and face grinding to enable programs to be built up quickly. It has an automatic, single-point, diamond wheel dressing. One diameter is ground first, deliberately slightly oversize. The automatic gauging checks the diameter and resets the wheel position to compensate for errors. Other diameters are then ground automatically without checking.
11. *Robot loading of grinder:* As shafts are mounted vertically on the pallet in supports, the robot picks up a shaft from the pallet by its end. To load the machine the grip has to be changed to nearer the center of gravity. Two extra fixed grippers are located at the grinder. The robot places the shaft in one set and picks up a grease gun to apply grease to the shaft ends. It then picks up the component at its C of G and loads the grinder. After the grinding the robot uses the second fixed gripper to regrip it prior to return to the pallet.
12. *Tool wear measurement:* The lathes are fitted with probes to measure tool wear and component diameter. One probe is fixed above the chuck and used for adjusting tool tip offsets. The other probe is mounted in the tool turret and is used to measure critical diameters. Offsets are adjusted automatically to compensate for tool wear, and indicate the need to change tools.
13. *Tool changing on lathes* is currently done manually. Most of the tools for all workpieces can be stored permanently in the lathe turrets. Block tooling is used. The system will be developed to have the robots change the block tooling automatically.
14. *Servicing:* There is some duplication in the servicing of the system, e.g., of lathes and computers. Maintenance engineers are on 24-h call to repair faults immediately.

3.4.2.8.2 British-built FMS of Cincinnati Milacron

One of the most recently developed FMS of Cincinnati Milacron was demonstrated at the MACH '84 machine tool exhibition at the NEC, Birmingham (June 19-29, 1984).

Thirteen different prismatic parts — some steel, some cast iron — will be manufactured by the Milacron FMS. The parts involve some 50 milling, 120 drilling, 23 reaming, 90 tapping, and 20 boring operations on the T-10 machining center.

System Elements. Utilizing an automatic, wire-guided vehicle for workpiece transportation, the system occupies some 200 m^2 (2000 sq.ft.) and comprises the following principal elements:

1. Milacron T-10 horizontal CNC machining center with ACRAMATIC 900 CNC and dual pallet shuttle
2. Horizontal automatic coordinate measuring machine (LK Metre Four)
3. Automatic washing station (CERA Ltd.) incorporating a Milacron T^3-726 programmable robot
4. Automatic, wire-guided vehicle (ACS)

5. Eight pallet loading/unloading stations buffer stores
6. ACRAMATIC flexible manufacturing cell controller

The system is capable of expansion up to eight docking points for a single automatic guided vehicle (AGV). These eight points could comprise, for example, two rotary pallet loading/unloading stations with the balance made up from any combination of CNC machining centers, washing machines, or inspection machines. The system can manage a total of 24 work pallets with routing of a single pallet to any five of the six processing stations in the cell.

T-10 Horizontal CNC Machining Center. The T-10 in the MACH '84 FMS is equipped with storage for up to 90 tools, a capacity found to be necessary for multishift, unmanned operation. The layout of the tool storage mechanism (which is fixed to the rear base of the machine and does not move with the column along the base) offers the possibility of tool delivery by guided vehicle if required. A dual work pallet shuttle unit is built into the T-10, providing a buffer which permits delivery and removal of pallets while the machining is in cycle. All pallets are coded to ensure complete integrity of the system, i.e., only the part of the program unique to a given pallet as it enters the T-10 (likewise the coordinate measuring machine and the robot within the washing station) can be brought into operation.

High Technology Features for Unmanned Operation. The T-10 incorporates a number of high technology features essential to its continuous and safe operation during unmanned or remote operating. These include a precision surface sensing probe and torque-controlled machining (TCM). This latter feature provides both machine tool protection and continuous adaptive control with tool protection/broken tool detection for every tool in the storage system. In addition, the ACRAMATIC 900 CNC software provides automatic tool setting/broken tool detection, sister tool replacement, and automatic tool recovery routines.

Automatic Work Transportation. Automatic transfer of work pallets between the loading/unloading stations and machines (Fig. 46) is provided by an automatic, wire-guided vehicle (ACS model TP218). The vehicle is a highly flexible means of transport allowing future expansion of the system (providing that the constraints on shop layout are minimum) and permitting the system to be introduced in a phased manner.

The journey route is selected by the FMCC while a separate vehicle controller ensures that the instructions are properly carried out. The docking procedure at each station is operated by a programmable logic controller which ensures that the correct docking procedure has been followed and reports back to the FMCC when docking has been completed.

The channels include routes to all four docking stations within the FMS and allow for charging of the AGV electric batteries during an idle shift or between journeys at a charging station.

Loading/Unloading Station. The loading/unloading unit is based on a standard Milacron 8-pallet rotary automatic work changer (AWC) and arranged with its own docking station. A programmable controller within the FMCC controls pallet selection and docking procedure.

Fig. 46. The transfer of a pallet from the automatically guided transfer car to the loading/unloading station of the automatic pallet store

A separate, manually controlled station is provided for the operator to index the AWC and to select any pallet as required. All pallets reside in their own home positions on the AWC and are identified by binary code. Upon dispatch or receipt, the pallet code is automatically read and checked against the control-specified pallet — again to ensure complete system integrity.
Following loading or unloading, the operator returns the pallet to the rotary AWC and enters the new status of that pallet into the FMCC. In this way, release of pallets into the system is fully controlled.

Acramatic FMCC. The Acramatic Flexible Manufacturing Cell Controller. (FMCC) synchronizes the actions of each piece of equipment at the system level and allows the operator to establish the work schedule (see Fig. 47). A route is established for each pallet by the operator via a VDU, with a maximum of five stops on route (at processing stations) in addition to start and finish points.
Operator loading/unloading activity is also entered via the keyboard and VDU, the primary display showing pallet status including (a) pallet number, (b) condition (finished, loaded, parked, inactive, in progress, defective), (c) identity (workpiece

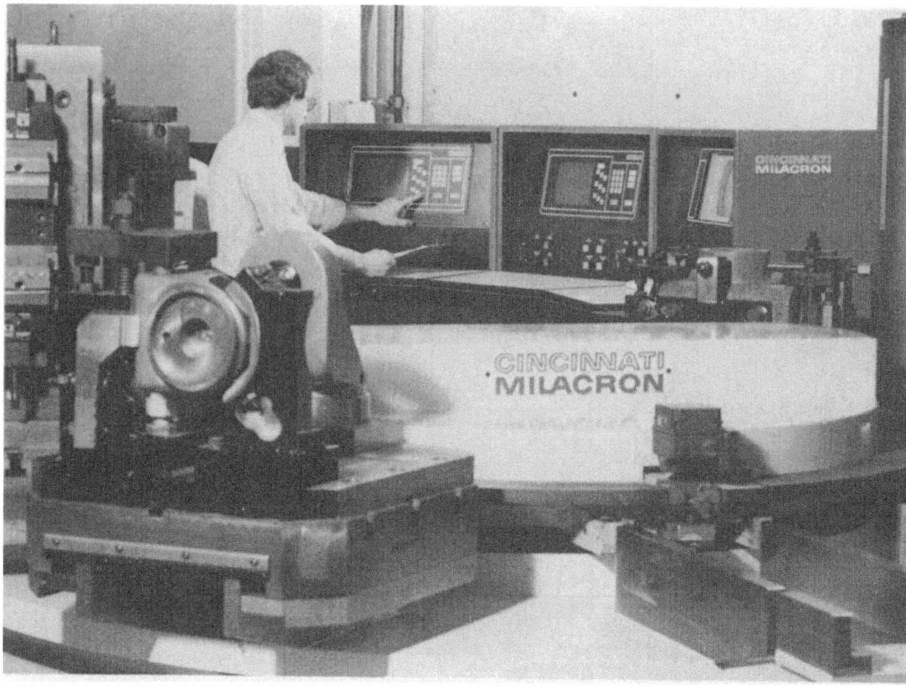

Fig. 47. The ACRAMATIC control system for flexible manufacturing cells

description or part number), and (d) priority (left unchanged or up-dated). Equipment and vehicle/cell status are also displayed. In addition, the FMCC controls the following functions:

1. System initiation (after a full shutdown)
2. System start/stop
3. Inclusion or exclusion of resources within the system
4. Selection of system mode or stand-alone mode
5. Single shuttle/dual shuttle mode
6. Normal system shutdown
7. Abnormal interruption

Program storage, data collection, and distribution is handled by an IBM Microcomputer incorporated in the FMCC. It includes a 10-megabyte sealed Winchester disk for data storage and a 360 K flexible disk drive unit. Upward communication with a mainframe computer is available through an optional RJE interface.

During machining center setup, it is not necessary for the operator to stand at the machine to operate the ACRAMATIC 900 CNC. Full remote display of individual CNC control status is available on the FMCC.

Automatic Washing Station. The FMS incorporates a Flexi-wash 700 R parts washing machine (CERA Ltd.) as an important element in the unmanned machining system. A wide variety of components can be accommodated in the cabinet,

which has a working envelope of 700 mm^3 (28 in^3) and is able to take a work load of 1 ton.
Within the cabinet is a Milacron T^3-726 6-axis programmable robot — suitably encased in a waterproof shroud — equipped with a CERA Flexi-jet microprocessor-controlled washing and drying jet.
A docking station placed at the front of the washing station enables pallets to be delivered into the cabinet through an automatically operated door onto an indexable work table. The washing process is carried out by a total of five Flexi-jets, four at fixed overhead positions above the work table and the fifth held by the T^3 robot. Programming of the robot arm is done by the use of a hand-held "teach" pendant.

Automatic Inspection. A Metre-Four coordinate measuring machine (LK T001 Company Ltd.) is integrated into the FMS and measures all surfaces machined within the system. It measures absolute component dimensions relative to design dimensions and datums, irrespective of pallet or fixture location.
Tolerance data included in the Metre-Four computer store enable the machine to accept or reject the part. Reject parts or out-of-tolerance conditions are signaled to the FMCC. Inspection data are displayed on the built-in VDU and can be printed out or transmitted to a remote host computer for permanent record.

3.4.2.9 FMS in Bulgaria and the USSR

3.4.2.9.1 Bulgaria

To date, four machine systems have been put into operation in Bulgarian industry. These include systems for rotationally symmetrical and prismatic workpieces. Two typical systems are characterized here.

SM-RSD-05: Machine System for Rotationally Symmetrical Workpieces

The system is designed for the full machining of parts turned in chuck — covers for direct-current motors made of an aluminum alloy with a steel socket.

D_{max} 400 mm
Max weight 10 kp (determined by the robot type)

SM-RSD-05 consists of four manufacturing cells for turning (ATM CE 062.10) and three manufacturing cells for boring (ATM RV 001 : 10.15), one pallet-turning station for each module, and a rack and rack-operating device for automated addressable transport.
SM-RSD-05 is a complex system with an extensively automated production process for machining. This has increased productivity by 25%-50% compared with a manufacturing section equipped with conventional machines for machining 163 410 workpieces per year. The manufacturing costs per part have been reduced from 2014 leva to 1793 leva. Depreciation of investment is 4.7 years.

PKD: Machine System for Prismatic Workpieces

The PKD consists of machine tools for the machining of prismatic components with maximum size of 1000 mm. It has five machining cells. Subsystem CM 100 has one machine tool (SIP 4000) for precision machining. There are two machines for the machining of prismatic components with maximum size of 400 mm, built entirely of Bulgarian machine tools with CNC control (MC 032). There are two each of subsystems CM 040 and RV 501.24. The warehouse comprises three rack blocks with three stacking lifts TC16G; the AGV transfer system comprises two types of automatic-guided vehicles (robocars). There are two of the first type with load capacity of 2500 kg, serving machining centers CM 100, and two of the second type with load capacity of 1600 kg, serving the rest of the machines. These robocars transfer the tool sets to the machining centers as well.
There are two tracing units. When tracing operations have been performed, automatic corrections are fed to the machining center. There are two measuring machines that check the final machining, and two washing machines. There is a tool shop for storage, preparation, and sharpening of the tool sets. The data control system has two levels. The lower level has four minicomputers to control the separate subsystems, and the upper level has one computer for operational planning and control on the production processes, with a link to the lower level and to the CAD/CAM configuration. The auxiliary subsystems consist of a swarf-removal area, a coolant and lubrication station, and a battery-charging shop.

3.4.2.9.2 USSR

In the USSR, the number of operators can be reduced only insignificantly by the use of individual NCM. A solution is provided by the complex automation of (a) the manufacturing of parts, (b) transport and storage, (c) production control, and (d) production-technological preparation/projecting.
FMS in the USSR are used for the automation of single-batch or small-batch production. FMS are defined as follows: a section replacing NCM, controlled by DNC, with a common transport and storage system and common planning of fixtures and tools.
The creation of FMS in the Soviet Union was based on the following starting conditions: Only 1.5%-6% of the time required to produce a workpiece is purely for the use of machine tools: the remaining 94%-98.5% is taken up by auxiliary activities, transport, and storage. The employment of FMS has increased the utilization of machines to 70%-98%.
The economic effectiveness (assumed to be 100%) rises by 55% due to the reduction of manufacturing costs, by 30% due to the quick reaction to market trends, and by 10% due to the increased flexibility in manufacture.
The n_L can be reduced to 4, even to 1, by the employment of FMS. Part assortments may comprise up to 10000 parts. The first Soviet FMS was created in 1972 under the direction of the Central Institute for Manufacturing Engineering (ENIMS).

According to predictions by MRJRA and CIRP, more than 50% of machine tools will be integrated in FMS by 1990. The main research on FMS has been done by ENIMS. The criteria for combination of FMS with existing systems are:
1. Existing NCM
2. Automatic transport and storage system
3. Central control computer
4. Simultaneous machining of different parts in the system

To date, 20 FMS have been set up. There are 14 FMS for rotary parts, six for complicated parts, such as double-curved surfaces, one for plane prismatic parts, and five for housing parts; 70% of the FMS in the Soviet Union are employed for the machining of rotary parts.
On an international scale, 70%-80% of FMS are used for other classes of parts; prismatic and rotary parts are machined in one system. Eighty percent of FMS operate according to the product principle, 20% according to the shop principle. According to the transport and storage system, FMS are subdivided on an international scale, with 5% being centralized workpiece store, 45% decentralized workpiece store, and 40% (six of these in the USSR) a combination. The centralized type can be further subdivided into (a) central store, (b) transport store, and (c) intermediate store. There are 14 FMS of the centralized type in the USSR, which is 5% of the total number in the world.

The Degree of FMS Automation. Systems with only a minimum of automated functions are also regarded as FMS. It is possible to integrate conventional machine tools with nonautomatic transport.
Automatic loading is desirable. This is the case with only 50% of FMS (rotary parts) because of a lack of flexibility of the control systems and the industrial robot (IR) grippers. The possibilities for IR employment for tool transport, magazine loading, and pallet loading with workpieces for a special FMS should be taken into account. The systems in the USSR have not yet been equipped with automated control and workpiece measurement at the machining locations.
FMS are used in series production. The assortment of parts machined in FMS ranges from 50 to 200 in the Federal Republic of Germany, from seven to 130 in Japan, from four to ten in the United States, and up to 1000 in the USSR. There is a worldwide trend toward complete machining of workpieces in such FMS.

Production-Technological Aspects of FMS. The basis of FMS is group technology. There are several systems for the automatic preparation of production in the USSR, e.g., ASTPP (including automatic classification, working out of the NC programs); these are only partly integrated in FMS, e.g., STPW and SOPP of ENIMS.

Production-Technological Parameters in FMS. The operations per part range from two to six for rotary parts, and from one to five for prismatic parts. There are two to four jigs and fixtures per rotary part and one to three for prismatic parts. The machining time per rotary part is 1.5-12 min, per prismatic part, 2-180 min.

Table 11. European FMS, not including those in the German Democratic Republic the Federal Republic of Germany and Czechoslovakia

			Bulgaria	Sweden		
Line		Current number	1	2	3	4
1		User		Volvo Lindesberg	Volvo Köping	ASEA Ludvika
2		Producer	Machino-exp. GmbH Sofia	Heller	Heller	Kearney & Trecker
3		Name of system	MODUL MC-RSD	Heller 2	Heller 2	unmanned machining center
4		Year put into operation		1976	1976	1978
5		Products	Machine tools	Cars	Cars	
6	Part assortment	Kind of part	Chucked parts	Housings	Housings	
7	Part assortment	Geometry	Rol	pl	pl	pl
8	Part assortment	Type variety			3	45
9	Part assortment	Lot sizes				
10	Part assortment	Max. size (mm)	200	700	700	1000
11	Part assortment	Max. weight (N)				
12	Manufacturing system	Manufacturing	subst.	suppl. subst.	suppl. subst.	suppl., subst.
13	Manufacturing system	Lathes	3			
14	Manufacturing system	Drilling machines				
15	Manufacturing system	Milling machines		1	1	
16	Manufacturing system	Drilling/milling machines		2	2	6
17	Manufacturing system	Grinding machines				
18	Manufacturing system	Others				
19	Manufacturing system	Total number of machines	3	3	3	6
20	Manufacturing system	Machine tool control	NC	NC	NC	NC
21	Manufacturing system	Tool number				320
22	Manufacturing system	Quality control	Manual	Manual	Manual	Manual
23	Material flow system	Conveyor	SE/Roll	Roll	Roll	
24	Material flow system	Part carrier	Pallet	Pallet	Pallet	Pallet
25	Material flow system	Number of part carriers				80
26	Material flow system	Transport loading	un	ge	ge	ge
27	Material flow system	Transport control	PC	RE	RE	M
28	Material flow system	Storage positioning	ze			de
29	Material flow system	Handling	SE	SE	SE	SE
30	Info.-system	DNC	x			x
31	Info.-system	On-line registration of manufacturing data				x
32	Info.-system	Optimization of dispatch				

Abbreviations: x = in application, (x) = in preparation, SE = special equipment, UF = under-floor chain conveyor, PR = process computer, RE = relay control, ge = tight, de = decentralized, ze = centralized, un = loose, K = crane, Roll = roller conveyor

Sweden		UK	USSR		Italy	Belgium	
5	6	7	8	9	10	11	12
Bofors Katskoga	Bugsy Transp. Mjölby	IBM Deptford	Enims Moskow	Orgstan-kinprom Inst. Moskow	Fiat Brescia	Cater-pillar	Cater-pillar
Kearney & Trecker	Yasuda Corp.	Molins Machine Co., Ltd.	Enims	Orgstan-kinprom	Comau	Hüller-Hille	Schar-mann
MM 200 S	Yasata YBM 90	Molins System 24	AU 1 ROTA	PRISMA	Flexible Linie	NC-Fertig.-linie	Rechner-geführtes FFS
1978	1979	1967	1972	1972	1979	1981	1981
	Fork-lift, trucks	Office machines	Machine tools	Machine tools	Cars	Dredgers	Dredgers
	Housings	Housings	Wheels	Housings	Housings	Lift-arms	
p	pl	pl	Ro1, Ro2	pl	pl	pl	pl
	15				6	4	5
			3—10	10—5000	220/d	1	
		300	320	750		2500	
				5000		15000	20000
	suppl. subst.	suppl. subst.	suppl. subst.	suppl. subst.	suppl. subst.	suppl. subst.	suppl. subst.
			6				
		2	1	1			
		3	3	4		2	
2	1	1		1	12	2	5
		1					
2	20	7	10	6	12	4	5
	NC	NC	NC	NC	NC	NC	NC
68	70	420	90	200		130	300
Manual	Manual	A/man.	Manual	Manual	Manual	Manual	Manual
		Roll	K	K/SW	Roll	SW	SW
Pallet	Pallet	Pallet	Magazine	Magazine	Pallet	Pallet	Pallet
			200			6	10
ge	ge	ge	un	un	ge	ge	ge
	PC	PR	PR	PR		PC	PC
	ze	ze, de	ze	ze	de		de
HA	SE	SE	SE	SE	SE		
x	x	x	x	x	x		x
		x					x
		M					A/M

Assortment. FMS in Japan allow for a maximum of 3600 parts, in the German Democratic Republic 2000, in the United States 1800, and in the USSR for 20000 rotary parts and 10000 prismatic parts. Lot sizes for prismatic parts run from four to 30, for rotary parts, from ten to 1500. The trend is $n_L \rightarrow 1$ for nonautomatic workpiece feeding; in the FMS ASW-2 the trend is $n_L \leqq 5$.
The time required for changing of rotary parts is 2-60 min, of prismatic parts, 5-70 min. Machine utilization runs from 75% to 95% for rotary parts and from 55% to 83% for prismatic parts.
The following feature of the adaptability (flexibility) of FMS is given: Period of time from the finishing of the drawing to the beginning of manufacture.
A large number of Soviet FMS operate on the two-shift system. Unattended operation during the night shift is not yet possible in the USSR because the level of automation is too low (no control of workpiece quality, no tool monitoring). Table 11 is a survey of FMS operating in Europe, not including these in the German Democratic Republic, the Federal Republic of Germany, or Czechoslovakia.

3.4.3 Manufacturing Cells

3.4.3.1 Historical Development

The first flexible manufacturing systems installed for batch manufacture were characterized by the use of NC machine tools and machining centers with specially developed material flow and information flow (cf. Sect. 3.4.2.1). The single workstation was not developed for application in an FMS environment. The rapid development of flexible automation, which might be characterized as familiar hardware forms controlled by the modular hardware and software of microcomputer-type controllers, has made it possible to develop a new generation of machine, sometimes called "unmanned station". This modern machine tool can be viewed as a complex example of flexible automation, and is called a manufacturing cell or machining cell.

3.4.3.2 NC Manufacturing Cells, Characteristics and Features

Manufacturing cells are characterized by the following features:

1. They fulfill some requirements of system solutions (adaptability in complex automated systems like FMS).
2. Solutions of workpiece handling by robots, overhead loaders or other kinds of material-handling equipment are part of the manufacturing cell.
3. Powerful machine tool controllers (CNC, see Sect. 3.4.1.1) now make possible the control of a large number of axes, vastly increased system memory, significantly more powerful interfacing, and program storage capacity.
4. The new controllers are the basis for the addition of new kinds of tool monitoring, of sensors for tool breakage detection, post-machining in situ gauging and adaptive control, coupled with the necessary software.

5. They facilitate a new kind of tool changing or tool-head changing equipment.
6. They offer possibilities for automized chucker changing.

This flexible automation allows for quick configuration of FMS, but manufacturing cells can also be used singly and are part of the general trend toward unattended, or unmanned manufacturing.

3.4.3.3 Manufacturing Cells for Rotary Workpieces

The turning cell has the following functional features:
1. Fully automatic finishing of chuck and shaft workpieces
2. Minimization of technological operations through an automatic process cycle
3. Continuous loading of the lathe with blanks and continuous removal of finished parts
4. Manufacture with limited manpower, even when entirely different parts are being produced
5. Changeover to a new part assortment with no problems
6. Automatic process monitoring through workpiece and tool checking, as well as checking of work handling
7. CNC systems for the handling and turning process, with program input directly at the flexible machining cell (FMC)
8. Possibilities for coupling with DNC

The main features characterizing a machining cell, i.e., automatic process monitoring and workpiece handling, will be explained more in detail below.
The manufacturing cells for turning can be characterized by the following general features:
1. Basic machine with an extendable control system (two supports, each controlled by two axes)
2. Automatic workpiece handling (portal, robot, special constructions)
3. Automatic workpiece measuring
4. Automatic tool monitoring
5. Automatic tool change and storage
6. Optical workpiece recognition
7. Automatic change of jaws
8. Complete machining in the form of drilling milling centers

For an example of a manufacturing cell for turning see Fig. 48.
Automatic process monitoring is effected by simultaneous collection and evaluation of process-typical measured variables. By sensors in the electronic section of the machine, and using specific methods of decision realized by a PC module, tool breakage and irregularities in the process can be signaled.
Correction is effected automatically. Hence, process monitoring comprises part monitoring and tool monitoring, while also including the measured variables of ultrasonic emission and main spindle power and feed power in the x and z directions, e.g., main and feed torque.

Fig. 48. A machining cell for turning with a portal-type robot, the MC DFS 2/2 from VEB "8. Mai", Karl-Marx-Stadt (WMW — Niles), German Democratic Republic

Part monitoring is done by means of a sensor arranged in the tool carrier in the same way as a tool. The length of traverse of the slide is used for internal part measurement. Measurement can be performed for outside diameters, inside diameters, and lengths through inductive transmission of measured variables. Variations in depth of cut caused by the blank are compensated for. Starting cut and run-out operations are skipped in the program. Evaluation is made on the basis of a sliding average value; there is a set point formation for each individual cut, and set point override is handled by means of a step switch. There is also periodical interrogation for process values.

Tool monitoring is done by means of a sensor with a feeler pin, the monitoring device being fastened to the spindle housing. The thermally neutral suspension of the sensor is done by means of compensation rods (see Fig. 49).
The feeler pin is suited to all tools; there is a reference point for zero positioning. The distance from the center line to the feeler pin is kept constant with compensation by thermal influences. The sensor is protected from dust, splash water, and chips by an automatic covering.

Material Handling. Owing to its portal shape, the IR 2/P-2 NC robot is ideal for the continuous loading of machine tools. Thanks to the special design of its coupling points (beams/supports) and its dimensions (clear width and clear height), it can be easily adapted to any given conditions of application (Fig. 50).

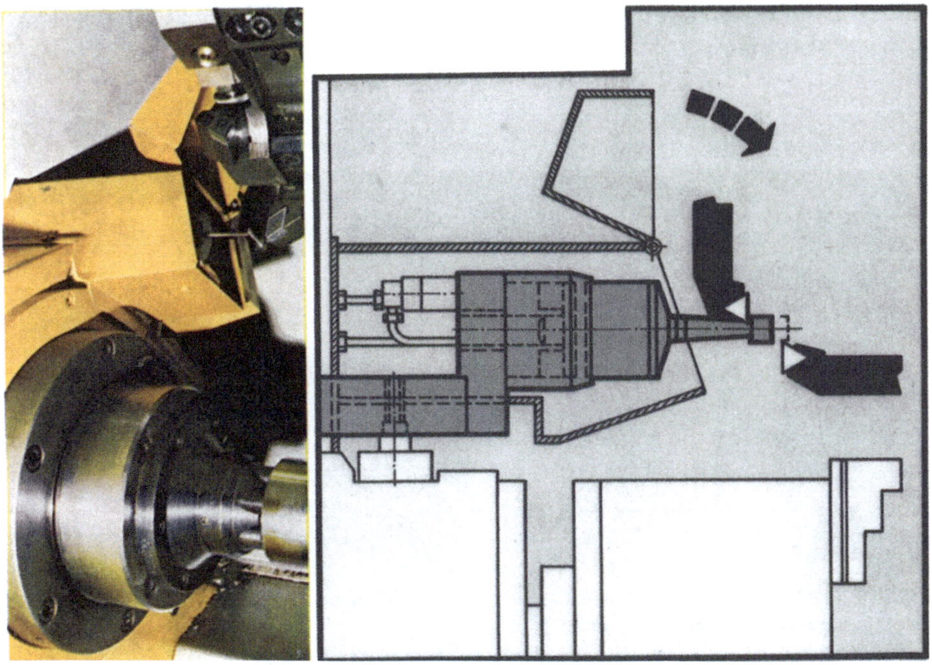

Fig. 49. Principle of tool monitoring

The DFS 2/2 consists of several units: (a) the portal, (b) the portal carriage, (c) the linear unit, (d) a wrist unit, and (e) a double gripper.
The portal carriage runs on V-ways (special slides). Its movement along the portal is freely programmable, as is the positioning of the linear unit, which is directly fastened to the portal carriage with its two-column circular guide and sliding screw drive.
The wrist unit with the hydraulically operated double grippers flanged to the linear unit enables gripper rotations to be performed around the C and D axes. The C axis enables the workpiece to be turned through 180° for two-side machining. The swiveling movement of 90° around D is required for work change. Changeover to chucking component grippers is ensured by the modular design. The four axes X, Z, C and D are NC controlled.

The *pallet station* consists of various units such as cassettes, a lifting mechanism, a pallet table, and Europallets. It stores the pallets carrying blanks or finished parts and moves the pallet into position for removal and depositing of parts by the portal robot.
One cassette is capable of accommodating four pallets. It can be loaded directly at the FMC or it can be moved to the cell in a loaded condition while the fourth pallet is in the execution plane.
The lifting mechanism, with its electrically controlled cycling facility, moves the finished-part pallet from the pallet table back to the cassette, and then transfers

a new, blank pallet to the table. Loading and unloading in the cassette are effected story by story, the cycling and lifting movement being integrated in the control of the portal robot.

Fig. 50. Material handling with the DFS 2/2

3.4.3.4 Manufacturing Cells for Prismatic Workpieces

3.4.3.4.1 Examples from the German Democratic Republic

The first machining cells for prismatic workpieces were developed and used in the late 1970s and early 1980s. The machining cells produced by VEB Wema Auerbach rank among the technically most developed solutions in the working range of 600x410x500 mm, i.e., for about 90% of prismatic workpieces.
These machining cells are produced in two basic varieties. A special further development for almost unattended operation is the machining cell Kompakt FC 400 P (Fig. 51), with the following essential features:

1. It is combined with a pallet magazine.
2. It has automatic pallet transfer from and to the machining cell in optional sequence without influencing the cutting operations.
3. There are no effects on clamping and machining operations due to separation of the pallet-changing mechanism and the manipulator.
4. Clamping and unclamping of the work in each of the eight magazine stations is independent of the machining cycle.
5. In-process resetting of the pallets takes place in the vacant stations of the annular magazine.
6. Various tailor-made pallets are available for constantly alternating and periodically repeated machining of small lot sizes without the need for resetting.
7. There is a simple, trouble-free action principle, using a single-arm manipulator.

Fig. 51. Machining cell Kompakt FC 400 P (German Democratic Republic)

Machining cells Kompakt FC 400 P and Kompakt FC 400 K are for the machining of plate- and box-shaped workpieces 400 mm^3 and utilize the most modern cutting and testing techniques.

All functional groups are fixed to a steel baseplate. Thus, the design is extremely compact and space saving and the machine can be installed easily. It can be transported without having to be disassembled and can be installed at the user in only a few hours.

The working area is completely enclosed, providing favorable conditions for both labor and environmental protection and for manufacturing accuracy.

Operation is practically unattended because of:

1. Vertical arrangement of the work carrier face, allowing for an ideal flow of chips, coolant flow, and transport
2. Automatic workpiece changing with a changing device which also serves as a tilting station

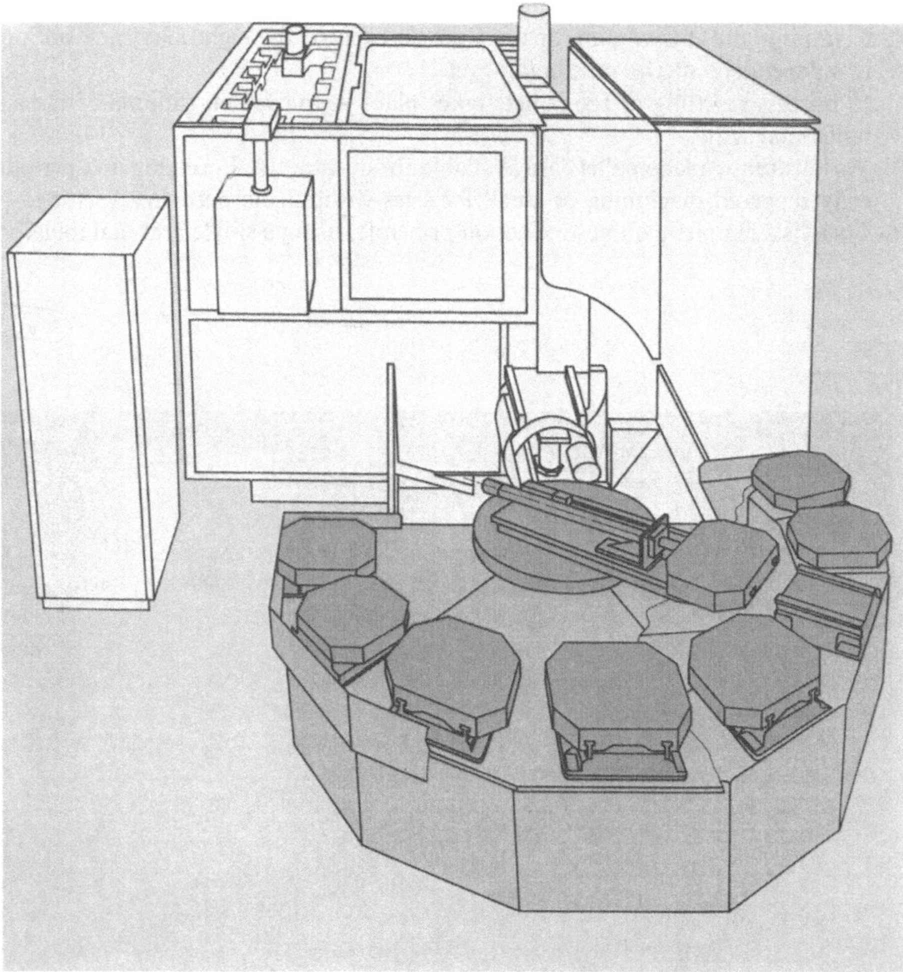

Fig. 52a. Workpiece pallet pol

3. A maximum of 36 storage locations, including the tool store
4. Process monitoring in the form of a system of testing for tool failure, an interchangeable sensor with a feeler pin, and supervision of tool life
5. Workpiece storage by means of a pallet pool (see Fig. 52)

The microelectronic manual input control type CNC-H 646, which is integrated in the machining cells, provides access to a wide field of technical-organizational conditions in the manufacture of small, medium and large series in a cost-saving way.
Employment is possible as an independent machining station or as a machining unit in manufacturing structures with a higher level of automation (FMS).
One generation of machining centers or machining cells for prismatic parts with which an adaptable modular concept has been translated into reality is produced by VEB Fritz Heckert, Karl-Marx-Stadt.
While the machining centers are produced in horizontal and vertical designs, the design of the machining cell is only horizontal. The sizes are 800x800 mm or 1000x1000 mm.
The modular concept is illustrated in Fig. 53 a-c.

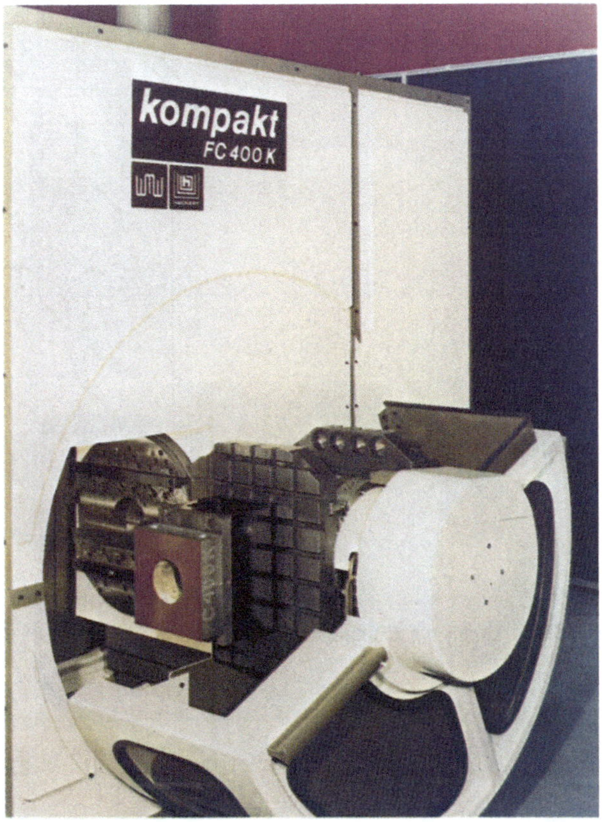

Fig. 52b. Workpiece pallets

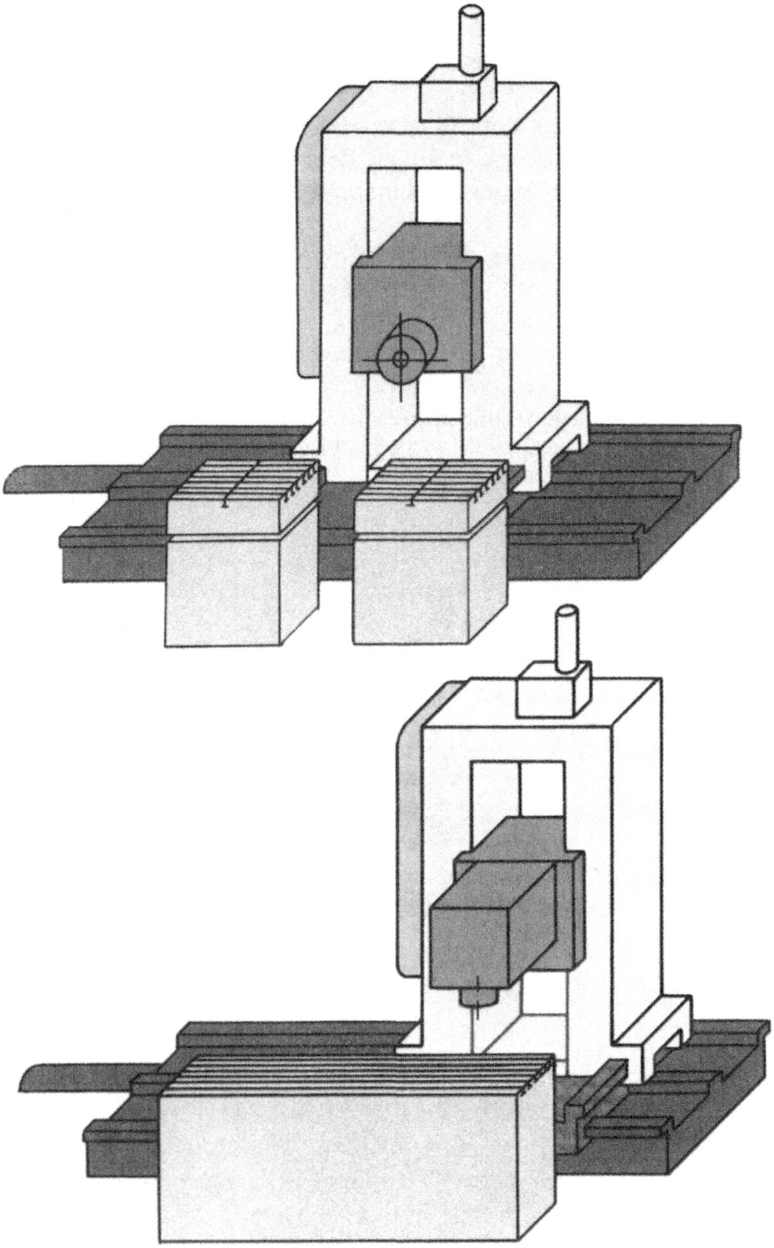

Fig. 53a. Modular machining center

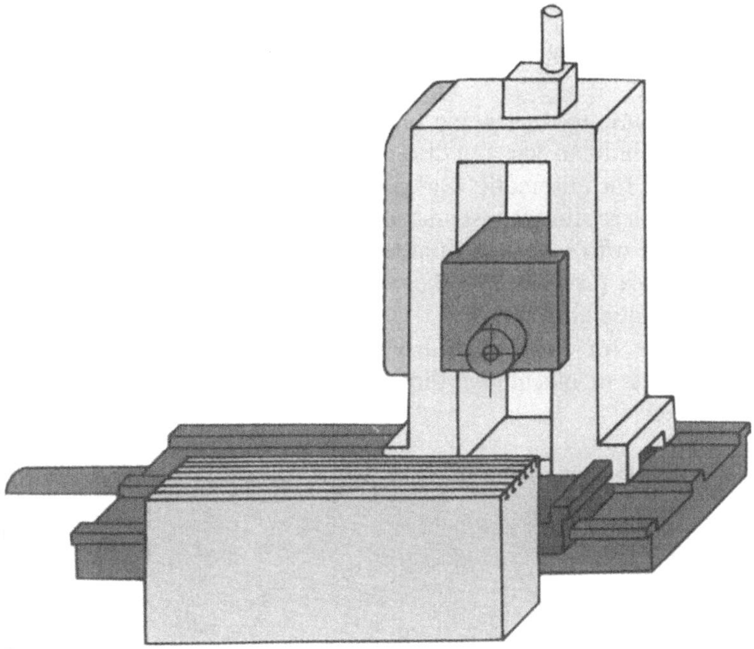

Fig. 53b. Modular machining center

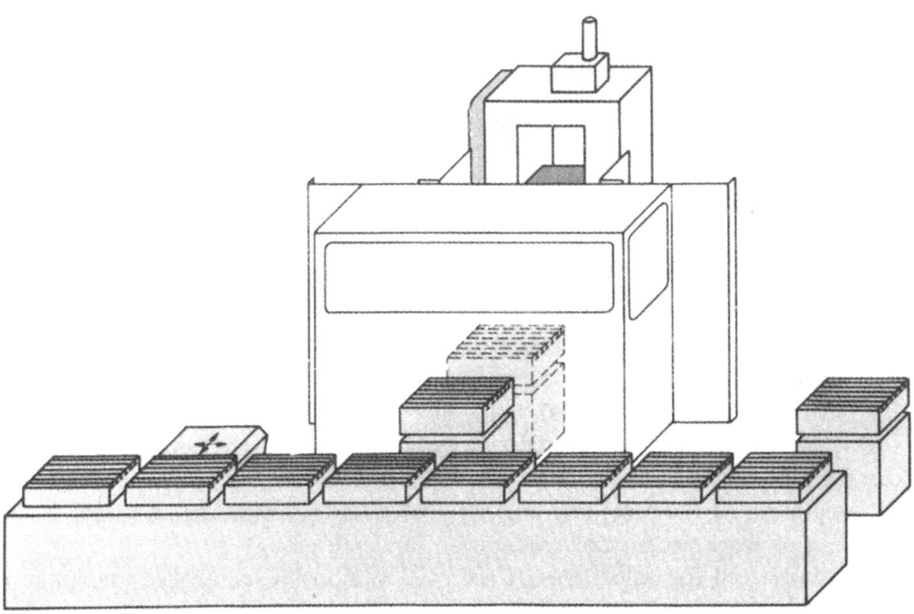

Fig. 53c. Modular machining center

Other essential features are:

— Attachment of free-programmable CNC controls as desired by customers
— Variable speed d.c. motor and recirculating ball screws for all coordinates of motion
— Automatic tool changing (40 or 60 tools) within 12 s
— Linear inductosyne as a direct measuring system with a resolution of 0.001 mm
— Device for automatic workpiece changing within 20 s
— Workpiece storage system consisting of six to eight storage locations for manufacture with a limited number of operators (see Fig. 54)
— Complex protection of the work area
— Automatic chip removal
— Systems for checking and monitoring for quality control, checking for completeness of machining, direct checking for tool failure, and active tool monitoring

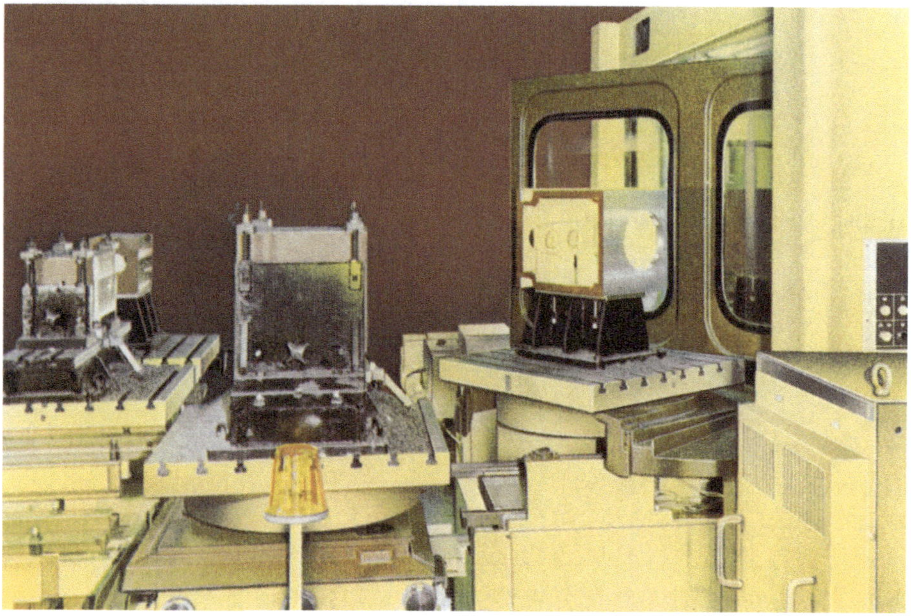

Fig. 54. Machining cell FCW 800/1000 with a workpiece storage system consisting of six to eight storage locations, assigned workpiece carriers, a clamping station, and a transfer mechanism

3.4.3.4.2 Example from the Federal Republic of Germany

The flexible duplex cell DFZ 400, made by the firm Fritz Werner of West Berlin, is a highly developed, novel solution. It was demonstrated for the first time at the European Machine Tool Exhibition in Paris in 1983 [91]. Its essential feature is the integration of the preceding auxiliary processes of manufacture (loading and reloading of workpieces, tool preparation, and presetting).

In the duplex cell for pallet sizes of 400x400 mm an NC-controlled manipulator in portal construction carries out all feeding or loading tasks within the manufacturing system (Fig. 55).

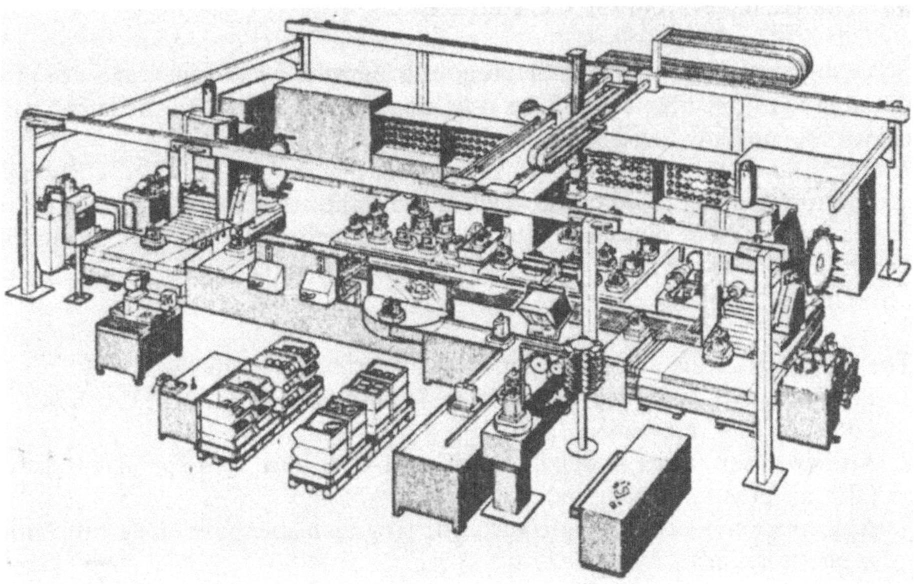

Fig. 55. Perspective representation of a flexible duplex cell (DFZ 400) for the automatic drilling and milling of medium-sized workpieces with an NC-controlled portal gripper to supply workpieces and tools (Fritz Werner, West Berlin)

These include supplying the two machining centers with workpieces and tools, placing workpieces and tools at the appropriate storage locations or at defined transfer locations, and feeding a washing machine which is integrated into the system. The workpieces are fixed on standardized carriers by means of universal clamping fixtures and are also transported with them.

To separate the operators from the machine with respect to time, and to produce different workpieces continuously and simultaneously there are extendable stores for the stockpiling of workpieces and tools. The control, coordination, and disposition of operational data and the administration of the store of workpieces and tools are performed by a cell computer. To make an operator-connected, automatically monitored operation possible, the machines are equipped with both a tool supervision system and a workpiece measuring system.

The automatic tool change during a change of orders and in case of tool wear makes continuous manufacturing without any interruption of the setting time possible. This considerably increases the degree of utilization of the machines. Moreover, production in smaller lot sizes is possible, so that the workshop stock can be considerably reduced.

The interconnection of two machining centers by a handling device allows for optimal utilization of personnel. As a rule, loading is carried out on the first and second shifts by one person each, who is occupied mainly with preparation of the workpiece carriers, workpiece changing, and replacing worn tools. On the third shift manufacture can then be continued without operators.

3.4.3.4.3 Examples from the United States

Some different kinds of manufacturing cells have been developed and are produced in the United States. *Turning cells* are a typical product of the largest machine tool company there, Cincinnati Milacron (Fig. 56).
An important feature is its high accuracy. Gear blanks from ten CINTURN cells are part print dimensioned to a ±0.001 in. tolerance. The Marposs gage is set to 75% of this ±0.001 dimension and will always compensate when this dimension begins to reach 0 on the plus side. This really means that the gage is working minus zero plus .00075 in.

The capability of the automatic parts-handling robot is as follows:
1. Body — servo-controlled, programmable positioning from 0° through 90° (.01° program resolution)
2. Arm extension (slide) — full 12 in. of servo-controlled, programmable stroke (.001 in. program resolution)
3. Arm rotation — 270° servo-controlled, programmable positioning (.01° program resolution)
4. Wrist — servo-controlled, programmable positioning from 0° through 270° (.01° program resolution)
5. Grippers — provide hydraulically actuated clamp/unclamp action for the parts

Fig. 56. Cincinnati Milacron turning cell

An interesting new solution is the two-machine robot cell from Cincinnati Milacron (Fig. 57). This flexible cell is another way of integrating dissimilar components into a single, unmanned part-processing unit — using off-the-shelf Milacron equipment. Included are a standard CNC step-grinding center, a turning center, and a T^3 robot arm.

Traditionally, this robot arm is mounted on a stand-alone base, but here it is attached to an overhead gantry, which demonstrates how versatile and modular this device really is. Mounting it to an overhead gantry opens up the floor area so that equipment can be positioned side-by-side, as shown in Fig. 57.

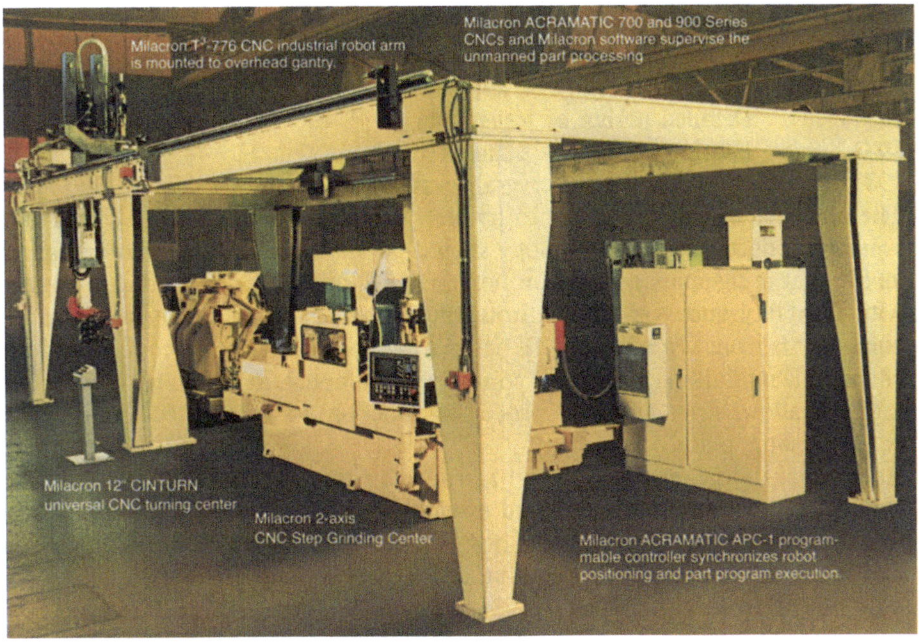

Fig. 57. Two-machine robot cell (Cincinnati Milacron)

Machines can be positioned face-to-face or back-to-back, as well, according to need.

Operation of the cell is unmanned, and handled by three Milacron computer controls. Both the grinding and turning centers are managed by a CNC. A third unit performs a dual role, directing the robot and managing the overall operation of the cell.

3.4.4 Flexible Manufacturing System Complex Provided with Laser (FMSC or FMC)

A large Japanese research and development project is one of the internationally outstanding activities in high technology for flexible manufacturing in small-batch production.

3.4.4.1 Historical Development

The flexible manufacturing system complex provided with laser, begun in 1977, is now in the final stage of an integrated operation test at the Tsukuba Test Plant (see Appendix II). A research organization composed of three national research institutes (Mechanical Engineering Laboratory — MEL, Elektrotechnical Laboratory — ETL, Government Industrial Research Institute Kyushu — GIRI Kyushu) and 20 leading companies in collaboration are working to pursue this research as a national project.

The long-term program developed as follows:

1977	Conceptual design
1978	Basic design
1979-1980	Performance test of elemental technologies
1980-1984	Design and study of application system
1981	Detailed design of test plant
1982	Fabrication of test plant (Tsukuba)
1983-	Overall test runs, overall evaluation

The objective of this project is to develop manufacturing technology for small-volume production. The technology should permit a level of automation equivalent to that in the mass production field, namely, an integrated production system with flexibility and very rapid throughput times for small-batch production of components from raw material. To achieve high productivity, the concept of interchangeable modules in an organic complex was adopted, developing a new technology that allows several types of processing functions to be executed simultaneously in the same place.

As product objectives three types of products were selected for the test plant; a headstock, requiring machining and assembly functions with high accuracy; a biaxial gear box; and a triaxial gear box, consisting of various part configurations and requiring complicated machining and assembly functions.

3.4.4.2 Concepts and Constituent Technologies

In the FMSC, greater productivity and flexibility are expected to be attained through two elemental system concepts: "Technology of Complex" and "Modular Construction" [127]. The basic idea of the technology of complex is to carry out simultaneously several different manufacturing processes in an integrated space. An enormous quantity of NC programming work is inevitable to meet higher product variation demands. This burden should be alleviated by auxiliary hardware devices. For such devices modular construction is the only answer. The following were chosen as constituent technologies and research and development were completed during the first 4 years of the program.

1. Metal forming

Objectives were to develop metal forming mechanisms and methods which could promptly and efficiently manufacture blank materials with minimum machining.

The following four machines have been developed;

a) Automatic open die forging machine: A 200-ton machine has been completed to forge large and medium-sized metal shafts. The aim is substantial reduction of forging time and a high degree of automatization; performance tests are well underway.
b) Hot isostatic press (HIP): The conventional HIP technique has been improved by preheating under atmospheric pressure, which can successfully reduce processing to 1/8-1/24 of the usual time. Cost reduction has been remarkable.
c) Disc-ring rolling machine: Various problems concerning accuracy, yield rate, etc. have been solved satisfactorily. The newly designed machine is successful in forming various material shapes.
d) Multi-ram precision forging machine: The machine can forge small parts, including stepped shafts, and similarly complicated parts with a high degree of accuracy. One of the features of this machine is the capability of applying combined stress to a blank from five directions by computer-controlled rams.

2. Machining

The purpose was to develop a multi-functioned machining system (machining complex), in which the processing functions of cutting, grinding, quenching, welding, and measurement of parts are integrated. In this system, modular units can be recombined according to required machining functions.

3. Automatic assembly

The automatization bottleneck of small-batch production systems is in the assembly process. A very useful NC assembly machine has been developed, in which various assembling functions are collectively integrated.

4. Product inspection

Efficient operational performance tests were needed. A multi-functioned inspection system complex has been developed for assembled products such as headstocks and gear boxes. Studies have been done on the automatization of product positioning, connection of a drive unit, sensor setting, and data sampling.

5. Laser processing

After completion of the first stage of development of two 5-kW CO_2 gas laser oscillators, one biaxial- and the other triaxial-orthogonal, 10-kW and 20-kW oscillators have been successfully developed in series. Two 300-W YAG and 200-W argon ion laser oscillators have also been developed. For the purposes of improving CO_2 laser performace, development and evaluation have been done of transparent optical parts and metal mirrors. Laser processing techniques such as heat treatment, welding, deburring, and chip cutting have been studied. The 10-kW oscillator is used in the Tsukuba Test Plant.

6. Design

In the FMSC program, the CAD system can make geometrical data modification, graphical simulation of assembly parts, interference checks, and drawings.

7. System control

To meet changes in product requirements and the subsequent reorganization of manufacturing processes a control system for the FMSC has been developed, in which the intelligence of each processing module is enhanced so as to realize high independent module control. These independent modules are arranged in a hierarchical structure, forming a distributed control system with simplified instructions. In order to realize this system configuration, novel noncontact transmission techniques for power and information have been developed.

3.4.4.3 Structure and Functions of Tsukuba FMSC Test Plant

The principal components of the Tsukuba FMSC test plant are (a) the machining complex, (b) the laser-processing subsystem, (c) the assembly complex, (d) the product inspection complex, and (e) a control system for these mechanisms. The layout of the Tsukuba FMSC test plant is shown in Fig. 58. The components of each mechanism are shown in Table 12.
The functions of the machining complex, assembly complex, and product inspection complex are shown in Tables 13, 14 and 15. The functions of the laser-processing subsystem are included in those of the machining complex.

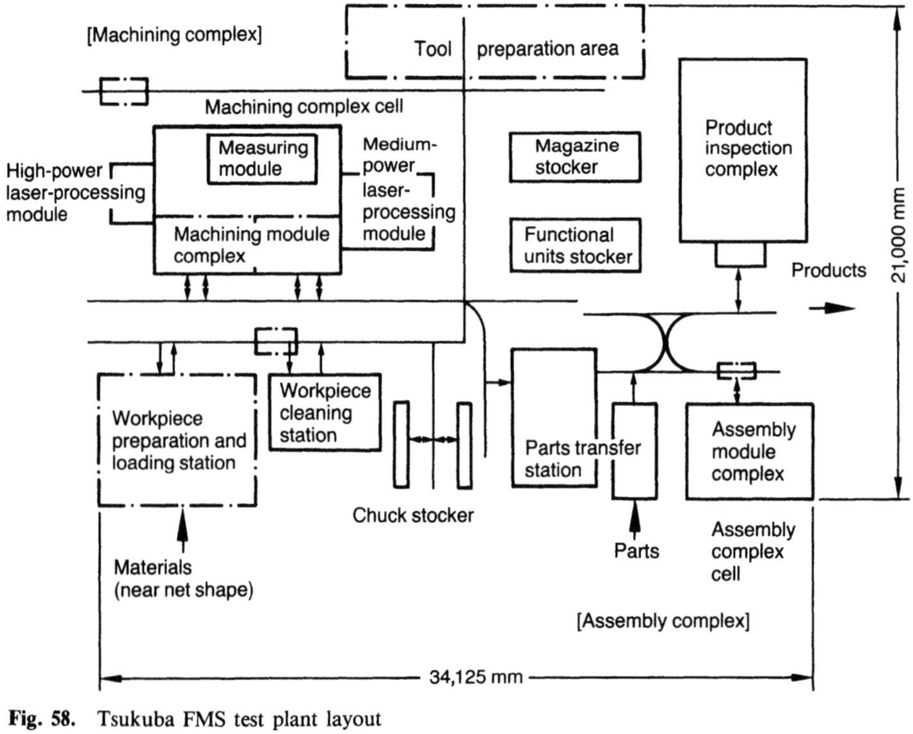

Fig. 58. Tsukuba FMS test plant layout

Table 12. Components of the Tsukuba FMSC test plant

Mechanism	Components	Quantity
Machining complex	Machining module complex (machining module, ACC, ATC)	2
	Functional head unit	7 types, 11 sets
	Measurement module	1 sets
	Workpiece preparation station (chuck, jig, centering device)	1 set
	Tool preparation station (tool presetter, tools)	1 set
	Cleaning station	1 set
	Machine parts transfer station	1 set
	Carrier (for functional head unit, magazine, and chuck)	1 set each
	Stocker (for functional head unit, magazine, and chuck)	1 each
	Incidental equipment (hydraulic unit, oil and coolant chilling unit, etc.)	1 set
Laser-processing subsystem	High-power laser-processing module	
	1) 10-kW CO_2 laser oscillator	1 set
	2) 5-kW processing module (on-line)	1 set
	3) 10-kW processing module (off-linie)	1 set
	Medium-power laser-processing module	
	1) 300-W YAG laser oscillator	1 set
	2) Processing unit	1 set
	3) Incidental equipment	1 set
Assembly complex	Assembly module complex	1 set
	Assembling parts transfer station	1 set
	Carrier (for cassette)	1 set
	Cassette stocker	
Product inspection complex	Inspection module complex	
	No. 1 Inspection unit (automatic sensor and probe changer)	1 set
	No. 2 Inspection unit (torque sensor)	1 set
	No. 3 Inspection unit (dynamic rotating accuracy)	1 set
	Carrier positioning unit	1 set
	Product transfer unit	1 set

Table 13. Functions of machining complex

Machining (main)	Cutting	(spindle system, feed system, AC servo)
	Gear cutting	
	Grinding	
	Special machining	(welding and hardening with laser)
	Complex control	
	Adaptive control	
Assembly (auxiliary)	Jointing of two workpieces for simultaneous machining	
Transfer	Indirect transfer	Automatic workpiece changer
	Direct transfer in machining module	
Storage	Functional head unit stocker	
	Chuck stocker	
	Magazine stocker	
Additional	Exchange	Automatic unit changer (functional head unit)
		Automatic magazine changer (magazine)
		Automatic chuck changer (chuck)
		Automatic changer (for tool, chuck, and measuring probe)

Table 14. Functions of assembly complex

Assembly (main)	Thread fastening Press fitting Inserting Holding Complex control	
Transfer	Transfer in mechanism (assembly cassette transportation)	
Storage	Procurement stocker Assembling jig/tool stocker	
Additional	Detection	Assembled part position Attitude
	Confirmation	Parts on cassette (assembled parts, assembling tools)
	Transfer	Industrial robot (assembled parts, assembling tools)
	Inspection	Press fitting, fastening torque, holding force
	Separation, integration, and exchange	Assembling jigs/tools

Table 15. Functions of product inspection complex

Inspection	Dimension Shape Run-out Vibration Noise Temperature Torque Operation efficiency Lost motion
Transfer	Transfer in mechanism (inspecting product transportation)
Storage	Inspecting jig Fixture Accessories (oil, belts) Various units
Additional	Preparation (cleaning, jig mounting) Product reloading (assembly — inspection) Loading (carrier — mechanism)

All functions are subjected to practical experiments at the Tsukuba FMSC Test Plant.

1. Total system

The following experiments and studies are performed:

a) Verification of the layout, functioning, and application of the system as regards the elemental technologies
b) Verification of flexibility and processing speed, which are the most important features of the FMSC for small-batch production

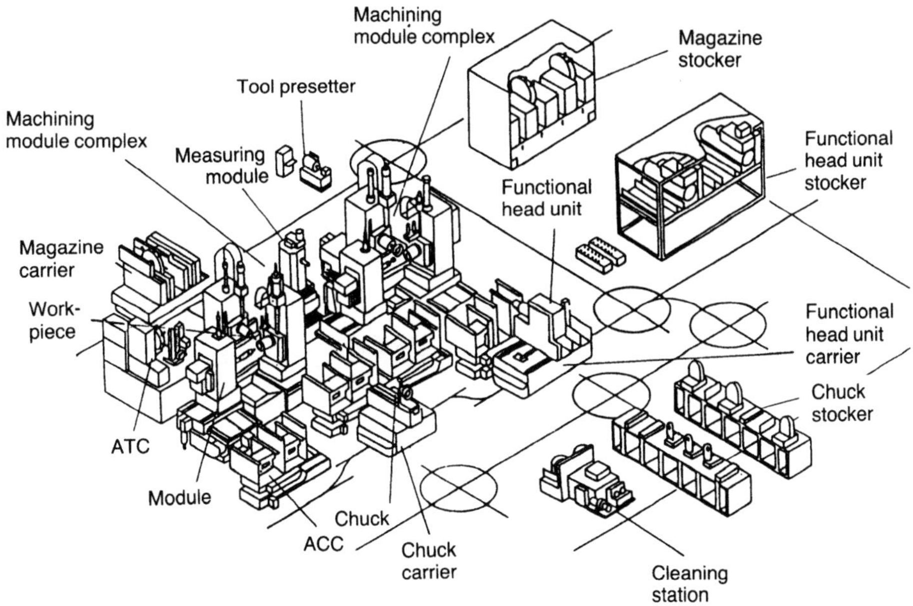

Fig. 59. Machining complex

2. Machining complex

The machining complex is to be a mechanism of many functions that can quickly process supplied materials. The structure of the machining complex is shown in Fig. 59. The following experiments are carried out to verify functioning, performance, and accuracy:

a) On complex machining such as cutting, grinding, laser processing, and partial assembly of round or square parts while utilizing the adaptive control function and structural accuracy compensation function
b) Measurement of parts not completely machined to determine a finish allowance; finishing, and measurement of the dimensions, shape, and surface roughness of machined parts in order to evaluate quality

3. Assembly complex

The aim of the assembly complex is to automatically supply the assembly space with assembly parts, standard goods, and procurement goods. The basic structure of the assembly complex is shown in Fig. 60. The following experiments are carried out to verify the performance and functioning of the assembly complex:

a) Automatic array and transfer of assembly parts of various types onto a cassette
b) Automatic array and transfer of the standard goods and procurement goods whose shapes have been recognized onto a cassette
c) Automatic assembly of the assembly parts, standard goods, and procurement goods supplied from the cassettes at one location

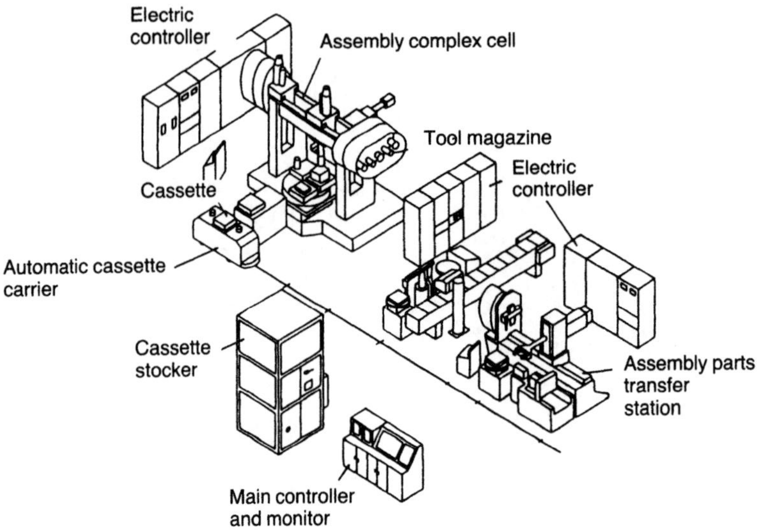

Fig. 60. Assembly complex

4. Product inspection complex

The aim of the product inspection complex is to inspect the quality of products manufactured at the machining complex and assembly complex. The product inspection complex is constructed as shown in Fig. 61. The following experiments are performed at the product inspection complex:

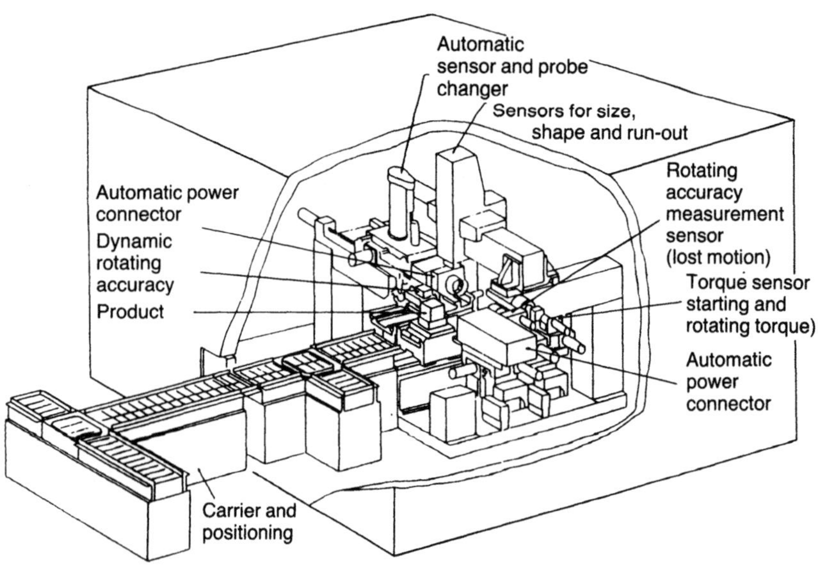

Fig. 61. Product inspection complex

a) Automatic and static inspection of the reference dimensions and shape accuracy of the manufactured products, and, by simulating the operative state of such products, automatic and dynamic operation performance
b) On the gear boxes, the output of reference data, according to diagnosis of the cause of vibration and noise if the measurements of these exceed the reference value

5. Laser-processing subsystem

The laser-processing subsystem consists of a 5-kW CO_2 gas laser (high-power) processing unit and a 300-W YAG laser (medium-power) processing module.
The aim of high-power laser processing is to perform surface treatment or welding of the machined parts in the on-machine state during machining at the machining complex.
In an experiment of complex treatment with cutting, the quality of processing, the system adaptability (such as the influence) of the cutting unit on the operating rate and the safety of the laser unit are verified. For such purpose, the laser processing unit is installed near the machining complex. In addition, a process such as cutting or welding is performed in the off-line state.
The aim of medium-power laser processing, where the processing head is built into the machining complex, is to deburr and cut chips during boring.

6. Production control

The chief aim of this FMSC project is the development of hardware. However, automatic design, machining data check, and the abnormality monitor were studied and developed as technical subjects related to production control, for the purpose of application of the complex manufacturing system in the future.
a) In the study of automatic design, the possibility of drawing and assembly of redesigned product objectives of the test plant are checked, and an experiment is performed on the assumption that such work will be applied.
b) In the study of the machining data check system, the format and mechanical collision of the computer-determined data changed due to redesign are checked, and an experiment of correction of the machining data by means of an editing function is carried out.
c) In the study of the abnormality monitor, the function of an external trouble diagnosis is checked; this monitors the temperature and its variation near the cutting point of the module complex by means of an infrared camera and detects abnormality. An experiment is also done on the function of the central monitor panel regarding abnormality of the mechanisms and processes in order to cope with automatic and continuous operation.

7. Control system

A hierarchical dispersion control method was adopted for flexible and speedy control of the various mechanisms. The highest unit controls filing, commands, and feedback information for the entire manufacturing system. Then follow a control

for the machining complex, one for the assembly complex, and one for the product inspection complex, and below these various types of management devices and control devices for individual purposes.
In the machining complex especially, area functions must be changed, or new ones invented, because the functional units are freely selected, replaced, and combined with each other to cope with the many different types of machined parts. To this end, the machining controller has an independent control unit for any axis, and an exchangeable functional unit contains a unit controller for any axis to cope with a change in function. Each mechanism is tested for functioning within the entire system at the Tsukuba test plant.
In short, the things verified at the Tsukuba FMSC test plant are (a) productivity, (b) flexibility, (c) adaptability and conformity, (d) facility of operation (system and equipment), (e) reliability, (f) the influence of disturbance, and (g) safety.

3.5 Direct Numerical Control (DNC) and Hierarchical Computer Control

DNC is a system of direct computerized control for several manufacturing, processing, and measuring machines and for workpiece and tool manipulation and transport and storage systems which are equipped with binary control systems by a digital computer. The digital computer is suitable for the storage, collection and distribution of instructions concerning sequence of actions, paths, and process state from and/or for the binary control systems. Communication can be performed bidirectionally in competitive operation according to programmed priorities between the control systems and the digital computer.

3.5.1 Stages of Development

In the course of 17 years of development as computer NC integration, DNC has gained from the progress with both kinds of digital equipment (see Table 16). The development of DNC was initiated by the possibility of controlling machine tools from a central computer [144]. The time-sharing capability of digital computers and the low degree of resistance of the paper-carried information input in the rough workshop climate for NC machines, as well as the cumbersome program handling constituted the motivation for DNC. A characteristic feature of these systems is the time- and format-adequate distribution of the programs stored in the computer to the machine tools.
The general structure of DNC systems comprises a central computer, one or several coupling elements, and several NC machines with an attached DNC system. Various forms of execution and arrangements of the transfer channels are used between the individual units of equipment (Fig. 62). The direct computerized guidance of NC machines includes basic and extension functions. Whereas the basic function represents the purpose of the first DNC systems, the extension function is already a result of first steps in the development of DNC (Fig. 63).

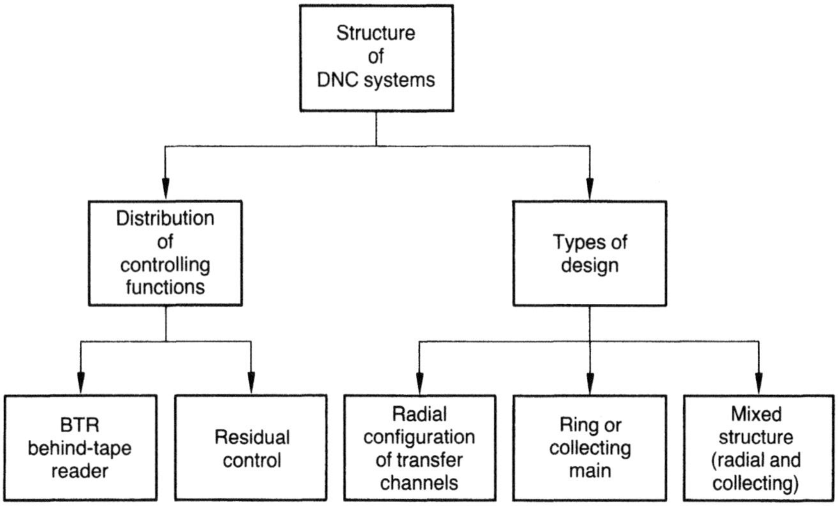

Fig. 62. Structure of DNC systems

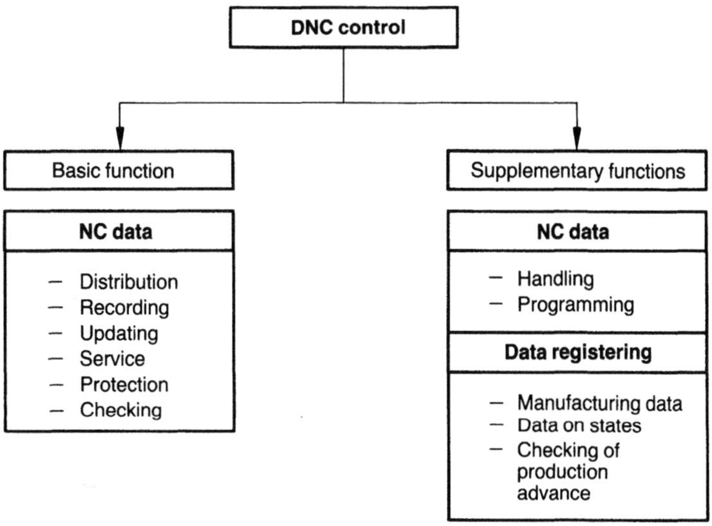

Fig. 63. DNC of several NC work stations

For the broad effect of the extension function unidirectional data transfer from the digital computer to the NC machines was insufficient. Bidirectional information transfer [65] can be executed between the devices of the DNC basic structure with format- and time-adequate data distribution via a communication software module or bidirectional hardware equipment which is capable of communicating. Since the early 1970s, DNC systems with this functional content have prepared essential prospects of development for workshop-wide process automation and automated, almost unattended manufacture. One more step toward bidirectional information transfer without paper between technical preparation and the workshop

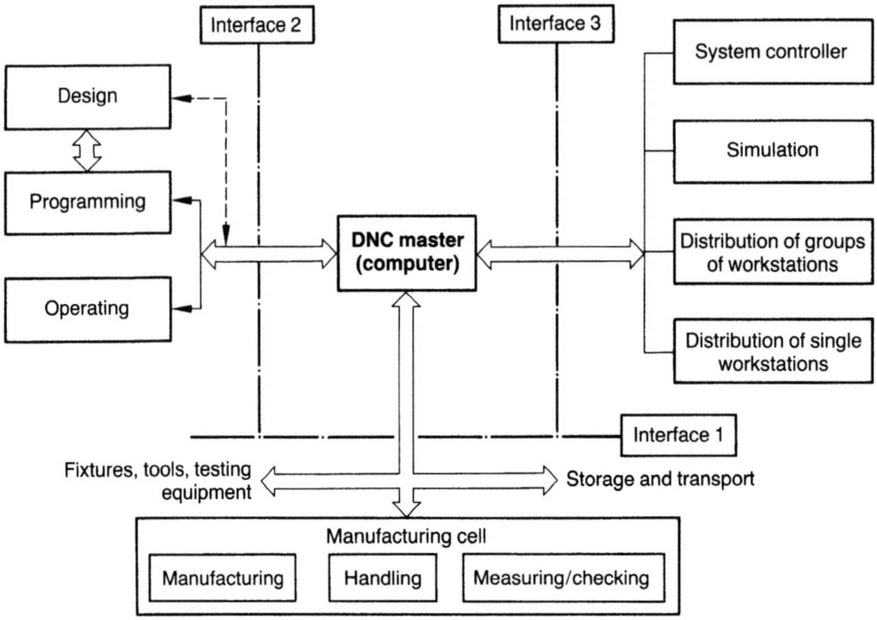

Fig. 64. Forms of communication with the DNC computer across interfaces

has been taken by adapting the hardware and software structure of DNC systems to the advanced developments of NC programming methods. This demands a variety of forms of communication with the DNC computer across interfaces S1, S2, and S3 (Fig. 64).

The data transfer is classified as:

— S1, for communication with the NC machines
— S2, for communication with programming systems
— S3, for communication with differentiated functional complexes for the control of systems and manufacturing

The data-transfer mode at interface S1 is determined by DNC-operated NC machines. It is connected in real-time for NC machines and characterized only by the communication interface; it is free of real-time for CNC machines. The possibility of storing machining programs in CNC control units permits the transfer of whole NC programs. Hence, the supply of information does not have to keep up with the manufacturing process and shows a basic difference to the real-time program supply. This considerably relieved the software of real-time interruptions and made possible the inclusion of functions with routines which are unfavorable for interruptions via interfaces S1, S2, and S3. Thus, new prospects are opened up for the continuous flow of information on preparatory tasks and for the monitoring of actions which are closely related to processes. This provides a decisive precondition for the implementation of computerized programming [227].

The programming of NC programs for NC machines of the DNC system can be executed via interface S2. Dialogue programming on the display screen for the generation of source programs is possible without any restrictions. In the DNC computer the data are converted into control programs for NC machines by the programming software and postprocessors for data transfer via interface S1. Interface S3 permits the integration of new equipment which is adequate to the increasing complexity of functions in flexible manufacturing systems. DNC does not serve the supply of these units of equipment with programs, but it does serve the recording of system state quantities and the derivation of control functions for devices which are closely related to the machining process, the NC machines, and the transfer of information to superior control computers of factories.

The hardware and software structure of present DNC systems is shown in Fig. 65. They are characterized by the following general features:

1. The DNC basic function is moving more and more to the background with respect to its importance.
2. Since real-time data supply of NC machining stations has ceased, the priority of the basic function over the extension function no longer exists.
3. The computer-internal representation of NC programs in the form of NC control programs is decreasing in favor of more general forms of representation as source programs.
4. The extension functions are increasing in number and influencing the basic functions.
5. The interpretation of the extension functions is more and more becoming the starting point for generating the basic functions.
6. The degree of consistent information processing is continuously increasing, aimed at multiple use of already generated datasets.

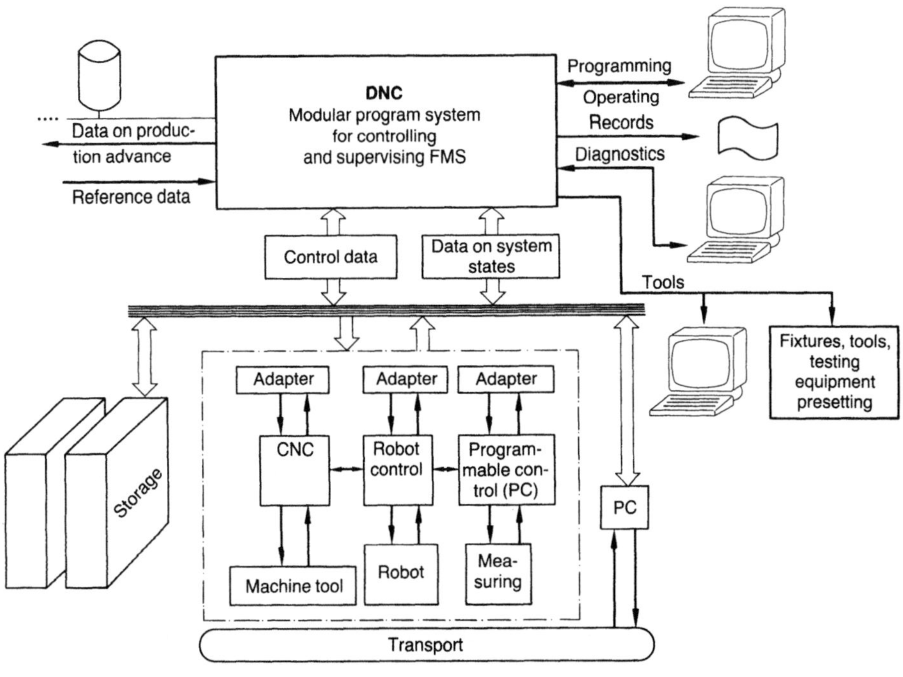

Fig. 65. Structure of hardware and software in present DNC systems

These general features lead to the following developments for DNC:

1. Complete build up of CAD/CAM systems, which are characterized by information transfer without any paper, from design and technology up to machining on CNC machines by the use of coupling modules and CLDATA-type computer-internal sets of information
2. A workshop communication system for production process control by the interconnection of system state information of the total manufacturing process and computer-aided instructions for planning and guidance [74]
3. Integration of process-accompanying monitoring systems as well as transport and storage equipment to raise the degree of automated, almost unattended manufacturing, leading to a high degree of self-containedness in flexible manufacturing systems

Even the real-time-guided NC machines showed essential qualitative gains [200]. Together with advances in sensorics, informatics, and the development of equipment, these trends are the driving forces in the development of DNC. The following selected examples illustrate the present state of the DNC technique.

3.5.2 Variants of the Realization of DNC Systems

3.5.2.1 DNC as Modules of a Complete CAD/CAM System

CAD/CAM systems originate with the primary aim of the computer-aided engineer's work — the designing of workpieces and pertinent production-technological documents. CAD/CAM systems improve the flexibility of production due to shorter design time, interactive design of alternative solutions, flexible and fast reaction to short-term changes, and reduction of error sources.

The degree of complete continuous information transfer without paper is determined largely by what basic software is used for the production of drawings. The system for the description and representation of the geometry must contain the interfaces for interaction as well as an interface for the CAD/CAM system. The connection to CAM is executed preferentially in such forms of representation as CLDATA or CADCPL [4]. These coupling modules permit problem-free takeover of the interactively generated workpiece data and supplementary information on blank dimensions, tools, jigs, and fixtures, as well as means for testing. They provide the portability of CAD information through a preprocessor into a processor for NC programming. At this interface, a sliding transition from the CAD or CAM system to the DNC system takes place (Fig. 66). These DNC systems have their connectors to CAD in the dialogue-aided programming of the source program. NC editors, NC programming systems with postprocessors, or programming systems with separate descriptions of geometry and technology for source programs are used [91]. The third program type improves the possibility of automation and flexibility for different machining techniques. Generation fitted to the requirements of the machine tool and the manufacturing process is carried out for the selected technique automatically with access to technology data files which were established earlier.

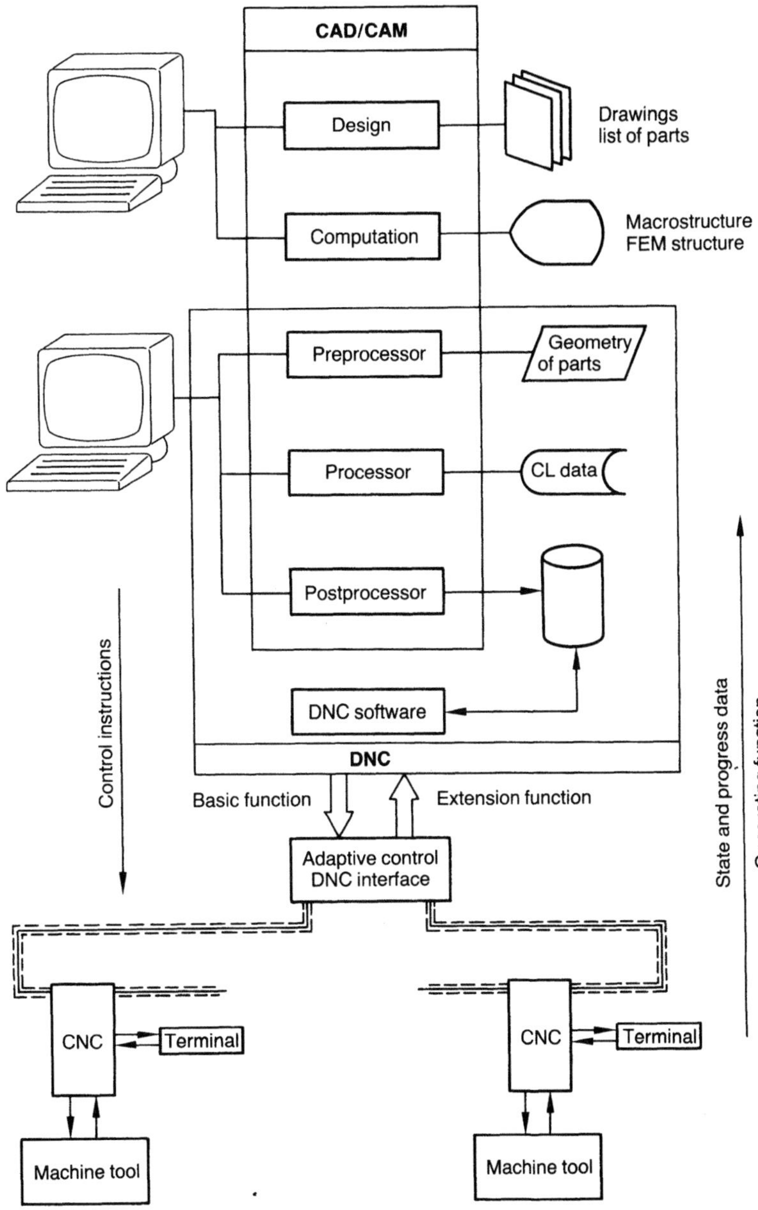

Fig. 66. Transition from CAD/CAM system to DNC system

DNC systems as components of CAD/CAM systems allow for multiple use of already-generated geometry data files. The machining programs are supplied with geometry data via coupling modules. The subprogams are generated independently of jobs and machining processes.

3.5.2.2 DNC Systems as Workshop Communication Systems

Automatic manufacturing in FMS requires communication with the equipment for part machining, part transport, and quality. Moreover, the various kinds of equipment must be coordinated hierarchically by programs of manufacturing control and disposition, and their combined action must be organized. Between this equipment and the functional complexes of the manufacturing control there must be effective hardware and software architecture which is able to communicate. The functions can be recognized in Figure 67. They have a modular design and are capable of communicating with each other. In the German Democratic Republic, the "modular program system" for FMS has been developed for this function [164]. Within such a system, DNC has to fulfill a central task. Figure 67 illustrates the entire set of equipment required for DNC communication. Real-time communication is superfluous due to the CNC and PC systems which are capable of being stored. Communication between the control units and the computers of the system is executed according to the Binary Asynchronous Communication (BAC) procedure. Competitive operation is the main feature. Interruptions by the hierarchically superior computers have priority. The following features of the BAC procedure are most significant:

1. Systematic perfection of the qualitative basic and extension functions of the classical DNC for all units of equipment in the manufacturing and assembly processes
2. Modular supplementing of DNC basic functions through additional functions of programming and interfaces to create CAD systems

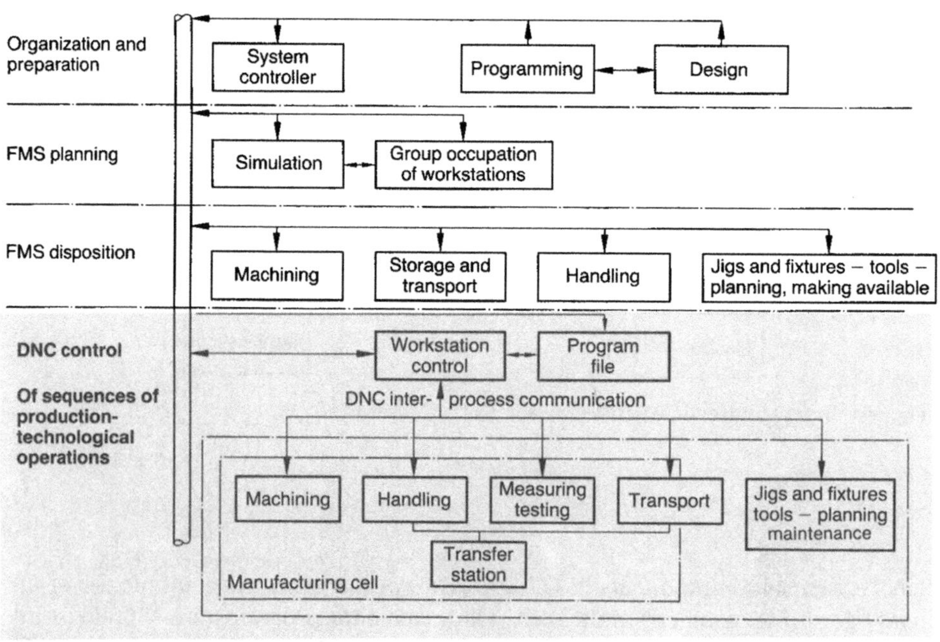

Fig. 67. Hierarchical structure of computer support in the factory

3. Expansion of the basic and extension functions of DNC to communication between the manufacturing or disposition computer and the machine control units, manual work stations, or operating and managing personnel
   With the BAC procedure in FMS, the unidirectionally guided machine controls on the interruption basis in the real-time operation are replaced by bidirectional DNC interprocess communication. Both principles jointly comprise control instructions, operational, state, and progress data, management information, and diagnostic functions [201]. With the BAC procedure direct computer control of machines doing mechanical work has developed further into a workshop communication system.

3.5.2.3 DNC Systems for Process Monitoring and for a Guiding-Center Function

The organization of an FMS is performed by a guidance computer or a process-control system. It has to comprise the program administration and distribution to the equipment for machining, transport, and handling as a DNC basic function. In addition, the acquisition, evaluation, and recording of production and machine data must be carried out. The state data are recorded by the polling technique in a real-time-shifting way in connection with the process. This DNC extension function was already known as interruption organization with classical DNC systems [201]. For the performance as a guiding-center, statements on quality, wear, and disturbances, as well as investigations into the capability of being processed are required, in addition to a statement on the operational status of the system components in the FMS [75]. A suitable evaluation of the recorded measured values is required in control stations to ensure a high level of quality.
Two variants can be realized through DNC:
1. Direct workpiece measurement at an automatic measuring station by continuous checking of the quality of measured values, deviations from the true shape, and trend determination [144]
2. Indirect measurement of quality by automatic recording of measured values of the used tools in the tool-storage unit or turret [11]

Both variants are effective with DNC, and in their evaluation they lead to new correction values for CNC machine control or to the automatic replacing of the tool in the machine.
Investigations into the capability of machines are important functions of guiding centers. Their results are decision aids for disposition and planning. The advanced CNC controls are equipped with a self-diagnostic system for error state determination. Error states of control are recognized by the BAC procedure [164] of the DNC computer. In polling, the DNC computer examines the error system register of the machine control system and introduces an error indication at the control station. At present, diagnosis must be carried out manually. On the other hand, the functional complexes of the workstation occupation, the workplace group occupation, and disposition are informed automatically of failure by the DNC computer. Job data and job stocks can be newly disposed by simulation.

For the long-term optimization of manufacturing the densification of recorded data and statistical evaluation are a good basis for learning from past errors, for deriving important conclusions concerning the prophylactic analysis of disturbances, and for maintenance strategy. Control stations for FMS serve to provide high flexibility, low volume of stocks, optimal productivity, reduction of costs, and constant quality. The most important information basis, on the other hand, is constituted by process-realistic, updated data. They can be made available to an FMS through the DNC extension function, making DNC very important for good organization in an FMS.

3.5.3 Summary and Prospects

DNC is a history-making numerical control technique (Fig. 68). The extension function gains in importance with the stepwise detachment from the real-time guidance of numerical machines doing mechanical work, and with the transition to transfer procedures in the competitive operation between control units and the computer. The basic function is reduced (by the capability of the machine control systems for workpiece program storage) from program distribution, which has to keep pace with the manufacturing process on a batchwise basis, to a single transfer at the job start. By program filing in machine-central, computer-internal data formats, the basic function opens up good prospects with a double function for the DNC principle:

1. Machine-independent and neutral program administration permits the complete connection to CAD/CAM systems via a standardized interface, avoids data redundance, and offers production-technological flexibility for metamorphic FMS.
2. The extension function can be used as a situation-dependent generator for the basic function, making possible a process- and utilization-dependent FMS organization.

These statements elucidate the process of development of DNC. The extension function, orginally developed as a secondary phenomenon of the basic function, more and more performs the process-guiding function in FMS. It can do this by process monitoring all the equipment of the FMS to which programs are supplied in the basic function. Machine-central program filing makes situation-dependent machine occupation in the FMS possible without any delay, and thus guarantees a high degree of flexibility [171]. It is of special importance for future transfer lines with a metamorphic structure capable of job-dependent resetting and changeover. DNC is a technique for the informational interconnection of equipment in preparation for manufacturing; it is capable of keeping pace with progress in the flexibility of manufacture and preparation and constitutes an important step on the way to an automated factory that uses no paper.

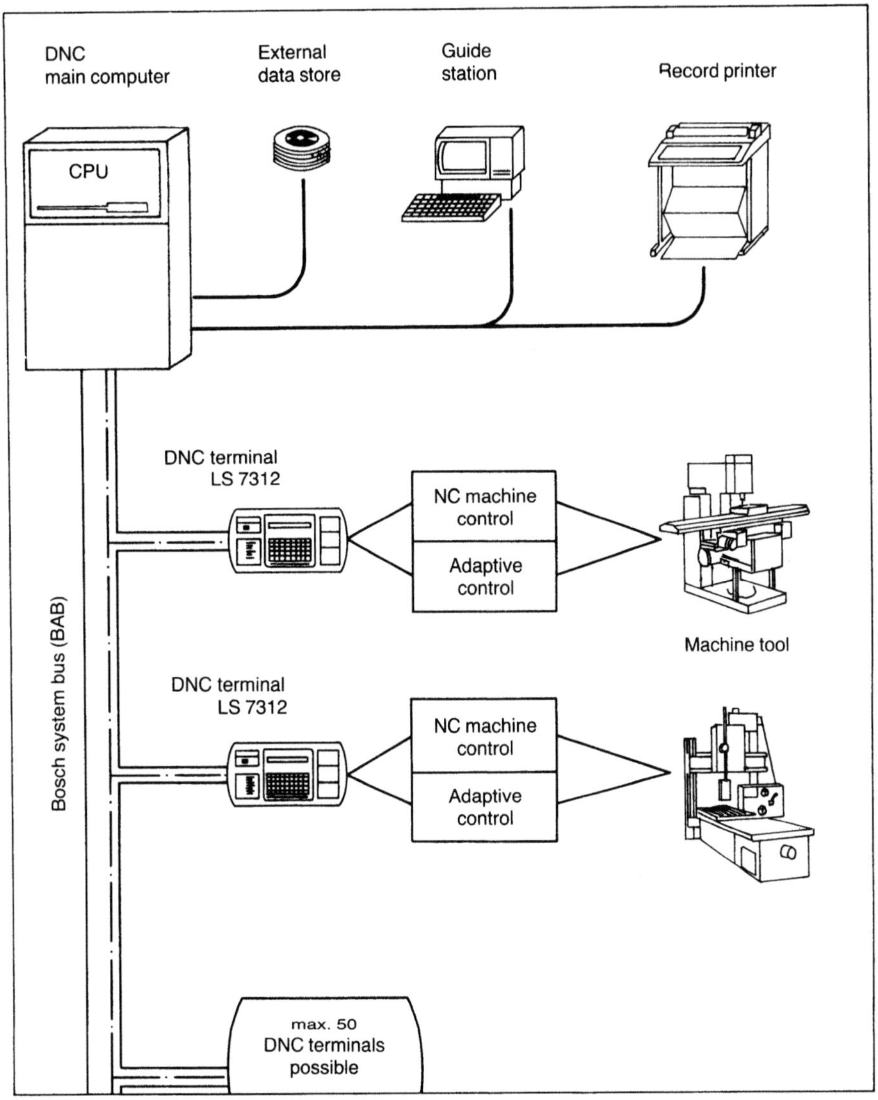

Fig. 68. DNC guidance system LS 7200 in bus design

Table 16. Important stages in the development of DNC

	Event	Designation	Company	General characteristics
1966	PS 1016427	Operation of numerically controlled machine tools with the aid of data processing plants	Molins — ICL	All numerically controlled machine tools are activated according to *one* NC program which is stored in the computer
1966	OS 1538595	Numerically operated programming arrangement for machine tools	IBM Stuttgart	General division of tasks and functions between computer and NC for course and fine interpolation
1967	OS 1588083	Device for controlling several machine tools from one center	Allen — Bradley	Organization of NC data storage and data distribution
1967	OS 1602895	Automatic machine tool system	Molins Machine Co.	NC data distribution and collection of data for SYSTEM 24
1968	OS 1900833	Arrangement for simultaneous numerical control of several machine tools	IBM Armonk	Multiplex channel makes it possible to produce different parts on the individual machines with a whole program put in
1969	OS 1901905	Monitoring system for numerical control systems in time multiplex operation	Bunker Ramo Corp.	Collection of process and state data on the basis of the machine tools with central control by the computer
1969	OS 19095453	Monitoring unit for a simultaneously operated NC system	Bunker Ramo Corp.	Process monitoring on the basis of centrally controlled machine tools by means of specific computer-internal distribution of data to the machine tools
1969	OS 1908757	Unit for the direct control of machine tools by computers	Sundstrand Corp.	Data flow in alternating directions on the basis of "all-purpose connection in both directions"
1969	AS 1966794	Unit for the numerical control of machine tools by means of a central data processing plant	Sundstrand Corp.	NC program change in dialogue with the data processing plant; editing of new programs with translation into machine language
1971	PS 97963	Circuit arrangement for the numerical control of work stations, especially of machine tools, by a central data processing plant	FZW "Fritz Heck."	CONCEMA-DNC with a ring-line structure and a computer external address distribution

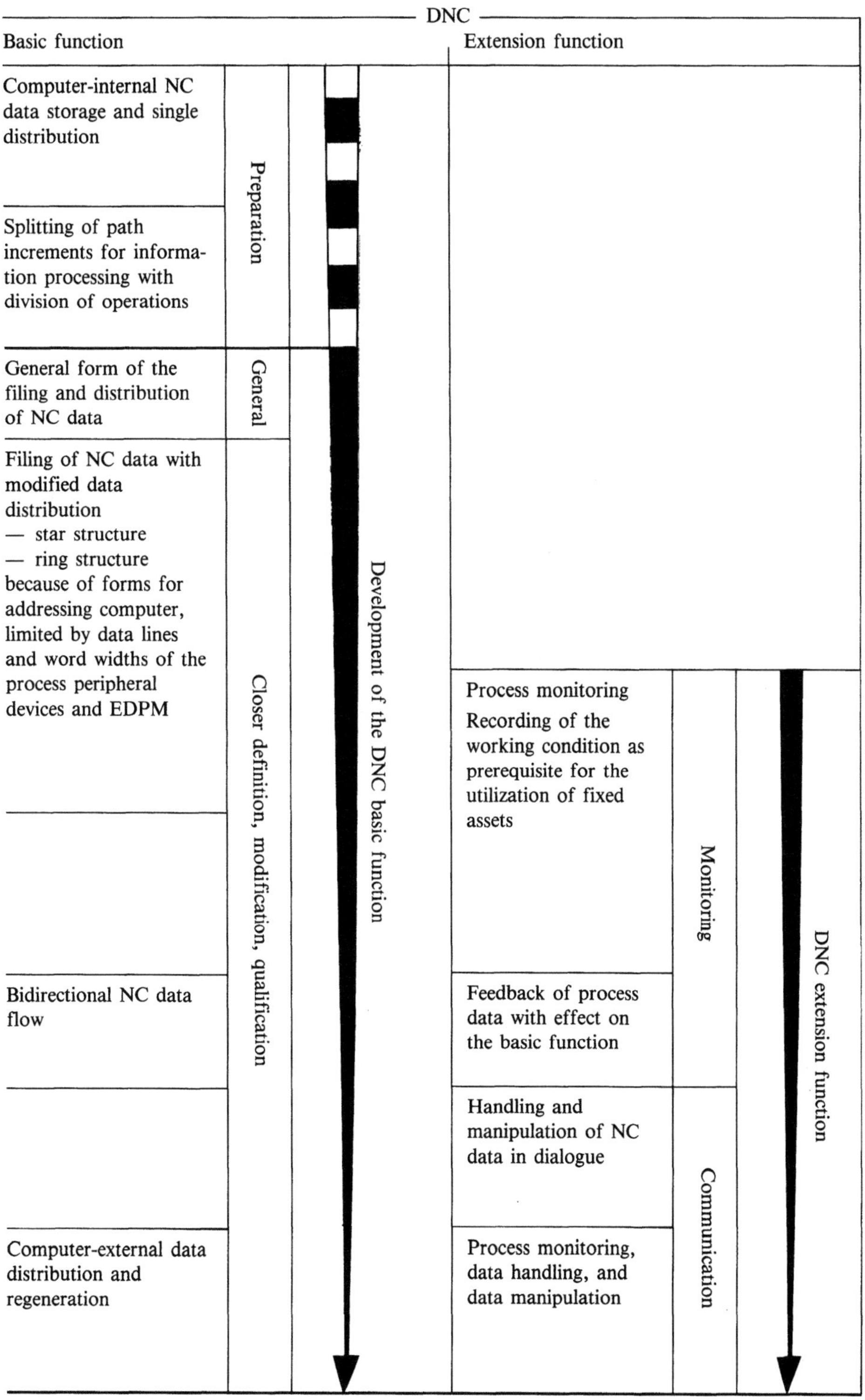
DNC
Basic function
Extension function
Computer-internal NC data storage and single distribution
Splitting of path increments for information processing with division of operations
Preparation
General form of the filing and distribution of NC data
General
Filing of NC data with modified data distribution
— star structure
— ring structure
because of forms for addressing computer, limited by data lines and word widths of the process peripheral devices and EDPM
Bidirectional NC data flow
Computer-external data distribution and regeneration
Closer definition, modification, qualification
Development of the DNC basic function
Process monitoring
Recording of the working condition as prerequisite for the utilization of fixed assets
Feedback of process data with effect on the basic function
Monitoring
Handling and manipulation of NC data in dialogue
Process monitoring, data handling, and data manipulation
Communication
DNC extension function

Table 16. Continuation

	Event	Designation	Company	General characteristics
1972	AS 2244170	Program control for DNC-aided control of several machine tools at one production plant	Toyoda Koki	Data are interpolated by the computer and made available to the servo drives of the machine tools with adequate clock pulses and frequency
1975	OS 2500086	Diagnostic connection system for computer-controlled machine tools	K & T	Monitoring of operational data during run and comparison with desired data; signaling of ranges of changes and prediction of abnormal states
1977	OS 2714267 OS 2714268	Circuit system for control of machine tools; circuit arrangement	IBM Armonk	Reciprocal data transfer through two unidirectional data buses, as organized through the central processor and macrofunctions
1979	OS 2904080	Control system for machine tools	Toshiba Kikai	Tool monitoring and control-internal tool change; the STARTISTOP operation is controlled on a superior level and serves process monitoring
1979	OS 2932790	Numerical program control for machine tools	Siemens	Input of NC data is performed in the source program form, and they are computer-internally stored; processing in the machine is organized by subprograms
1981	OS 2950979	Numerical control system	Allen-Bradley	Diagnosis with interface processor and front-panel processor in the main store; diagnosis via terminal in dialogue with the computer
1981	OS 3146342 A1	NC data-division and data-processing units for NC machine tools	Hitachi Ltd.	New computer-internal representation of NC data in processor transfer form; fixed program assignment executed by postprocessing at job start

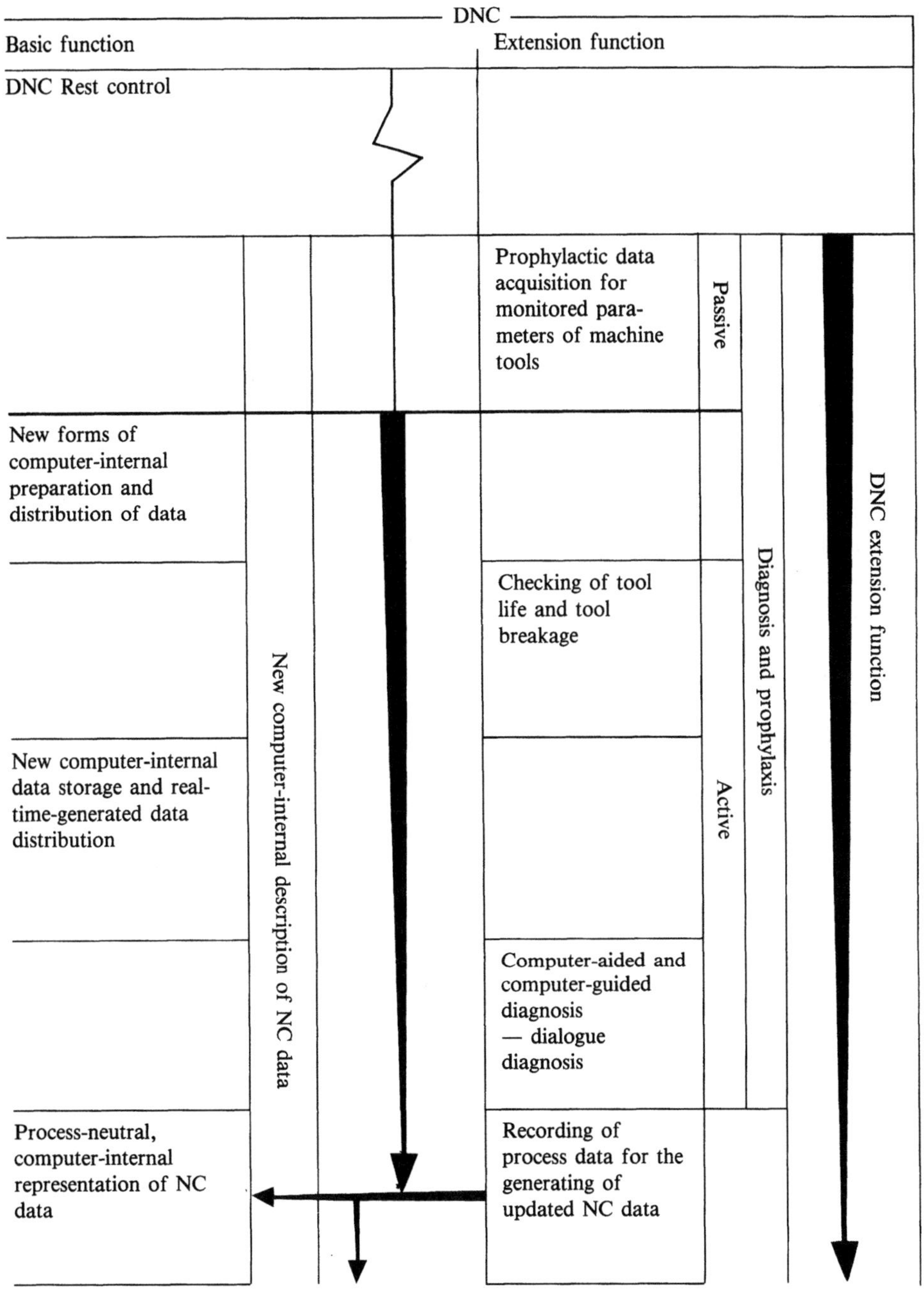
DNC
Basic function
Extension function
DNC Rest control
Prophylactic data acquisition for monitored para-meters of machine tools
Passive
New forms of computer-internal preparation and distribution of data
Checking of tool life and tool breakage
New computer-internal data storage and real-time-generated data distribution
Computer-aided and computer-guided diagnosis — dialogue diagnosis
Active
Diagnosis and prophylaxis
DNC extension function
New computer-internal description of NC data
Process-neutral, computer-internal representation of NC data
Recording of process data for the generating of updated NC data

4 Methods of Decision-Making for Investments

Due to the high investment value of highly productive NC manufacturing systems it became necessary to develop scientific and computer-aided methods of decision-finding for optimum employment.

The following initial conditions have to be taken into account:
- The investment costs run between several hundred thousand and millions of marks or other currency units.
- On an international scale, several hundred different control types in connection with different kinds of machine tools are developed per year and offered with a large number of modifications of equipment.
- The starting point for all decisions is the given assortment of manufacturing tasks.

On principle, in the field of production-technological preparation of production decision aids are required for the following ranges of tasks:
a) for long-term planning of investments,
b) for efficient utilization of the manufacturing system employed, and
c) for further technical development based on experience derived from practice and overall technical development (see Fig. 69).

While points b and c above are covered in Sect. 5, some principles of decision-making in the employment of highly productive manufacturing systems are outlined here.

A number of decision aids have been developed which can be subdivided into two main categories:
1. Analytical methods of computation for individual machines or equipment variants
2. Simulation methods for complex system solutions (FMS) and dynamic sequences of processes or techniques

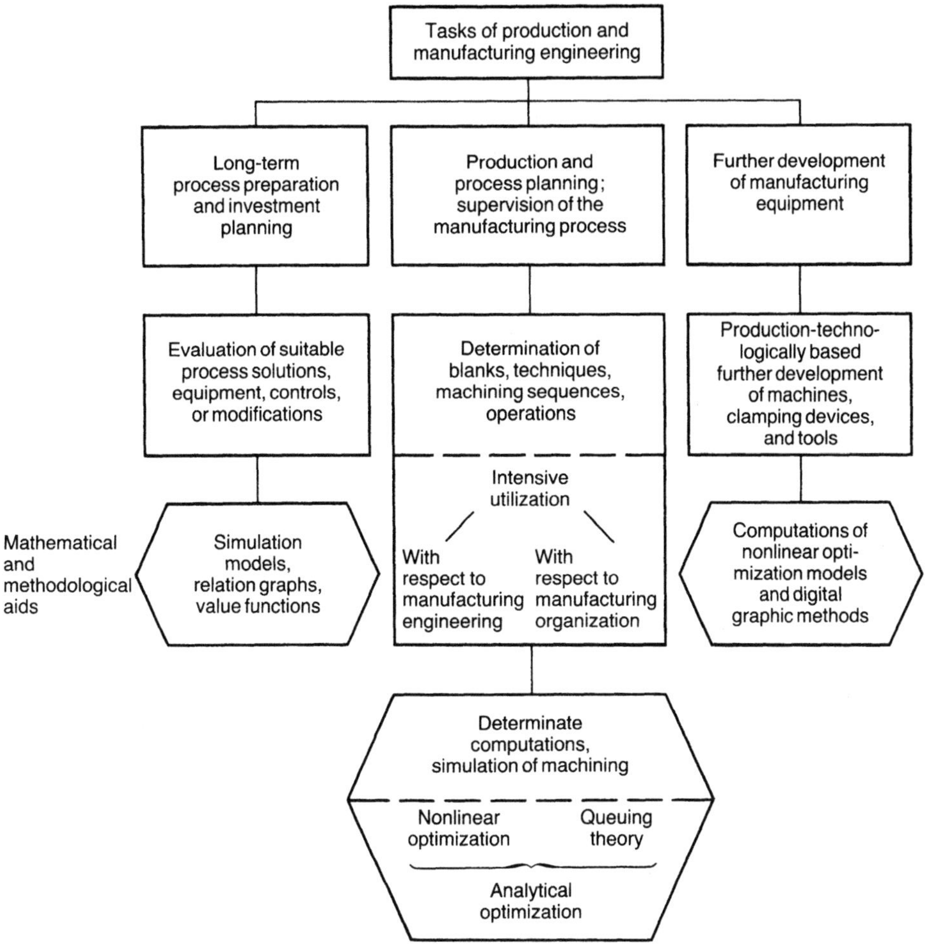

Fig. 69. Important tasks in manufacturing process design and suitable methodological aids

4.1 Analytical Methods of Computation for the Production-Technological Preparation of Employment

In this field, some methodological aids have been developed during the past few years. For problems connected with decision-making mostly enterprise-specific solutions are available. Therefore, no methods with general validity on an international scale can be presented. The formulation of problems and some essential principles for decision-making will be discussed as examples.

Variants of the component	Technical assignment			Economic estimation		
	Parameters for decision-making	Possible process solutions from technical viewpoint	Optimum process solution	Economic decision-making based on operating time	Economic decision-making based on costs	Optimum process solution
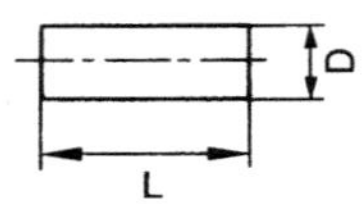	Diameter D length L batch size quality	– Lead screw, lathe – Copying lathe – NC lathe	Lead screw lathe →	Because of the extreme simplicity of components an economic estimation is not necessary		Economic application of lead screw lathes is guaranteed for all cases
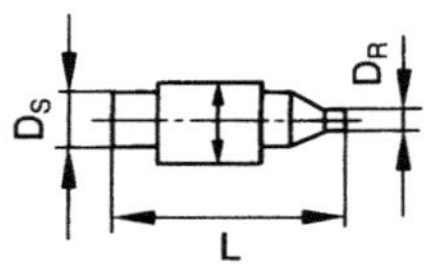	Diameters D, D_R, D_S length L taper curve radius R batch size quality	– Lathe with lead and/or feed screw – Copying – NC lathe	No unambiguous decision possible →	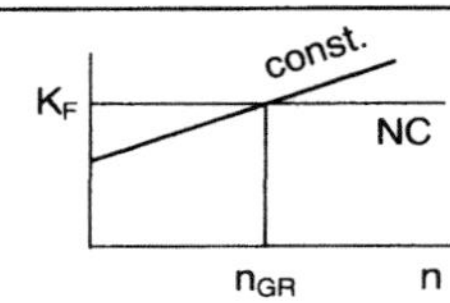Consumption of time	n_{GR} : limit of batch size	Up to n_{GR} application of NCM is recommended above the limit of the batch size har use a copying lathe
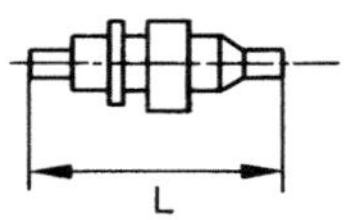	Diameters D, D_R, D_S D_E, D_C length L, L_E, L_G taper curve radius batch size quality	– Lathe with lead and feed screw – Copying lathe – NC lathe	NCM →	An economic estimation is not necessary because the components are very complicated, and NC manufacturing is the most economical (check the utilization of the machines)		NCM usable for all cases (verification of efficiency is needed)

Fig. 70. Selection of machine tools for long rotary parts from a technical-economic viewpoint

4.1.1 Formulation of Problems

As pointed out in the introduction, the production-technological preparation of employment focuses on a technical-economic coordination of the given individual parts. For simple parts with low machining costs, a small number of different shapes, and low demands for accuracy, it is clear that — from an economical viewpoint — conventional machining has to be used.
In contrast, with increasing machining costs, a higher degree of complexity, and higher demands concerning accuracy and surface, this decision is not unambiguous in all cases. This is explained by the example of rotationally symmetrical parts in Fig. 70. If one proceeds from the numbers of parts ordered which are typical of general machine building and lot sizes (100-500 parts/year = ca. 20-50 parts/lot), one arrives at the decision spectrum represented in Fig. 70.
It is found that at a high level of complexity no unambiguous statements are possible without more detailed considerations and calculations. These necessary technical-economic investigations will be characterized in more detail below, special attention being paid to integration into the overall decision-making process.

4.1.2 The General Sequence of Decision-Making

The sequence of decision-making on a production-technological basis is illustrated by Fig. 71. The exact determination of the whole part assortment to be considered is the first step. Here one proceeds from a classification of parts and a combination of suitable groups of parts.
Since as the result of the classification up to several hundred types of parts can be covered for which no more detailed investigation is possible for reasons of capacity, one is confined to suitable representatives. It is important that, with respect

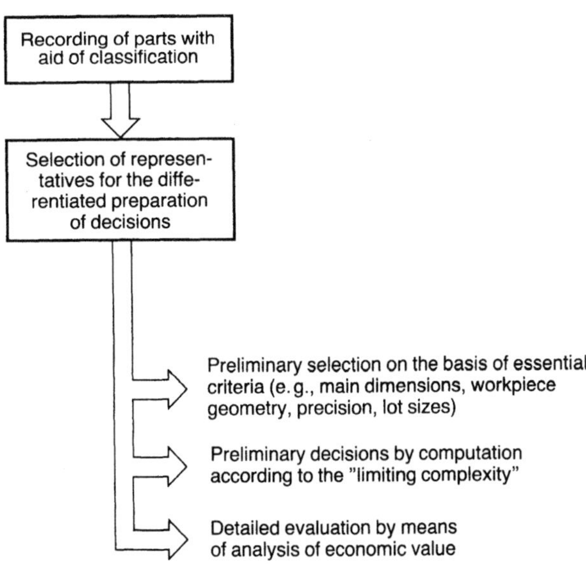

Fig. 71. Production-technological preparation of highly productive instruments

to complexity and all the other characteristic features, really representative parts are selected, i.e., the parts should be selected in such a way that the lower, intermediate, and upper limiting ranges are included for the most important features, such as dimensions, workpiece geometry, and lot sizes. The subsequent decision steps are based on this one.

4.1.3 Preliminary Decision on Machine Selection Based on Degree of Complexity

The general suitability for NC manufacture as compared with conventional machining on nonautomated machines can be determined on the basis of characteristic features of the workpiece. Deciding according to a specific degree of complexity has proved to be efficient. A differentiated approach is necessary for the more important cutting processes (see Fig. 72).

Manufacturing process	Turning	Milling	Drilling
Functional equation	$DC_{TU} = \sum_{j=1}^{w} (a_1 + b_1 + c_1)_i$	$DC_{MI} + \sum_{i=1}^{w} (b_2 + c_2 + f + g + h)_i$	$DC_{DR} = \sum_{j=1}^{w} (a_3 + b_3 + c_3)_i$
Charac-teristics	w = number of form elements i = for-statement	w = number of surfaces to be machined i = for-statement	w = number of holes i = for-statement
Evaluation factors	a_1 = geometrical shape b_1 = surface finish c_1 = dimensional accuracy of the form element	d = mode of clamping e = rigidity b_2 = surface finish c_2 = dimensional accuracy f = degree of confinedness g = form milling operations h = number of cuts	a_3 = geometrical shape b_3 = dimensional accuracy c_3 = surface finish

Fig. 72. Degrees of complexity for determining the suitability of workpieces for NC machining

4.1.3.1 Determination of the Degree of Complexity for Turning

The degree of complexity for turning can be determined by means of Eq. (1):

$$DC_{TU} = \sum_{i=1}^{w} (a_i + b_i + c_i)_i$$

w = number of shape elements
i = for-statement (consecutive index)

Factors:
a_i — factor for evaluating the geomtric shape of the element of shape
b_i — factor for evaluating the surface quality of the element of shape
c_i — factor for evaluating the dimensional accuracy of the element of shape

A summary of the evaluations for the individual factors is given in Tables 17-19.

Table 17. Factors for evaluating the geometric shape of the workpiece

Element of shape	Evaluation factor
Undercut Land Radius	$a_1 = 1$
Cylinder Plane surface Recess	$a_2 = 2$
Cone Thread	$a_1 = 3$
Curved surfaces	$a_1 = 4$

Table 18. Factors for evaluating the surface quality of the workpiece

Symbols	Peak-to-valley-roughness (μm)	Evaluation factor
~	>160	$b_1 = 0$
	41-160	$b_1 = 1$
	10- 40	$b_1 = 2$
	< 10	$b_1 = 2$

(The evaluation factor $b_1 = 0$ is assigned to the shape elements of undercut, land, radius.)

Table 19. Factors for evaluating the dimensional accuracy of the workpiece

Nominal dimension (mm)	Angle	Tolerance (μm)	Evaluation factor
10^1	1°	Untoleranced dimension	$c_1 = 0$
10^{-1}	1'	101—200	$c_1 = 1$
10^{-2}	1''	40—100	$c_1 = 2$
10^{-3}		<50	$c_1 = 3$

4.1.3.2 Determination of the Degree of Complexity for Milling

For the determination of the degree of complexity for milling on milling or drilling machines Eq. (2):

$$DC_{MI} = d \times e + \sum_{i=1}^{w} (b_2 + c_2 + f + g + h)_i$$

w = number of machining points (surfaces on the workpiece to be machined)
i = for statement (consecutive index)

Factors:
d — factor for evaluating kind of clamping
e — factor for evaluating rigidity
b_2 — factor for evaluating dimensional accuracy
c_2 — factor for evaluating surface quality
f — factor for evaluating degree of limitation
g — factor for evaluating form milling operations
h — factor for evaluating number of cuts

A summary of the evaluations for the individual factors is shown in Tables 20 and 21.

Table 20. Weighting factors for milling (extract)

d-Clamping device	Weighting factor
Vise	d = 2
Fixture	d = 3
Table	d = 4
e-Rigidity of the material	Weighting factor
Normal	e = 1
Risk of distortion	e = 1.25
b_2 Accuracy	Weighting factor
Without tolerance	$b_2 = 0$
0.2 mm ≧ tolerance > 0.1 mm	$b_2 = 1$
tolerance ≦ 0.1 mm	$b_2 = 1.5$
c_2 Surface roughness	Weighting factor
~ R_t or R_z > 160 μm	$c_2 = 0$
R_t or R_z = 41—160 μm	$c_2 = 1$
R_t or R_z = 6.3—40 μm	$c_2 = 2$

Table 21. Weighting factors for holes (extract) a_3—geometric shape

Primary criterion	Secondary criteria	Tool number	Weighting factor a_3
1. Through holes	Ø ≦ 25 mm (steel) in solid material without counterbored	1	1
	Ø = 32 mm (gray cast iron) in solid material without counterbored	1	1
	Ø ≦ 25 mm (st) in solid counterbored	2	2
	Ø ≦ 32 mm (gci) in solid counterbored	2	2
	Ø > 25 mm (st) hole precast	1	1
	Ø > 32 mm (gci) hole precast	1	1
	Ø > 25 mm (st) hole precast counterbored	2	2
	Ø > 32 mm (gci) hole precast counterbored	2	2
	Ø > 25 mm (st) in solid	2	2
	Ø > 32 mm (gci) in solid	2	2
	Ø > 25 mm (st) in solid	3	3
	Ø > 32 mm (gci) counterbored	3	3
2. Stepped holes	Ø ≦ 25 mm (st) in solid	1	2
	Ø ≦ 32 mm (gci) in solid	1	2
	Ø > 25 mm (st) hole precast	2	3
	Ø > 32 mm (gci) hole precast	2	3
3. Blind holes	Ø ≦ 25 mm (st) without counterbored	1	2
	Ø ≦ 32 mm (gci) without counterbored	1	2
	Ø ≦ 25 mm (st) counterbored	2	3
	Ø ≦ 32 mm (gci) counterbored	2	3
	Ø > 25 mm (st) without counterbored	2	3
	Ø > 32 mm (gci) without counterbored	2	3
	Ø > 25 mm (st) counterbored	3	4
	Ø > 32 mm (gci) counterbored		

4.1.3.3 Determination of the Degree of Complexity for Drilling

Equation (3) is used to determine the degree of complexity for drilling:

$$DC_{DR} = \sum_{i=1}^{w} (a_3 + b_3 + c_3)$$

w = number of machining points on the workpiece (holes)
i = for-statement (consecutive index)

Factors:

a_3 — factor for evaluating geometric shape
b_3 — factor for evaluating dimensional accuracy
c_3 — factor for evaluating surface quality

4.1.4 Technological-Economic Value Analysis for the Differentiated Evaluation

For all decision aids described so far the analysis started from more or less comparative considerations of the characteristic features of the parts assortment and the machine data. In consideration of the extraordinarily high costs involved with the high-capacity machines, a detailed method of evaluation must also be available.
Moreover, it should be taken into account that the machines are offered with basic equipment and with various kinds of accessory equipment; the decision on the production-technological and economic justification for using them must be made by the user. Therefore, there must be a method that proceeds primarily from given features of the workpiece assortment or a detailed quotation of price to permit both an overall evaluation and a differentiated assessment.

4.1.4.1 Buildup of the Target Complex

The basic aim of the evaluation model is the optimal adaptation of the numerically controlled manufacturing system to the given manufacturing task. In general, maximum utilization of the manufacturing equipment is in the foreground for the buildup of the target complex. A distinction must always be made between utilization of time and utilization according to functions (technologically).
While utilization of time must be clearly determined prior to investment and secured under operational conditions by the required organization of production, functional utilization is the main point of the production-technological evaluation. The general structure of the target complex is illustrated in Fig. 73.

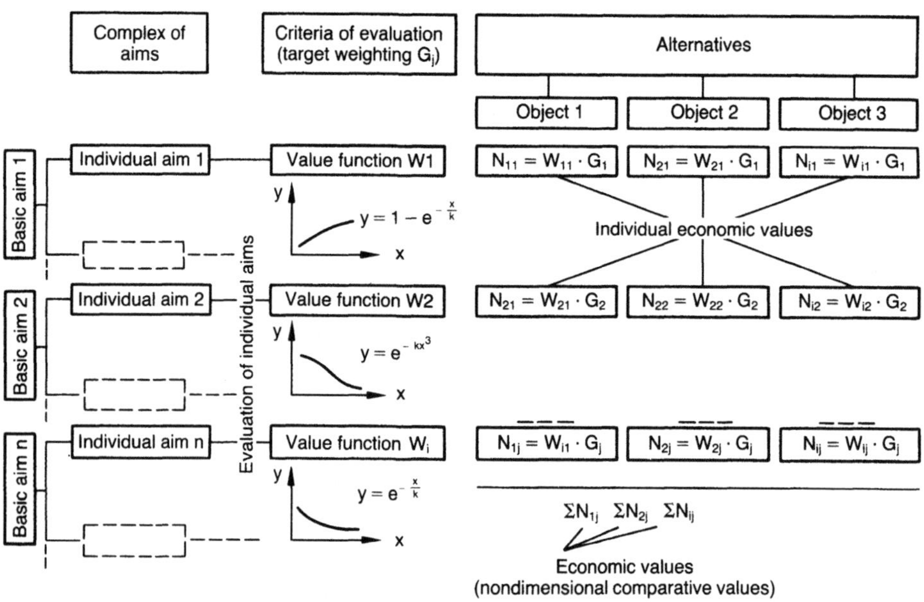

Fig. 73. Elementary process of the analysis of the economic value

By the formulation of specific targets and pertinent target criteria a systematic evaluation is achieved; it can be obtained both manually and in a computer-aided manner.

4.1.4.2 Weighting of Targets, Determining of Value Functions and Use Values

With the aid of the function value W_{ij}, the influence of the given aim z_j of an i-th variant on the increase of productivity or the reduction of manufacturing time is expressed. The influence of the costs arising for a subtarget is covered by the pertinent weighting G_j. In a concrete sense, this demand means the differentiated recording of the cost elements of the basic and accessory components.
If one succeeds in approximating an exact determination of the value function W_{ij} in a problem-related way, the subjective factor of this method will be largely eliminated. By multiplication of target value (function value) W_{ij} by weight G_j, an individual use value is determined for each of the individual targets to be evaluated. Therefore, a numerically controlled machine can be evaluated on the basis of an examination of the individual use values or by the combination of the total use-value numbers.
Use-value analysis has proved to be a comprehensive setup for complex evaluation problems. It is a method for systematic decision-making in complex evaluation problems, the absolute economic gain not having to be determined primarily.
The general sequence of operations in use-value analysis is presented in Fig. 74, the important stages being:

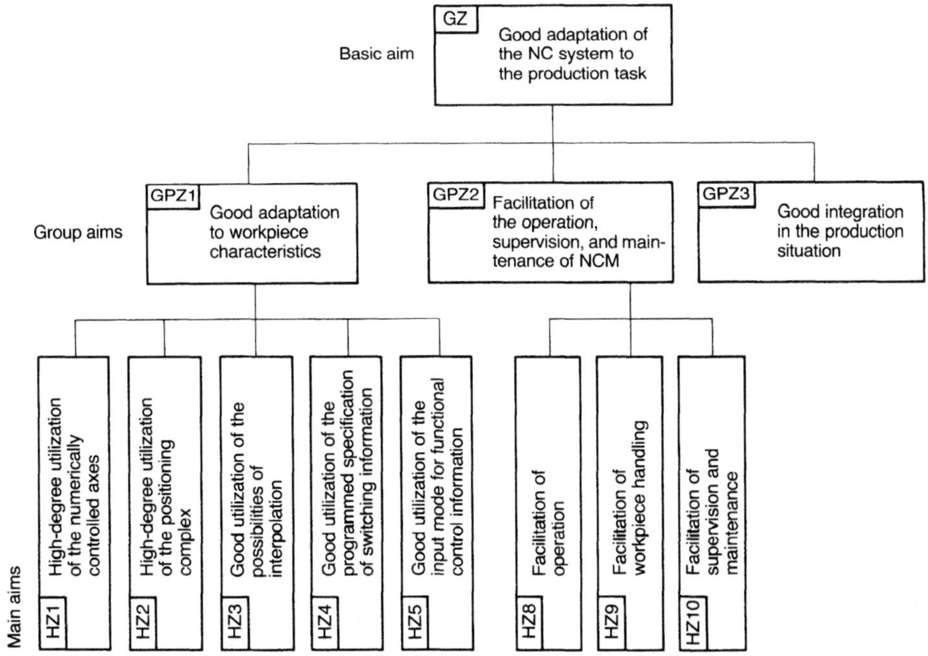

Fig. 74. Example of a complex of aims

1. Buildup of specific target complexes consisting of basic and individual targets (if necessary, a more detailed subdivision)
2. Determination of pertinent value functions for each individual target
3. Computation of individual use values
4. Combining to an overall use value for a particular object

4.1.5 Machine Selection on the Basis of Determination of Economy

Exact determination of economy is required for the final selection. Any statement on efficiency is based on the calculation of the prime cost differences between the conventional manufacture and the new one proposed.
The prime cost difference for the manufacture of a particular workpiece is expressed by Eq. (4):

$$K_S = K_{S_1} - K_{S_2}$$

where K_S = prime cost difference
K_{S_1} = old prime cost
K_{S_2} = new prime cost

The problem is that, in general, no exact time values are available from production-technological detailed planning. Usually, this is solved by (a) using selected representative parts as a basis for decision-making or (b) carrying out an approximate time determination by means of special analogy methods integrated with methods of group technology.

4.2 Methods of Simulation

The impact of the computer on the technological aspects of manufacturing has been discussed in preceding sections. In this section the use of the computer as a means of investigating the manufacturing function itself will be considered, not as a component of the system but as a tool for discovery. This approach views the manufacturing process as a system, and attempts to achieve productivity and efficiency improvements through a better understanding of the manner in which these components are selected and controlled, and of their interactions. A brief description of the simulation process is given, followed by a discussion of the need for simulation from the manufacturing problem perspective, examples of the use of simulation in FMS design by various groups, and a projection of probable development directions.
This section is restricted to discrete event simulation with the emphasis on computer-aided manufacturing, also known as flexible automation. With fixed automation the operations to be performed on a workpiece are set and the same sequence

of operations occurs for every workpiece. In flexible automation, the sequence of operations depends upon the workpiece being processed. The system must have information identifying the individual workpieces to invoke the proper operations and sequences; thus, this type of automation can be called "data driven". There are also uses of continuous and hybrid simulation in manufacturing; however, they are not sufficient in number or well enough documented to warrant their inclusion in this brief summary.

4.2.1 Discrete Event Simulation

Simulation depends upon the concept of a model which is an abstraction of the system under study, in the same sense that a play is an abstraction of real life. The objective is to capture the essence without the detail. There can be many different abstractions of a given system, both by level of abstraction and by content, the correct choice depending upon the modeler's objectives.

In discrete event simulation one develops a model by describing the elements, or components, of the system, i.e., the things of which the system is the aggregate, and the manner in which these elements interact. These elements are generally known as *entities.* In a job shop the entities of interest might be machines, operators, and workpieces. If the system under study were a traffic system, the entities might be cars, drivers, roads, and intersections. It is generally useful to give characteristics, called *attributes,* to these entities to distinguish between them. In the job shop the workpieces might have attributes such as due date, weight, and length. The operator attributes might be skill level, assigned machine, pay rate, and name. Entities are usually physical but can be logical as well. For instance, a machine group can be viewed as a logical collection of the machines in it; first-come-first-serve is a logical entity for choosing the next workpiece to be processed on a machine; and the assignment of an operator to a machine can also be viewed as a logical entity with the attributes of machine and operator.

Using the basic concepts of entities and attributes it is possible to describe a model giving a picture, or snapshot, of the system at a point in time. The total set of information needed to describe this picture of the model is known as the *system status.* Simulation is the process of moving the model through time by continually updating the system status, much as one might make a movie by combining a series of snapshots. For instance, if the status of the operator is idle, an assignment to a task may be made and the operator's status changed to busy. To do this, one must determine the rules by which the attribute changes are to be made, i.e., the logical relationships which must exist before a status change can be implemented. For example, when an idle operator, a free machine, and a waiting workpiece are simultaneously available, an assignment (i.e., a status change) can take place. It should be clear that a thorough understanding of the manufacturing system is required to develop the model and the rules for system change. Without this knowledge base the prospects of success are slim and chance dependent. There are many ways in which these rules for status change can be described. Techniques based on directed graphs or Petrie nets appear to hold great promise and are currently gaining wide acceptance.

Perhaps the most widely used directed-graph approach is activity cycles, which follow the entities through time. They recognize two types of states, active and idle, for entities and require that entities alternate between them. The behavior of each entity class is described by these alternating active (activity) and idle (queue) states. The rules for changing system states, i.e., the simulation, can be directly derived from the aggregation of these behavioral patterns, called an *activity cycle diagram* (ACD). The ACD contains both the structure and logic needed to derive the rules for simulation. Using the basic principles of ACD, at least two English systems, HOCUS and CAPS, have been developed for automating this process. Both systems accept ACDs as input and generate computer code implementing the logic of the ACD for simulation purposes.

By recording each change of system status one obtains the complete history of the system. This provides the information needed for system analysis and evaluation. It cannot be overemphasized that this historical record is but *one* sample of the system under study. In some cases a single sample is a sufficient basis for understanding. In most cases involving stochastic variables, multiple samples (i.e., simulation runs) are required to reduce experimental error. The ability to repeat model runs enables the analyst to reduce the experimental error to desired levels at the cost of simulation time. One can also increase the level of detail in the simulation to obtain a better representation of the system, as well as to divide the processes into their more basic elements. For instance, one could describe the processing of a workpiece on a machine by assigning it to the machine for a duration equal to a single attribute of the workpiece, RUNTIME. Or one could give the workpiece attributes of SETUPTIME, METALREMOVED, and ENERGYREQD, and the machine attributes of HORSEPOWER and REMOVALRATE, and determine the length of time required to process the workpiece as a function of the attributes of both. The latter method would be expected to provide a more accurate representation and to require a better understanding of the process, but it would require more effort in model development and substantially more data.

4.2.2 Need for Simulation in Manufacturing

Products manufactured in very high volume are efficiently handled with mass production/assembly line techniques. The high volume justifies large capital expenditure for equipment designed as product specific. Product variation and even engineering design changes are often difficult and expensive to incorporate.

At the opposite end of the discrete manufacturing spectrum is the classic job shop, where variation in product is a fact of life; batch sizes are small, work arrives constantly, and there is no commonality of routing. Capital investment is in the process, not the product. Job shop systems are designed to accommodate high product variability, paying the price in low efficiency.

In most industrialized nations the majority of discrete manufactured parts fall between the process-oriented job shop and product-oriented dedicated line, in an area often called "mid-volume, mid-variety". The item is produced in batches to spread the fixed costs of setup over a number of workpieces. Batch sizes are limited by the desire to keep work-in-process and finished inventories low, to meet due dates on shop orders, and to keep work centers from long periods of dedication to a single product. It has been estimated that over 75% of machined parts are made in batches of 50 or less.

All types of manufacturing situations offer ample oppportunity for the useful application of simulation, but the job shop appears to be particularly rich in simulation potential, particularly because of the emphasis that has been placed on automation in this area. The problem is roughly that of automating the job shop with the objective of achieving a balance between efficiency in product-oriented production and the variety of the job shop, and simulation appears to be one of the few tools available.

The operation for job shop is, in theory, rather simple: at any point in time to assign the best batch of work to any machine that becomes idle. The difficulty is that the information necessary to operate and control a job shop is of sufficient magnitude to exceed the comprehension of all but the ablest human minds. For instance, in a job shop with 50 machines, an average processing time of 20 min, and a backlog of 6 weeks' work, one might find 16000 workpieces from which to choose the best batch to be processed. Each of these batches could have information on due date, customer, priority, batch number, number of parts, assembly identification, operation identification, physical location, and raw material requirements which might have a bearing on choice of the best next batch to be processed. When one adds to the batch information the requirements for customers, part routings, work centers, and financial information, the data base necessary may exceed 10^{10} bits. It is little wonder that managers have resorted to infinite loading of machines when confronted with such a mass of data.

Data-driven, or flexible automation is based upon the development of automated procedures, or algorithms, an example being the job shop, i.e., making all of the required decisions, such as the choice of the next batch. The operation of the system is a function of the aggregation of *all* of the instances of *all* of the algorithms, i.e., the algorithms do not operate independently but in conjunction with the others in a dynamic environment. Thus, it is not just a question of finding a good procedure for choosing the "best" batch in isolation, but a problem confounded by all of the other choices. Under these circumstances, few people have the insight to make proper choices without extensive testing and modification. Testing in the actual system is usually so expensive in both time and money as to be impractical. Many system designers and builders have turned to simulation for this testing as the only reasonable approach. A model of the system is developed, the logic of the algorithms implemented, simulation runs made, results analyzed, modifications made, and the process repeated until satisfactory results are achieved. The final algorithms then become the basis for actual system implementation.

The question is often asked, "Can I use simulation for my problem?". This is the wrong question, since the response is always affirmative. The proper question is, "*Should* I use simulation?"

4.2.3 An Example of the Use of Activity Cycles in Computer-Aided Manufacturing

The fundamentals of activity cycles were described in Sect. 4.2.1. In this section an example of the use of activity cycles is given, starting with the description of the system and the problem. The activity cycle is described and its use in developing the simulation and attaching the problem is shown.

The system under study is the preliminary design of a small (four machining centers) flexible manufacturing system with the automated transfer of palletized workpieces. The problem was to determine whether automated scheduling of the system was feasible. The workpieces arrived at the load-unload area in batches of ten. The four machines were served by a single track cart. Each machine was on a separate spur with a single position for a pallet to wait processing, the machine bed, and a single position for a processed workpiece to wait for the cart to transport it. The physical layout of the system is shown in Fig. 75. The circles in the figure indicate the physical positions for pallets in the system. All are shown except the single position on the cart; thus, it is physically possible to have 16 pallets in the system, but logically impossible to have more than 14 if pallets are to move between locations. One of the subsidiary questions was how many pallets were to be used.

The cycle of the workpiece starts when it is loaded onto an empty pallet by the operator. Assume that any workpiece can be processed on any machine. (The machines are indentical; tooling restrictions make them unique, but this will be discussed later.) When a machine has a free entry position and the cart is available, the workpiece is transported to the entry position and unloaded, freeing the cart. When the machine is free, the workpiece is automatically moved onto it, releasing the entry position for the next workpiece. The workpiece is processed on the machine. When completed, it is moved to the exit position if it is available; otherwise, the workpiece and machine must wait until the cart picks up the workpiece occupying the exit position. The cart picks up the workpiece from the exit position when one of the three load-unload stations is free. The workpiece is moved onto the

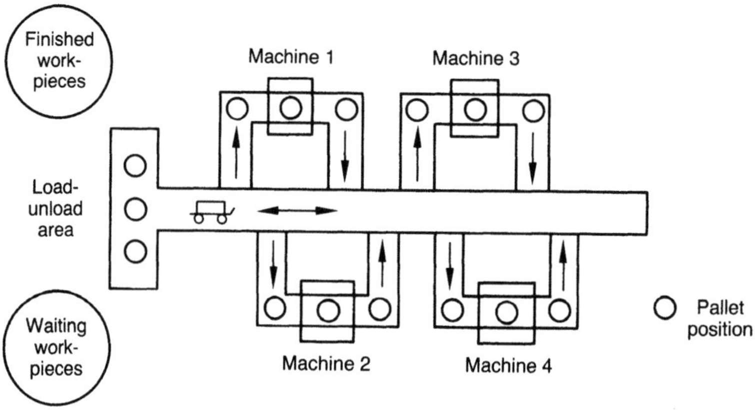

Fig. 75. Four-machine FMS physical layout

station by the operator, freeing the cart. The operator determines whether the workpiece (a) is completed, or (b) requires reclamping and refixturing and more machining. In the study it was assumed that the workpiece required one operation, i.e., to pass through the system, the idea being that problems with the control of this simplified system would strongly suggest that the more complex real world would have worse problems.

The operator unloads the workpiece from the pallet and places it on the transporter of finished workpieces, leaving the pallet empty and ready for the next workpiece. When ten workpieces have been completed, the operator processes them, i.e., they are shipped. The first simulations were based upon the assumption that the system was infinitely loaded, i.e., there was never a lack of a system workload.

The ACD for the system is given in Fig. 76. The entities of interest are workpiece, operator, load station, pallet, entry, exit, machine, and track. Note that the cart is not included in this list, as its behavior is indistinguishable from that of the track. In the same sense, once a workpiece is on a pallet, their behavior patterns or cycles are indistinguishable, so it is necessary to follow only the pallet through the system.

If we follow a workpiece through Fig. 76, it begins in a POOL. When ten are available (signified by the small 10 between POOL and COLLECT) and the operator is in IDLE, the activity COLLECT can take place, with the result that the workpieces are moved to FRESH. When a processed workpiece-pallet combination is available on a load station (i.e., in OLD) and the operator is in IDLE, EXCHANGE can take place with the result that the finished workpiece is taken off the pallet and placed in MADE; the next workpiece is then removed from FRESH and placed on the pallet. This combination is still on the load station and must wait there until the track is CLEAR and an entry is in FREE, at which time the activity MOVE FROM LOAD STATION to machine spur can begin. When this is completed, the track is returned to CLEAR; the entry goes to FULL (together with the pallet to an unnamed queue) and the load station to OPEN. When the machine is available in VACANT, the pallet begins MOVE TO MACHINE STATION. Upon completion of this move the entry is returned to FREE and the machine and pallet immediately begin PROCESS, as this is a bound activity. When PROCESS is finished and an exit is in EMPTY, MOVE FROM MACHINE STATION takes place, with the result that the machine goes to VACANT and the exit to OCCUPIED, together with the pallet to an unnamed queue. The pallet waits on the exit until the track is in CLEAR and a load station is in OPEN. Then the activity MOVE BACK TO LOAD STATION begins. Upon completion, the track returns to CLEAR, the exit to EMPTY, and the load station to OLD, together with the pallet to an unnamed queue. When the operator is in IDLE and a new workpiece is in FRESH, the workpiece cycle is ready to begin again with EXCHANGE.

In following a workpiece through its cycle, we have followed the cycles of all of the other entities as well. For instance, a machine goes successively through queues VACANT, WAITING, and DONE, with the obvious alternating activities. Note that there are only two possible points in the system, as drawn, for exercising management discretion — in the choice of activities for the operator and the track, as shown by the multiple exits from the queues CLEAR and IDLE. It is frequently very convenient to know the points where logical control options can be exercised. Note in Fig. 76 that "MATCH" shows up twice, associated with the queues FULL and VACANT, and again with EMPTY and DONE. The entry and exit positions are associated uniquely with the machines. Thus, a workpiece on the third entry must be processed on the third machine. The match of attributes of the entry in FULL with the machine in VACANT ensures this. The same holds for the exit in EMPTY and the machine in DONE.

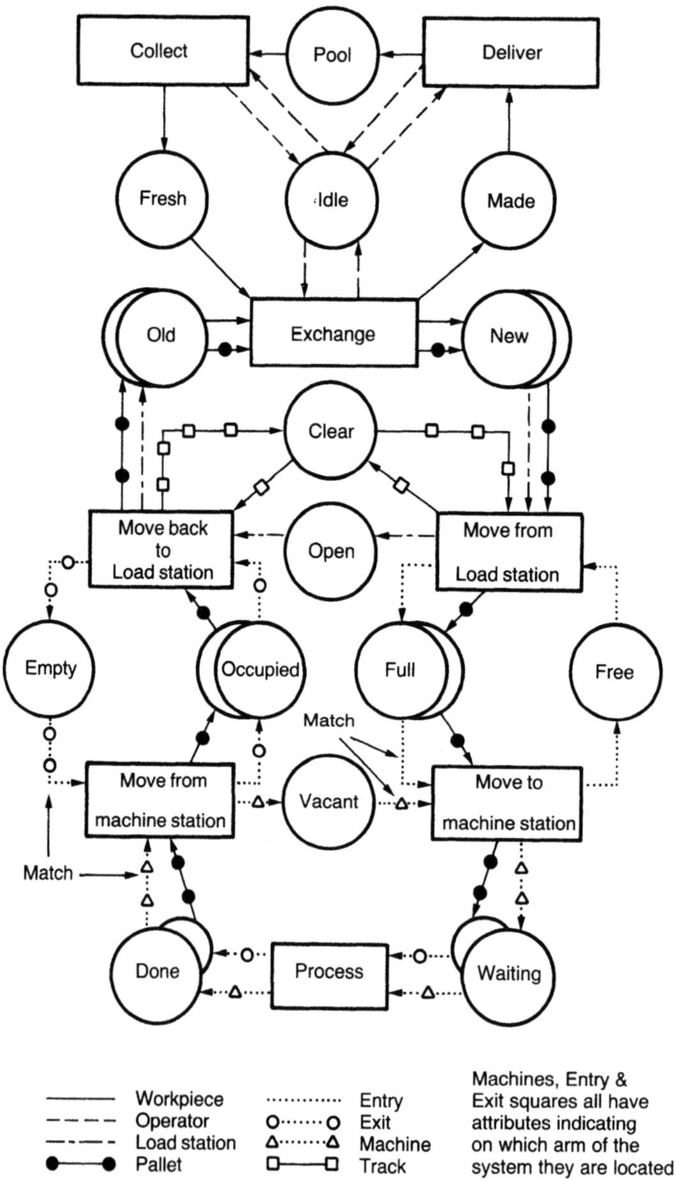

Fig. 76. ACD for automated production system

The ACD was input to the CAPS interactive program and the simulation program produced in ECSL. The program was compiled and executed for 10000 production time units for about 2000 different workpieces. Reports were generated every 2500 time units. There were ten pallets in the system. During the 10000 time units, DELIVER took place 182 times, i.e., 1820 workpieces were completed. The theoretical maximum is 1906, giving 95.5% as a performance ratio.
The workpieces actually had to be processed on a specific machine, due to tooling, although the workload was considered to be evenly balanced between the four ma-

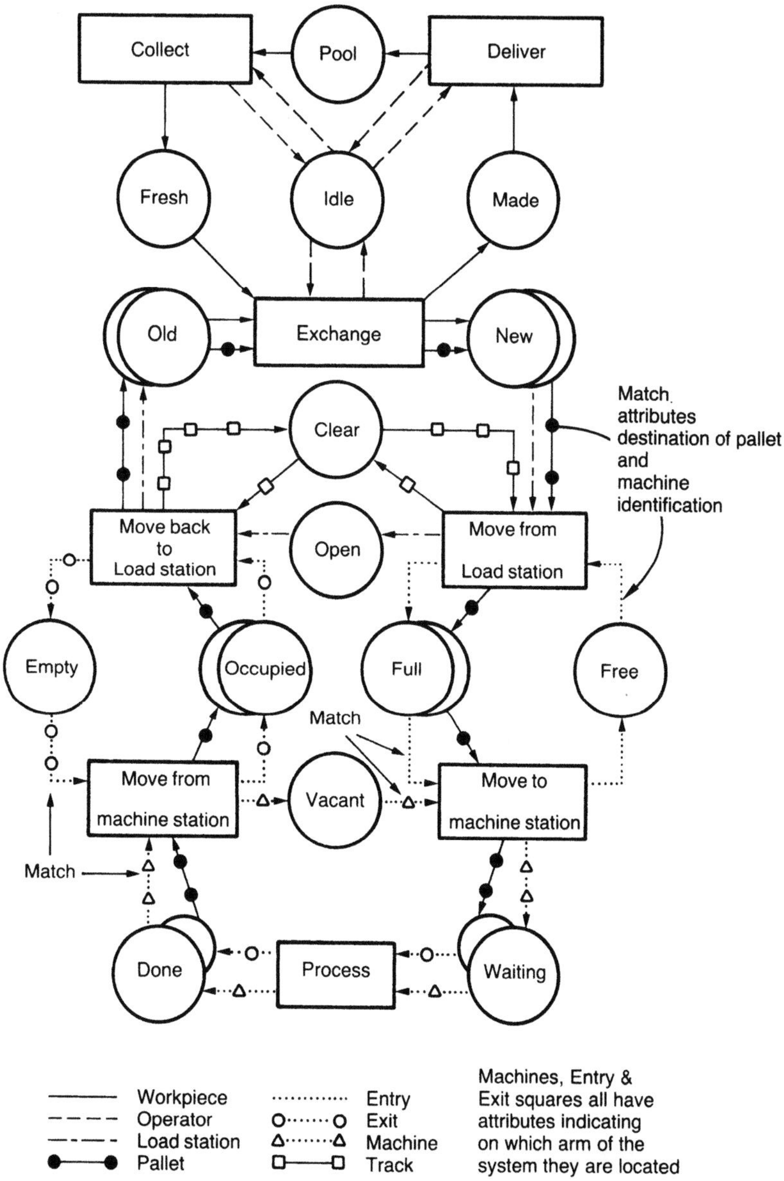

Fig. 77. Augmented ACD

chines. The model was updated to choose a required machine from a uniform distribution during the COLLECT activity for each entering workpiece. The ACD in Fig. 77 reflects the change to assign the workpiece to the required machine. This was accomplished by MATCH of the required machine of the pallet-workpiece combination with the machine number. Previously, a pallet could be sent to any machine with a free entry position (see Fig. 76).

With no change in run parameters, the augmented model was processed. It soon became apparent that no activities took place after 2500, i.e., the system reached deadlock. Since there were only ten pallets in the system, this had not been considered to be a problem. A postmortem dump of the program, a feature of ECSL, showed that three machines were idle, i.e., in VACANT, and the fourth in DONE, waiting to move a workpiece to the exit position; this workpiece could not move, as all three of the load-unload positions were occupied by pallets waiting for the blocked machine.

The next procedure attempted to alleviate the deadlock problem by requiring that the entry position to the desired machine be in FREE before a workpiece was chosen. This is shown on the redrawn ACD in Fig. 78, where an entry is required with a MATCH for EXCHANGE. Runs of this procedure prevented the deadlock at a slight expense in lost machine utilization. These losses occurred for various reasons, all having to do with the delays in loading workpieces for machines which needed them. For instance, when there were no workpieces for a machine in the entire system and then two arrived together, the second could not be loaded for the machine until the first had been loaded, transported to the entry, and shuttled onto the machine, freeing the entry. During this period, other activities might further delay the loading. The end result was that the machines, which were the limiting factor, lost processing opportunities and incurred unnecessary delays.

The next procedure eliminated this problem by permitting a maximum of two workpieces to be on their way to a machine, or at the entry of the machine, or on the bed of the machine. In the example given, this would allow both of the workpieces to be simultaneously loaded for transport to the machine. System performance was thus improved.

At this point it was noted that all of the scheduling activity had been negative in nature, i.e., the procedures were designed to prevent system deadlock, not to improve system performance. Procedures were incorporated to select the machine which had the largest current backlog of work and a place for another workpiece, then to select a workpiece for that machine. These procedures were implemented with the result that 182 deliveries were made, the same number as in the first test with no distinctions between machines. It was concluded that it would be possible to determine scheduling procedures which would overcome tooling restrictions and that ten pallets were sufficient to maintain machine utilization.

4.2.4 The Use of Simulation in Data-Driven Automation

In the foregoing section we considered the need for simulation as a methodology for developing the algorithms for the control of flexible automation and the design of systems, and we gave an example of the former. In this section the uses of

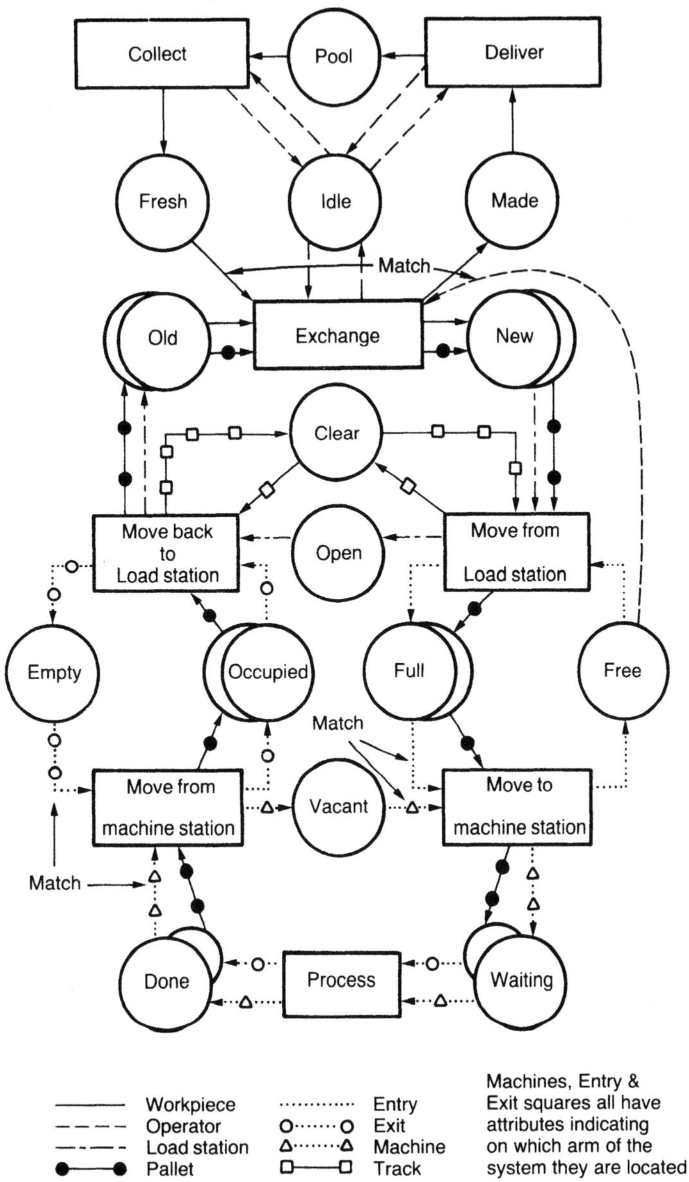

Fig. 78. ACD with early matching

simulation are considered and examples given. The typical uses of data-driven automation are in system design and control. Issues which might be addressed are:

1. System configuration
   a) Choice of work stations, type, and number
   b) Choice of material handling system
   c) Physical layout
   d) System and work station buffering

   e) Work station complexity — multi-spindle tradeoffs
   f) Sensitivity to workload fluctuations
   g) Sensitivity to workload mix
2. System control
   a) Assignment of workpieces to stations
   b) Allocation of material handling system capacity
   c) Sequencing of workpieces within a station
   d) Alleviation of material handling system congestion
   e) Timing and sequence of dispatch

Simulation is often used throughout the entire life cycle of these data-driven systems: before designing or acquiring a new system, to improve the performance of an existing system, and before expanding or otherwise modifying an existing system to handle new products or equipment. Since simulation mimics the behaviors of the actual systems, it can be used for training and research purposes as well.

4.2.5 Survey of Use in Manufacturing

Since the late 1960s, FMS vendors have used simulation for in-house design and control purposes. Many vendors have supported manufacturers acquiring their FMSs with simulation studies for system fine tuning, modification, or evaluation of proposed workload changes. An exhaustive list of the use of simulation in manufacturing, even if restricted to data-driven automation, is beyond on the scope of this section. Here a sampling of the uses of simulation is given, divided according to vendors, manufacturers, and research groups.

4.2.5.1 Vendors

Kearney & Trecker initially developed a specific simulation model to use when they designed their System Gemini. Then in 1971, in conjunction with the University of Wisconsin — Milwaukee, they developed a generalized FMS simulation model in SIMSCRIPT, one of the more popular simulation languages. This model evolved over a period of some 10 years to handle a wide variety of FMS situations. It provides the detail necessary to answer almost any system design or operation problem, and therefore requires extensive data input to fully describe the system and its operating environment. This model has been an integral part of feasibility studies, system proposals for both new and existing systems, detail system design evaluations, and the training of operators and managers.

Cincinnati Milacron is a major vendor of an FMS known as Variable Mission Manufacturing System (VMM). They have used ECSL and FORTRAN in developing a variety of simulators and production planning aids. Stressing the great importance they place on simulation, Cincinnati Milacron provides a macro-level simulator for the central computer of their VMM system so that purchasers can easily benefit from future simulation use.

4.2.5.2 Manufacturers

The *Caterpillar Tractor Company* uses GPSS-H, a simulation language for IBM computers, to perform simulations of many of their machining complexes. They have over 12 highly specialized computer-controlled manufacturing systems and find simulation to be particularly important in handling system exceptions, such as tool changes due to breakage and schedule modification. They also simulate other manufacturing systems and operations, including plant layout.

The *Hughes Aircraft Company* (HAC) has recently installed an FMS system in a California plant. HAC and the Charles Strak Draper Laboratory have developed a computer-based planning system called FLEXPLAN, which is used as a decision-support system for this FMS. Simulation is used primarily to predict how the FMS will respond to user decisions in handling daily operations and exceptions.

The *Garrett Turbine Engine Company* is currently using simulation for an FMS. They have developed a specific, detailed FMS simulation model to analyze their system, using SLAM. Garrett has found that the use of simulation improves interaction, communication, and understanding between their manufacturing, systems, and software personnel. In addition, they believe this experience will simplify FMS design and its integration.

The *Eaton Corporation* uses simulation in many areas, writing models in a variety of simulation languages. In its Eaton-Kenway Division, simulation is used in the design of every automatic storage and retrieval system that it proposes, to determine the physical configuration as well as the details of material handling devices. Model output is stored and an off-line computer displays the simulated system activity. Eaton also used simulation in studying the automation of a sheet-metal processing shop. System design, shop floor control, and dispatching were included in the study, done in conjunction with the University of Wisconsin — Milwaukee. The resulting system has reduced throughput times and transport requirements. Eaton has developed a number of other simulation models to analyze a wide variety of manufacturing problems. The largest of these projects was for a real-time Flexible Manufacturing Tooling Center which was determined, through simulation, to be unfeasible, thereby saving a large expenditure of capital.

4.2.5.3 Research Groups

The *National Bureau of Standards* of the United States is developing an experimental FMS. A special form of stimulation, emulation, is being used in the development of that facility. Simulation is used as a tool for computer-aided design, inspection, robot control, and vision.

The *Toyahashi University of Technology* reports that the Japanese have been using FMS technology since 1968 when the DNC system of Japan's National Railway was realized for electric locomotive replacement parts. Not all Japanese FMSs are simulated, and when simulation is used specific models are often the choice. The Japanese use simulation for equipment planning on high-volume production systems. Historical records are used to generate stochastic input for these simulation systems. Even in the high-volume area, the tendency is to develop specific models for each individual production system.

The *Computer and Automation Institute* of Budapest, Hungary, has developed a simulation model, written in SIMULA, to assess FMS performance for different scheduling strategies and product mixes. They report that similar models exist in Czechoslovakia, the German Democratic Republic, and the USSR. The Budapest Institute has developed a graphic simulator for robotic programming. This simulator is used in conjunction with a graphics package that constructs 2-D and 3-D images. A graphic description of the robot and its environment is constructed first, and the simulation package is then used to design and test the robot control by dynamically displaying its motions. This avoids subjecting the actual system hardware to potentially catastrophic failures.

The *Research Center of the Machine Tool Industry* in Karl-Marx-Stadt, German Democratic Republic, has developed a multilevel planning and control system, MAOSIM, for Mathematical Analytic Optimal Simulation. At each level of detail, analytical techniques are used to efficiently allocate and balance the resources and to generate alternative plans. Then simulation is used to test control strategies and to compare system performance capabilities with actual demand. MAOSIM also walks the user through all steps in building a model and analyzing all the different factors of the system to achieve desired system performance.

The *Illinois Institute of Technology* has been active in the field of simulation for manufacturing. Their efforts have produced an FMS simulator, a generalized assembly-line simulator, a specialized manufacturing simulator, and a computer-aided batch scheduling system, as well as a computer-aided manufacturing planner. Their Manufacturing Simulator is designed to deal with situations where resources may be shared, such as setup men or fork-lift trucks.

The *University of Wisconsin — Milwaukee* assisted in the development of the previously discussed Kearney and Trecker model, which they now use for research purposes. In addition, they have developed several generalized models of FMSs, transfer lines, and robotic cells. These models, written in ECSL, access the general adequacy of components in various manufacturing systems, using a minimal amount of data input. They are used primarily during system feasibility and initial design studies before detailed system data are available.

McMaster University of Hamilton, Ontario, Canada developed a general-purpose FMS simulator called FMSSIM in FORTRAN. This model operates on a variety of material handling systems, different configurations, and control rules which the user can select. It also allows the user to define his own control rules. This system is used mainly for research to identify trends in the behavior of FMS and to obtain some general guidelines for their design and operation. Besides the typical simulation reporting, FMSSIM also provides graphical animation of the manufacturing system; that is, it displays the movement of the parts through the system on a refreshed CRT.

The *CMP* Group has developed an FMS simulator that operates at the detailed level to follow individual workpieces and transporters through the system. The simulator can produce, as a by-product, a data file containing significant system status changes. This data file can then be used by a microcomputer program which produces an animated color display of the FMS under study. Transporters move about the shop with workpieces, delivering them to the appropriate machine, if available, or to the shuttle of the machine. The *University of Berlin* developed a similar program, although it did not use a microcomputer. At the *Gesamthochschule Kassel,* Professor Adolf Reinhardt is extending this scheme of portraying the historical output of a simulation by allowing users to pick system components for an FMS from predetermined sets. These components are positioned on the screen and the model is developed directly from them.

The *C. S. Draper Laboratory* have geared much of their general research and development in the manufacturing systems area to assist the users of FMS. They stress the importance of simulation to the success of any FMS in a manual written to help users considering an FMS acquisition. In addition, they added user enhancements to a generalized FORTRAN-based FMS simulator. They are also developing a minicomputer-based simulator to help users in operating FMS systems. Draper uses CAPS/ECSL for model development due to its lower development time and cost.

4.2.6 Future Trends

Simulation has now become a tool accepted by vendors, manufacturers, and researchers involved with data-driven automation. Most vendors of FMS now use simulation in their internal work, and several offer customers either simulation services or packages. It is probable that more vendors will use simulation and that simulation will be used in a wider scope of activities by vendors. Undoubtedly, more manufacturers will recognize the potential value of simulation in investigating their complex, dynamic problems and will increase their use of it as well. Those not having in-house simulation capabilities will use vendor-supplied models or services, or will turn to consultants. Initially, studies will center on specific problems of their systems, but as their confidence grows with familiarity, they will find applications involving other systems and eventually conduct studies of wider scope.

The use of the microcomputer, rather than a larger CPU, for the simulation of data-driven automation is well underway in Great Britain and the United States. This is a trend that is likely to grow rapidly as microcomputers become even more widely available and as knowledge about their power spreads. The availability of a full simulation package, such as CAPS/ECSL, has certainly enhanced their growth. It would be surprising if other developers did not make their systems available on micros in the near future.

The use of simulation to study the physical layout of robotic applications is relatively new and will grow. Interference avoidance packages will be supplied by researchers, vendors, and consultants, with manufacturers eventually developing their own capabilities. The sophistication of these programs will grow as they gain acceptance. One can anticipate the day when the capabilities of these avoidance models are linked to more traditional simulation models, so that the cell designer can first find feasible movement patterns, given the robot, tools, and parts, then determine the dynamic characteristics of these patterns to obtain better, if not optimal, cell designs.

The use of animated display of simulation results will increase. More simulation packages will provide a means of generating off-line displays, giving users a motion picture of their system. Eventually, generalized interactive models will be available, permitting the user to stop model execution, incorporate changes, and continue model execution. This approach makes the user an integral part of the experiment, however, and could easily lead to confounding and incomplete documentation of data and runs. In fact, perhaps the most promising aspect of this approach is its use in training. The instantaneous visual feedback would appear to be an ideal method of training for the design, management control, and operation of data-driven automation. In a like manner, one would expect the use of simulation of cutter paths to be visualized to assist NC programmers and/or process planners.

Perhaps the most interesting trend, and the one with the greatest probable impact, is the use of directed graphs as a means for specifying simulations. The University of Birmingham with its CAPS/ECSL system was the first to gain worldwide acceptance of this approach. Currently, most countries heavily involved in data-driven automation have found directed graphs — or their logical extension, activity cycles — to be a powerful approach to the analysis and simulation of these systems. It is all but certain that the use of directed graphs will find wider acceptance in the future. Other simulation systems, such as CAPS/ECSL, will be developed and other approaches using direct graphs, such as MAOSIM, will continue to appeal. It is likely that some simulation languages will find it convenient to incorporate directed graph input as a front end to their existing simulation structure.

5 Computer-Aided Production Planning

5.1 Scope and Main Fields

This section is confined to the problems of discrete manufacturing. For a systematic approach to production planning and control it is useful to subdivide the entire field into three main areas (Fig. 79):

1. Input information and influences of manufacturing
2. Production planning
3. Output information for the realization and control of manufacturing (including feedback from manufacturing itself)

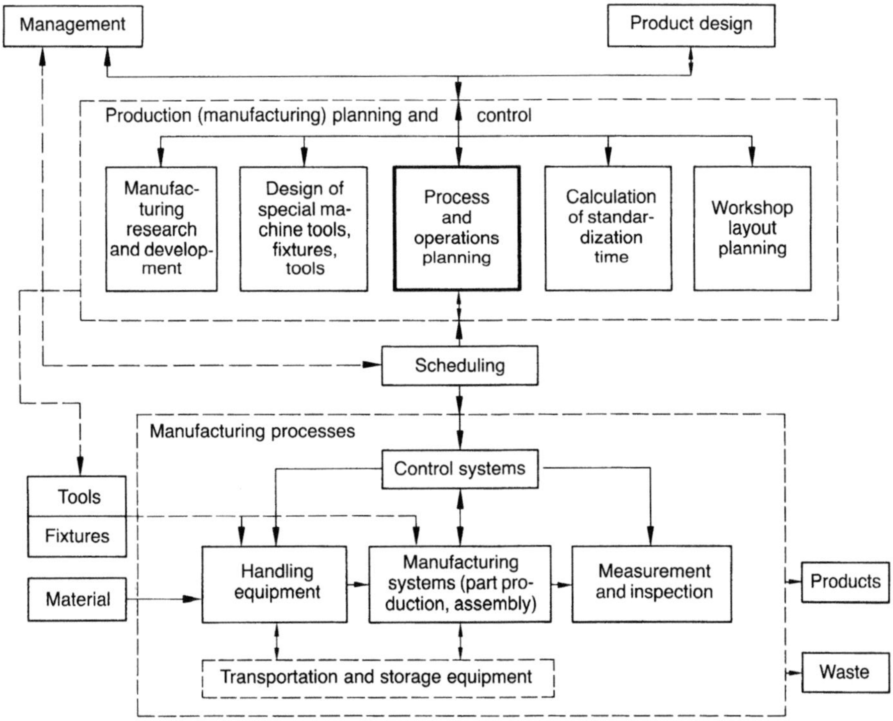

Fig. 79. Main tasks in production planning

5.1.1 Input Information for Production Planning

The input information for production planning is coincident with the output information of the design department and other planning departments. Design and functional information covers, for instance, shape, size, shape elements, accuracy, and surface quality. Input information of other factory spheres is especially economic data (lot sizes, limitation of costs) and planning data (assortment plan, time schedule).
Some methodological and technical aids are used in production planning, in general, such as methods for part classification and part retrival (part-group formation, definition of similar parts) and storage for workpiece data and further information on the manufacturing process.

5.1.2 Production Planning

The range of tasks in production planning covers several essential areas; the work division and structures in the factories are not uniform. The essential areas are:

1. Preparation of manufacturing processes, with the tasks of working out manufacturing documents (operational sheets, NC punch tapes), including the permanent tasks of alteration and operative-technological organization of production
2. Technological design for reconstruction of equipment and effective application of technical capacity
3. Manufacturing research and investigation speeding up the further development of manufacturing processes; manufacturing labs for rationalization and automation
4. Technological design and manufacturing of special machine tools, fixtures, special tools, and all kinds of technical aids for rationalization
5. Technological normalization of work, for the further development of factory time standards, and carrying out of time studies and work-scientific analyses.

Tasks 2-5 are based on the solution of task 1 with the use of methodological aids. Information stores such as data banks and suitable programs are necessary, in general, for all problems mentioned. Such data bases are necessary for machine tools, tools, materials and raw materials, recommended values for machining data and technological times, fixtures, and other aids. The classification and retrieval of all kinds of data is realized by suitable coding systems, which are usually included in preplanned programming systems.
Furthermore, specific data-bank executives (CAD/CAM-DB controls) are necessary for productive solutions. The efforts at unifying CAD/CAM-DB controls imply the possibilities for implementing programming languages, including different kinds of communication and library file maintenance.
The first-mentioned working area for the preparation of technological documents contains, beginning with information on the manufacturing problem, all tasks involved in the determination of the entire manufacturing process and the exact determination of the machining process.

Most of the primary data needed for the realization of manufacture are based on this area. These are, in particular, the operating sheets or manufacturing sheets (differences in the degree of detail of manufacturing information), working recommendations, information carriers for automatic machines (e.g., control punch tapes for NC machine tools, cam-adjustment plans, tool-adjustment plans), material sheets, and wage sheets.
An essential portion of the fixed data very often have to be changed; therefore, a maximum of flexibility in the entire working area is necessary. This can be ensured with a powerful data base with universal dialogue possibilities. Various national programming systems are available for integrated production planning, which can fulfill the requirements mentioned above in different ways (see also Sect. 5.3).
The international standardization of essential components of such systems as input languages, data-bank systems, library program, and computer systems including operating systems and programming languages used (e.g., FORTRAN) could be useful for further development. But this task is to be understood only as a long-term effort. The program packages for the preparation of entire manufacturing processes consist of a number of single programs for the different technological operations.
All types of programs are realized by two different principles:
a) the principle of variants (based on unified processes), and
b) the principle of generation (solves problems in all cases).

In some cases both solutions are used together. The principle of variants based mostly on factory-unified manufacturing processes is not primarily suitable for higher (national or international) degrees of standardization.
The principle of generation, including the computer-aided preparation of technological operational sheets, can be standardized in the same way as the process-programming systems mentioned above.
The programming systems for the part-programming of NC machine tools perform a special task in the determination of manufacturing operations (see Sect. 5.4).

5.1.3 Output Information and Working Areas Based on It

The output information in the field of production planning is coincident with the primary data for the entire factory organization. It is used for the technological and time control of production sequences.
Before production it is possible to simulate the manufacturing and production sequences on the basis of the given information. The simulation of manufacturing processes is especially important for the design of highly productive, and therefore expensive, automatic manufacturing complexes. Universal simulation programs (e.g., GPSS, SIMULA, SIMDIS) are available, which are able to modify the application of manufacturing tasks (cf. Sect. 4.2).
A good precondition for the international standardization of simulation technique is the universal application of these methods and programs.

The output information from production planning can be used in an off-line or an on-line mode. The off-line mode has recently become predominant for the entire factory organization. The on-line mode is used for more highly developed manufacturing structures, especially in connection with DNC solutions for NC machine tools and manufacturing systems (cf. Sect. 3.5).
A number of different documents (order proofs, operation sheets, working instructions NC punch tapes, tool lists, tool setting plans, material lists) are used for the whole off-line information offering, which is standardized in the majority of factories and nationally to a certain extent. Engineering and technological database systems are widely applied in the entire CAD/CAM field. Solutions are now for the following fields:

1. Data-bank controls for maintenance of technical documentation
2. Structure of the data-bank
3. Link between data-bank and manufacturing process

Different technical and methodological aids are used to supervise the production processes and the time scheduling in accordance to the information determined. At present, these tasks and problem solutions have not been prepared for international standardization.

5.2 Unification and Group Technology

Characteristic of the metalworking industry are versatile product assortments and manufacturing in small series. Therefore, methods of unification are very important.
The most important aim of the unification of manufacturing processes is the improvement of the series character of production. Hence, unification methods are a means for bringing mass-production economics to batch manufacturing.
The scientific fundamentals for the unification of manufacturing processes and part assortments proceed from the "unification of production-technological process", after Sokolovski [209]. Of special importance are the more advanced investigations by Mitrofanov [163], which led to "group technology".
The possibilities offered by computers were decisive for the uniform, general utilization of these methods with a broad effect. The principles of unification are applied mainly to:

1. Classification and analysis
2. Unified process design (unification of production-technological processes)
3. Group technology

The unification principles contribute especially to efficient process design (see Sect. 5.3).

5.2.1 Classification

The classification methods are used for the sequencing and analyzing of objects or object data. The following aims of classification are essential in manufacturing engineering:
1. Re-use of already designed individual parts and components for newly developed products
2. Unification of product design, manufacturing processes, and other elements of production
3. Concentration and specialization of production
4. Carrying out analyses and making available statistical information (e.g., analysis of features according to frequency distribution)

Several national and company standards are given for classification methods in parts arrangement. National and especially international standardization in this field would be useful, but is very difficult to realize. The reasons for this are (a) the large number of well-adapted classification systems in companies, and (b) the fact that every kind of classification is useful, and the advantage of using an international or national unified classification system is not so great.
In this field the MICLASS System is a development of international importance[1]. The MICLASS System is characterized here briefly, as an example of many classifications.
Figure 80 illustrates the interconnected elements of the MICLASS system and their functions. Although it can be used manually, MICLASS is essentially a computerized system tailored to meet the needs of design and manufacturing.

The system has four major parts:
1. The classification number, which makes it possible to classify parts by their engineering and manufacturing characteristics
2. The data base, which contains design and manufacturing information for the whole company
3. Retrieval programs, which make it possible to retrieve drawings, route sheets, manufacturing instructions, etc.
4. The analysis programs, which are used for design standardization, optimization of machine tool use, and manufacturing routines

These principles are illustrated in Fig. 81. In some industrial countries, such as the USSR and the German Democratic Republic, the principles of classification are regulated by national standards. For instance, in the GDR the following are compulsory:

— General terms for classification	TGL 28 259
— Classifiers for individual parts	TGL 28 260

[1] MICLASS TNO, Organization for Industrial Research, Netherlands, USA

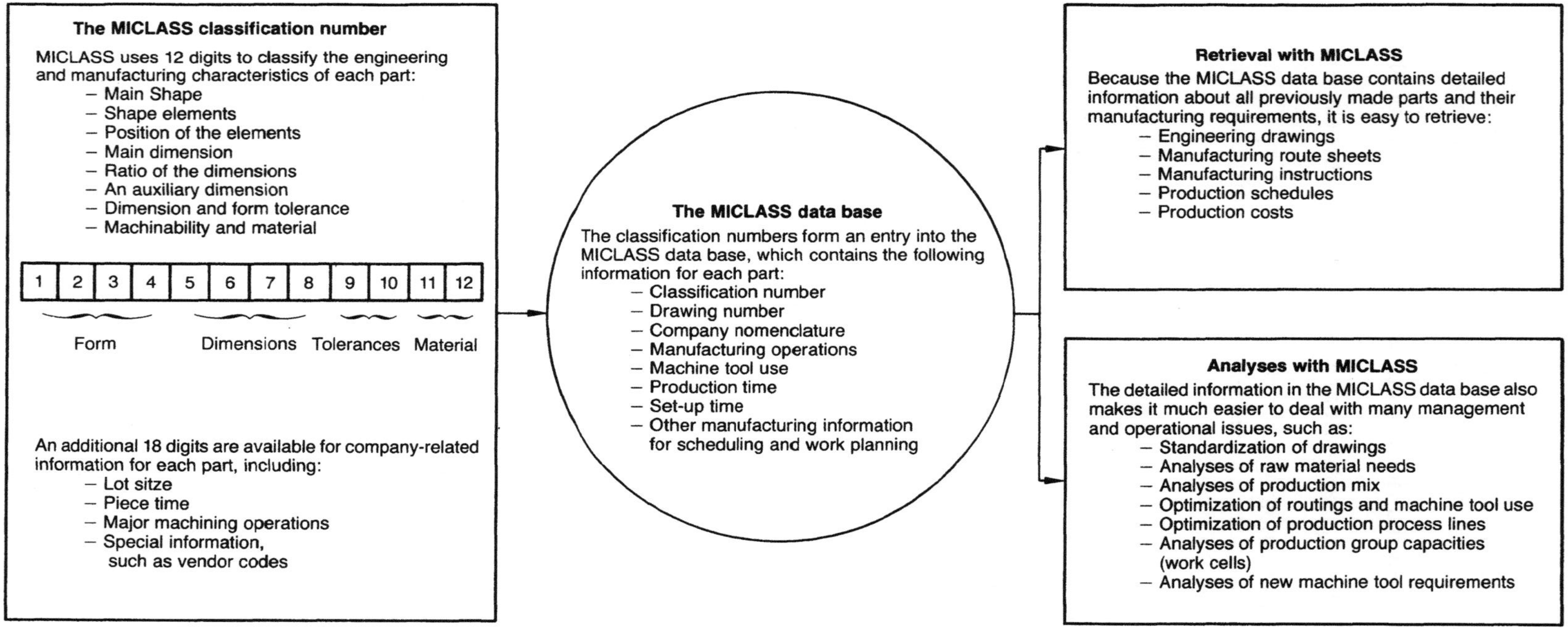

Fig. 80. MICLASS as an example of part classification

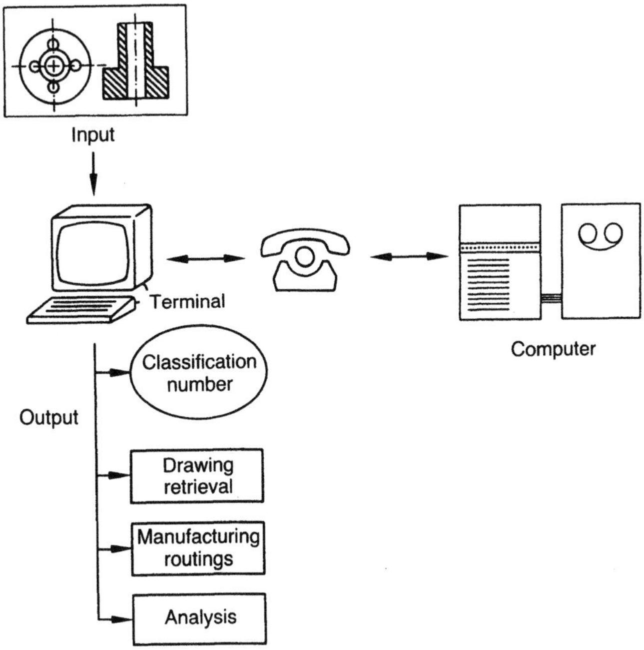

Fig. 81. Interactive computer-assisted part-classification

5.2.2 Group Technology

Group technology includes the combination of objects of labor into groups which can be manufactured at the same specialized manufacturing plant with a minimum expenditure for change at the machining station. Group technology may concern individual operations, process stages, or the entire manufacturing process; the greatest effects are achieved when the entire process is included.

With the development and application of group processes, good preconditions are created for high-level forms of object-specialized manufacture (transfer lines with opportunity for part changing, as well as flexible manufacturing systems).

5.3 Computer-Aided Process Planning

5.3.1 General Requirements

The working out of processes and operations is the main component of the production-technological work in the technological preparation of production. Because of the complexity and the various aspects of the tasks, effective, practice-adequate methods of rationalization were developed and introduced into industrial practice only in the late 1970s.

The following aspects and features of these systems are important:
- Supplementary to the main function — the computer-aided working out of production-technological documents (work-plan master cards, instructions for work, job documents — integration of other tasks)
- Efficient control of the operation-connected updating service by effective data access
- Communication with operation-connected production planning and machine occupation
- Application of dialogue-oriented forms of communication by means of interactive display-screen technique (see Sect. 7.2.6)
- The possibility of the communication with design preparation via suitable geometry stores (see Sect. 7.2.7)

These features and requirements are provided and met in different ways by the internationally known systems. For instance, functions for process planning are integrated in CAD solutions (cf. Sect. 7, Integration of CAD/CAM).
Some systems are being developed in the German Democratic Republic which will meet the complex demands mentioned above. One of these, the program system RATIBERT, will be described in detail.

5.3.2 The Program System RATIBERT

From a basic system for the computer-aided treatment of the work-plan master card up to a complex information processing system which integrates production-technological preparation in the enterprise reproduction process, the program system RATIBERT can be used at all conveivable intermediate stages according to the progress of development.

The entire program system can be applied as follows in both input and output of punch-card automated processing cycles and display-screen dialogue:
1. Working out of sequences of operations and production-technological source programs on the basis of
   - Dummy and type processes
   - Sequencing graphs
   - Generative process programs
   - Similarity techniques
2. Treatment of the work-plan master cards and operational instructions
3. Job-controlled and group-process-controlled generation of documents for production
4. Determination of price and cost for every new part and in case of change concerning production technology, design, and blanks
5. Computer-internal supply of data and automatic updating service for the enterprise data-bank
6. Mechanical provision of data on savings in time and material for planning and accounting
7. Determination of optimal lot sizes and transport techniques

8. Computerized finding of repetition parts, creation of similar parts, and geometrical (or detail-geometrical) analysis of parts
   — Automated preparation of offer technologies for the sales engineer with all economic parameters within the framework of the contract negotiations
   — Economic evaluation of variations in design and draft
   — Supply of important enterprise data in the display-screen dialogue (assortment plan, main plan of dates, material use, workplace use, indication of finished production, etc.)
   — General computer-aided production-technological updating service for all stored data stocks
   — Execution of any scientific-technological computations
9. Processing of production-technological operation programs for
   — Automatic programming of NC machines [142]
   — Preliminary computation of time for conventional center-lathe operation
   — A large number of conventional operations (peripheral milling, flat grinding, internal grinding, heat treatment)

The working programs can be adapted to any user conditions for
— Machine tools
— Clamping accessories
— Tools
— Machining data
— Machine setting values
— Multi-part machining
— Time determination
— Selection of means for measuring and testing

The program complex of the production-technological updating service through display-screen operation ensured for the first time the daily updating of all production-technological and economic-organizational data (when punch cards are used they are generally updated every week). With display-screen dialogue, moreover, production is continuously supervised in the following way: In cases of deviation from the specifications of the production-technological documents (e.g., machine displacement, change of dimensions, blank dimensions) new and exact production-technological parameters can be determined within seconds.
Moreover, the components of the RATIBERT program system (PS) permit short-term computer-aided introduction of new job documents. RATIBERT may be regarded as a general framing system for further development toward fully automated computer-aided technical preparation and execution of production.
The following basic functions of the data-handling process are supported by modules of PS RATIBERT:
— Loading (establishing) of data stocks
— Replacing, clearing, blocking, checking, selecting, compiling, restructuring and adding of data stocks, data records and data
— Output (printout or display screen) of data stocks or records completely, as well as with the use of variable group keys

— Direct access to data arrays, extraction of data, combination and averaging of data sets, and conversion and supply of data for computations
— Searches in data stocks and data records
— Rearranging and securing of data

Data stock designates a group of data records with the same structure or different structures, which establishes a definable relation to objects. In a borderline case, a data stock may also consist of a data record. In the storage structure only one address is assigned to the data stock, so that it can be located by one access only. As *data stocks,* not only the work-plan master cards and lists of parts (consisting of records of different structures) are imaged in the data and storage structure, but also data collections, e.g., texts, workplace data, material data, normative rules. *Data records* constitute a quantity of data and relationships which can be imaged and interpreted in different structures. Depending on the relation to objects, a key or addresses are assigned to data records. They serve for direct or group-related access. The keys or addresses may be used in different data records as references for the later substitution of data.

5.3.3 Examples of Computer-Aided Process-Planning Systems[1]

AMAPS

The advanced manufacturing, accounting, and production system (AMAPS) is an on-line, functionally complete, and fully integrated manufacturing and control system. AMAPS provides information that enables management to plan and control manufacturing operations with precision and thus make the most productive use of a company's manufacturing resources. A special option enables manufacturers to incorporate multiple plants into a single data base. AMAPS is a flexible system that utilizes control file parameters to enable its users to establish management policies and procedures unique to their particular companies and industries. The system is innovative and easy to use and includes such user-friendly features as on-line MRP-type analysis and optional real-time data base updating directly from inquiry screens.
AMAPS operates on IBM 43XX, 3033 under DOS, OS, ADABAS, DATACOM, IDMS, TOTAL, IMS, DL/1; on HP3000 under IMAGE; and on WANG and PRIME minicomputers under ANS COBOL. It was developed by the Comserv Corporation. The software has modules for the following: bill of material, material control, material requirements planning, purchasing control, process and routing, shop floor control, capacity requirements planning, standard costing, master production scheduling, cost management, and lot traceability.

ATS

The automated time standards (ATS) module was developed by the McDonnell-Douglas Automation Company, under the sponsorship of the CAM-I Process

[1] From the CAD/CAM Guidebook (1983) USA. Gettelmann, Cincinnati

Planning Program. ATS provides a means for proving the feasibility of automating the time-estimating process for manufacturing operations. It is a prototype software applications package which permits user-defined estimating sequences to be automatically completed with user-written calculation routines. ATS can be used by itself, or in conjunction with the CAPP system. When used with CAPP, ATS permits information in a CAPP-stored process plan to be used as input to complete an estimating plan.
Time-estimate values can likewise be merged from an ATS-stored time estimate into a CAPP-stored process plan. The Hazeltine 2000 terminal is used for ATS, with conversion routines available for the Hazeltine 1510, IMB 3270, and Lear Siegler ADM-1A and ADM-2. Documentation for the ATS module consists of a 9-track magnetic tape along with hard-copy documentation consisting of a user's manual and a program logic manual. ATS may be purchased from CAM-I, Inc. The software has a module for the function of time standards for process planning.

CADAM

The computer-graphics augmented design and manufacturing (CADAM) system is an interactive graphics system for computer-aided design and manufacturing. The system includes a design/drafting package, together with a number of aids for design analysis. It uses a central design data base for storing and retrieving drawings created by designers and draftsmen. This design data base enables users in both design and manufacturing to share geometric and alphanumeric data. It facilitates the communication of design information between design and manufacturing. Geometric construction is based on descriptive geometry. The graphic's display can be zoomed in and out to get the desired resolution. Views can be merged, stacked, rotated, and juxtaposed. With the aid of 3-D-mesh generation facilities, designers may construct finite element models. CADAM was developed by the Lockheed Corporation and is marketed by the I.B.M. Corporation. The software has modules for the functions of automated design drafting, design analysis and drawing retrieval.

CAPP

CAPP is a prototype software development that provides a data management framework designed to assist the functions of process planning in the manufacture of discrete parts. The system enables a process planner to automatically access standard process plan specification data in an interactive and dynamic manner. These standard process plans are directly related to the specific fabrication processes required to manufacture a family of parts. Once the standard process plan is retrieved, the planner can refine the plan to produce a similar part. The refined plan data can then be stored and become part of the computer data base for subsequent recall as required for production or update. CAPP 2.1A is directly implementable on an IBM 360/370 system under operating system OS/MVT, Release 21.7 with Time Sharing Option (TSO). CAPP 2.1A consists of a 9-track magnetic tape with supporting hard-copy documentation. The modifications are in hard-copy form only, with the exception of the IBM 3270 modification, which is available on 9-track magnetic tape with hard-copy documentation. A minicomputer version of CAPP is also available. CAPP was developed and is available from CAM-I, Inc. The software has a module for the process-planning function.

CUTPLAN

CUTPLAN is a computerized process-planning system which aids process planners and manufacturing engineers in developing the manufacturing information required to produce a given part. CUTPLAN has both variant and generative capabilities for the development of both routing sheet information and detailed operation information, such as cutting tool selection, feed/speed selection, and cut selection. Basic sets of machining data can be supplied by Metcut, or company-specific data may be incorporated in the system. CUTPLAN also has graphics capabilities for the development of manufacturing drawings such as setup, machining, and inspection instructions. Additional stand-alone modules for machinability data retrieval (CUTDATA), tool and cut selection (CUTTECH), and cost estimation (CUTQUOTE) are currently under development. CUTPLAN is available from Metcut Research Associates, Inc., Cincinnati, Ohio USA. The software has a module for the process planning function.

DAL

DAL is an interactive computer programming language that is an integral part of the Calma DDM (design, drafting, and manufacturing) CAD/CAM system. DAL includes all of the DDM graphic commands for geometric construction, plus standard mathematical functions, display controls, labeling and dimensioning, and branching and conditional execution. DAL has an interface to FORTRAN. Thus, the library of FORTRAN routines are available to DAL users. They may add their own FORTRAN routines if they wish. The numerical control applications package consists of software that communicates with DDM and the user through a library of DAL routines and support functions. Each DAL routine is tutorial, and each can be modified by a user to meet specific needs. DAL routines are supported by a HELP file that lists the major DAL routines and their functions. A number of DDM numerical control applications are written in DAL. Typical NC operations in DAL include: approach feed rate, cut approach point, cut approach vector, and set-up base rough cuts. DAL was developed by Calma, a subdivision of General Electric Co. The software has modules for these functions: interactive design, numerical control programming, automated design drafting, digitizing and scanning capabilities, engineering analysis, process planning, material requirements planning, production scheduling and control, inspection and quality control, inventory control.

DDM

Calma's 3-D mechanical design DDM system is a turnkey, stand-alone interactive graphic system. This versatile general-purpose manufacturing tool has a number of highly automated applications, such as automated design and drafting, finite element modeling, flat pattern development, family of parts, parts nesting, flame cutting, and numerical control. In performing the numerical control functions, the system operates with the same 3-D model data that were used to create the original part design geometry, to automatically or interactively generate tool paths. The system will handle such major numerical control areas as 2½- to 5-axis surface contouring, turning operations and profiling and pocketing to surfaces. The tool paths can be displayed on the graphics screen for visual verification and editing and for checks on dynamic and fixture clearances. When the tool path data has been verified it can be output as a CL file, APT source file, COMPACT II, or in any convenient form designated by the user. DDM was developed by Calma, a subdivison of General Electric Co. The software has modules for these functions: interactive design, numerical control programming, automated design drafting, digitizing and scanning capabilities, modelling and finite element analysis, process planning, material requirements planning, production scheduling and control.

5.4 NC Programming

In all spheres of production engineering in which electronic controls are increasingly used, the efficient generation of control data is important for the optimum utilization of the manufacturing equipment. Logically, the possibilities for programming numerically controlled machine tools are at the center of developmental research.
Proceeding from numerically controlled machines (see Sect. 3), more and more components such as controlled handling devices, NC measuring machines, and transport and storage systems are integrated in automated computer-aided programming. Programming systems for NC machine tools and machining systems are the most common application at present; therefore, mainly these plants will be discussed.

5.4.1 NC Programming Systems for Machine Tools

5.4.1.1 Historical Development and Basic Concepts

Immediately following the development of the first numerically controlled machine tool (see Sect. 3.1) the need was recognized for computer-aided generation of control information. In 1956-1957 the first programming system — APT (Automatically Programmed Tools) — was developed. APT was the basis for many modifications and succeeding developments in other countries.
All important programming systems have the same basic structure (see Fig. 82). Proceeding from this common basic structure the transfer of data from the processor to the postprocessor is executed according to an internationally standardized data structure, CLDATA 2 (Cutter Location DATA).

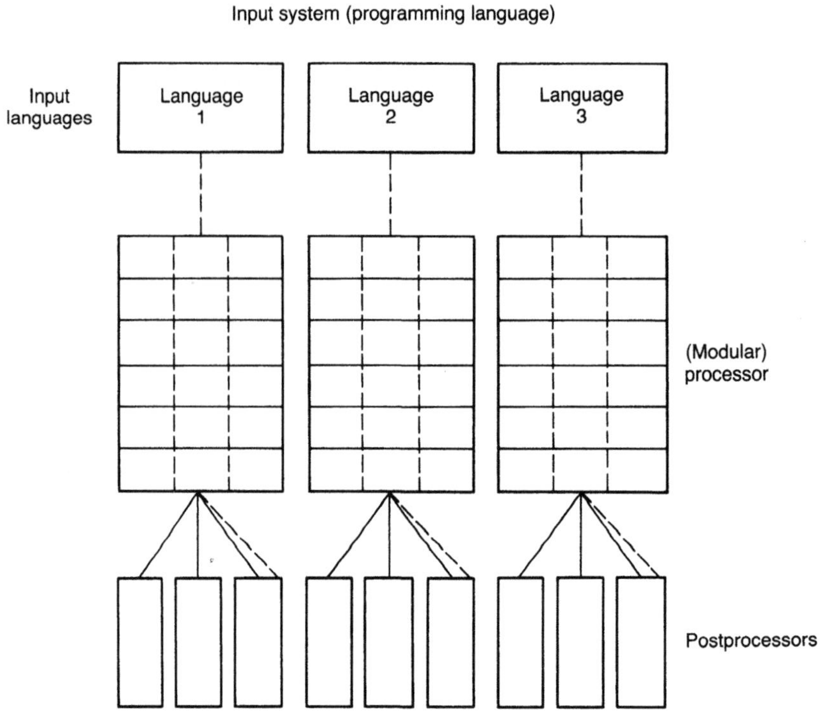

Fig. 82. Basic structure of an NC programming system

5.4.1.2 Comments on the APT Language

The APT language is the Nestor among NC programming languages. In the course of its long development it was repeatedly extended and realized in four main steps. For each of them innumerable implementations exist on very different types of computers. The language has a number of syntactic and semantic features which have become the maxims of many languages that were developed later, not only in the APT family [223].

An APT program constitutes a sequence of statements. Each of the statements is structured according to a framing syntax:

statement : : = [[mark] designation =]
noun [/parameter list]

Designated statements are connected with definitions of geometric elements or macros, to which symbols are assigned along with the designations. The parameter list serves for the closer definition of the type of statement determined by the noun. Statements in brackets, but without labels, are also permitted as parameters. That part of the language which is designed for describing the workpiece geometry within Cartesian coordinate systems is extraordinarily highly developed. Table 22 lists the kinds of definition possible.

Table 22. Variety of definitions for geometrical elements

Geometrical elements and transformations	Number of different possibilities
Points	10
Straight lines	15
Planes	7
Circles	10
Cylinders	2
Ellipses, hyperbolae	1 each
Cones	2
Curves of 2nd order (analytical)	3
Curves of 2nd order (approximated)	3
Vectors	10
Spheres	4
Product surfaces	5
Control areas (approximated)	4
Matrices of transformation and others	11

Two important reasons for the development of a large number of smaller programming systems inclusive of modified languages are the many possibilities of this language, which are not always required, and the demand for efficient large-scale computing facilities. At present there are more than 300 NC programming languages or specific modifications all over the world.

5.4.1.3 General Features of NC Programming Systems

Numerical control systems are currently classified under two basic functional types: positioning, also called discrete or point-to-point, and continuous path or contouring. Positioning systems are used to control machine tools — such as the drill and jig borer — that perform operations only at a specified point on a workpiece. In drilling, for example, the drill spindle is positioned at a single specific point; the proper drill size, speed, and feed are selected; the drill then advances to cut a hole to the specified depth, withdraws when drilling is completed, and is repositioned to the next point of work. The instructions required to operate the positioning system can usually be completed easily by a programmer with the aid of a manual desk calculator.

The contouring system is more complex, and is used when the route taken by the tool is of consequence to configuration throughout the entire tool movement. Contouring is used to control such machine tools as the lathe, the milling machine, and the machine center, which removes metal continuously over the surface of a workpiece, as in the milling of a cam. Contouring may also apply to flame cutting, sawing a continuous path, welding, and to such processes as the application of adhesives. The problem is to control the tool continuously, which requires frequent changes in movement along two or more machine axes simultaneously. During this time, the tool must be in continuous contact with the workpiece. A continuous-path system requires a much greater input of detailed instructional information than does a positioning system. Although direct programming with a desk calculator is possible in contouring, the time consumed in calculating the cutting tool path renders this approach relatively inefficient and impractical. An automatic data processing technique to program a contouring system is therefore virtually essential.

Through the use of a computer, further important advances have been made toward simpler, faster, and more automated preparation of the machine instructions and control tapes. This computer-assisted programming consists essentially of a specially devised symbolic language capable of transmitting a brief, simple description of the machining operations and the shape of the workpiece to a computer. The computer then calculates and translates the brief directions into a detailed instruction and the coded language of the control tape.

Computers are being used more and more for the complex programming needed for NC contouring machines. Simple manual programming remains satisfactory for NC positioning systems, although some authorities have argued that the day of the computerless NC user is fast disappearing.

Although a considerable amount of point-to-point programming is still being accomplished by manual methods, computer-assisted programming is a virtual requirement for complex contouring, and many users find it useful and economically desirable for point-to-point work as well. In addition, many industrial firms find they can increase utilization of the computers they are leasing or have purchased by applying the computers to numerical control functions.

The process of computer-assisted NC programming involves computers that cannot perform calculations other than those that human beings can do, in spite of the fact that the computers can do the calculations at exceptionally high speeds. The output of the process must be in the same form as would be generated manually, and it must be consistent with the format of the numerically controlled machine that will produce the part.

Table 23. Possibilities of geometric description

Geometric description	Control	Typical techniques
Single points	P, S, B	Drilling, milling, stamping, flame cutting
Point pattern	P, S	Drilling
Straight lines	S, B	Milling, grinding
Circles	B	Milling, profile grinding
Sequences of points (any curve elements)	B	Milling
Trains of lengths (rectangular)	S	Milling, turning
Elements of area (1st order)	S	Milling
Defined elements of area (2nd order)	S, B	2-D milling
Any forms of area	B	3- to 5-D milling
Translation (points, point patterns, cycles of line segments)	P, S, B	Drilling, milling, grinding
Translation of systems of coordinates	P, S	Drilling, milling, grinding, turning
Reflection	P, S, B	Contour milling
Turning (especially for systems of coordinates)	P, S	Drilling

P, point-to-point; *S*, straight line; *B*, continuous-path contouring

Supplementary to the prevailing classification into point-to-point and contouring, which is used at present in most controls, there are also wiring-programmed controls with straight-line controls, which have to be taken into account in a system for computer-aided programming. The possibilities of geometric description and typical fields of application are given in Table 23.

5.4.1.4 Processor-Postprocessor

The computer program consists of a processor and a postprocessor, which gives a general-purpose computer the "brain" that it needs to understand and execute the instructions of the part-program. The computer program (processor) accepts the part-program instructions and calculates the cutter path. The computer program also accepts additional instructions for control of feed rates, spindle speeds, coolant, etc., and passes these on to the postprocessor for output.
Because each machine tool and control system may differ, a means of communicating these differences to the computer is required. This task is performed by a supplementary computer program called a postprocessor. Its instructions are processed by the computer after the cutter path has been determined by the processor, hence the name "postprocessor". It is in this phase that the coordinate points of the cutter path, plus the part-programmer's instructions on feed rates, spindle speeds, and other auxiliary commands, are put out in the format of the machine tool.
The final computer output contains all the commands required to produce the part, either in the punched tape format of the machine tool or directly in 1-in. wide, eight-channel perforated tape.
Wherein, then, lies the advantage of computer-assisted programming? The advantage is that a part-programmer may give the machine tool its necessary instructions in a human language, and that he is thereafter relieved of the need to do arithmetic calculations. The result is a part-program produced in a fraction of the time otherwise required, and free of human error.
The problem of programming instructions can be explained as a function of computer language limitations. If it were feasible, the simplest arrangement would be for an operator to tell his NC machine in his own language what was required of it. Unfortunately, this is not practical for several reasons: the vocabulary is too large, the words are too long, and conventional language is often ambiguous. Extensive compromises must be made to communicate with a computer. The part-programming languages such as APT and AD-APT, therefore, have their own system of expression, containing a limited vocabulary of completely unambiguous words and a fixed maximum number of characters. The result is a form of Pidgin English, usually made up of phonetic contractions of multiple words. Hence, programming instructions can be given by the part-programmer in "English", which is then translated into digital form by the computer and produces the necessary instructions for the machine tool.
The general machining statements produced by the processor such as APT and AD-APT must be converted to instruction sets designed to fit the particular requirements of the machine tool/numerical control system. This is the function of

the postprocessor program. A separate postprocessor program is written for each machine tool/controller combination. The postprocessor has many functions, falling into the following areas:

1. Numerical-control system input media

The postprocessor must produce the machine tool instruction sets in the medium required by the numerical-control system. The most common input medium in use is 1-in., eight-channel punched paper tape. This has been adopted as the standard by the EIA[1]. However, there are older numerical systems still in use which require machine tool instruction sets on punched tape.

2. Control-tape coding

The postprocessor must produce machine tool instruction sets in the correct character coding required by the numerical-control system. The standard accepted by the EIA is binary-coded decimal character coding. There are still numerical-control systems in use which require a pure binary character coding.

3. Control-tape format

There are five control-tape formats currently in use: fixed sequential, block address, word address, tab sequential, and universal. Each requires the machine tool instruction sets to appear in a specified sequence and form.

4. Machine tool characteristic

Each machine tool manufacturer builds individual characteristics into their machine. The postprocessor must take into account the characteristics of the specific machine tool for which it was designed. Some of these are:

a) *Axis notation*. If the machine tool has a different notation for the principal axis than the EIA standard, programming must allow for this.
b) *Reference point*. Some machine tools have a fixed zero reference point on the machine tool worktable, while others have the capability to move the zero reference point through programming.
c) *Measuring*. Some machine tools measure absolute distances from a fixed reference point on the machine tool worktable. Others measure distances incrementally.
d) *Auxiliary commands*. There are many optional capabilities which can be added to numerically controlled machine tools. These options vary with both the capabilities of the machine tool and the desires of the customer. If they are added and are programmable by the part-programmer, they must be made a part of the postprocessor.
   A few of these are:
   — Tapping cycle (including overshoot determination)
   — Coolant control (ON, OFF, FLOOD, MIST, TYPE)

[1] EIA (Standard-Code RS 244) Electronic Industries-Association

— Automatic tool changing
— Feed-rate selection
— Spindle-speed selection
— Table rotation

5. Numerical-control system interpretation capability

The interpretation capability of the numerical system varies considerably. Depending upon the system, straight-line, circular, parabolic, or spline-curve interpretations are possible. Systems other than straight-line interpretation systems need only the starting and ending points and the type of curve to complete a machining operation. Straight-line systems must have the starting and ending points of line segments which together approximate the desired curve to be machined.

6. Time estimates

It is common practice to build into the postprocessor the ability to project a time estimation of the actual production run in producing a part. This is accomplished by counting the machine tool/numerical-control system cycles for each operation. These cycle counts then are multiplied by the basic cycle time to give a reasonable time estimate.

7. Reasonableness tests

These determine whether the particular machining program might exceed the capability of the given machine tool/numerical-control system. For example, they check to ensure that a point location off the worktable is not specified. These functions change a general-purpose machining statement into a specific machine tool instruction set for a given machine tool. Postprocessor programs are usually developed by the supplier of the specific numerical control/machine tool combination in use. They are not provided as a part of the computer numerical-control language.

5.4.1.5 Data Communication

As numerical-control parts to be programmed for point-to-point or contouring operations are becoming more complicated, the need for large-scale computers for processing is becoming critical. Often, the NC operation requires larger computers than the rest of the plant operation does. Also, even if an industry has a large-scale computer, it might not be able to process NC programs economically, either because of limitations of the in-house computer or because NC programs are not available or cannot be properly maintained.
NC data processing on large remote computers via a telecommunications system makes it possible for any size plant or operation to use various remotely located, large computer systems over standard existing telephone wires. Thus, a small plant equipped with a series of low-cost data communication links can, in fact, have as much computing power and versatility as any large plant in which the program-

mers are obligated to use the in-house computing facilities and therefore have less flexibility.

Until recently, the basic measure of data processing efficiency has been "throughput" — how efficiently the equipment could successfully complete a specific job. Now, however, the criterion is "turnaround" — how quickly the programmer can retrieve essential information. A variety of equipment for specific needs is available from most computer manufacturers and communications systems suppliers. Great care must be exercised in selecting the type of equipment to be used, in order to achieve compatibility and economy.

One of the reasons for using a remote system is the need for a wide variety of languages and postprocessors. Another reason is a combination of the distance to a computing center and the need of service that provides rapid turnaround as well as high throughput rates. Direct-access computing is accomplished by the installation of terminal equipment on the user's premises and linking it by data-transmission networks to a remote computer.

Time-sharing gives organizations real-time data service of any complexity, without the attendant full-time costs. The power of third-generation computers can be brought directly and immediately into the user's office or plant on an as-needed basis — and at minimal expense. The cost of operating the large-scale computers in a computing center is distributed among many users who otherwise might operate with less efficient systems.

A good computer-assisted system will allow programmers to improve their productivity while increasing the quality of their efforts. The biggest question is which computer-assisted system is best suited to a company's needs. Items to consider include ease of use, availability or turnaround, cost-effectiveness, and, most importantly, continuing support and maintenance from the vendor. Typically, the general machining statement produced by the processor, such as APT or AD-APT, allows programmers to create source in a mode that is universal to the machines within a company. The input source statements will be similar for all controllers and machine tools, and the computer will substitute the specific codes for individual controllers. The software package that performs the conversion for a specific machine tool/controller combination is the postprocessor. The quality of a particular vendor's postprocessor should be checked carefully when a company is evaluating systems.

Of the available systems, the one likely to fulfill most needs is the APT system (Automatically Programmed Tools). The basic characteristics of the APT philosophy have been taken over by nearly all subsequent programming languages, e.g., the concept of separating the software into a processor and postprocessor, or using free-format input through words linked according to syntax rules. Further, APT has undergone continuous development, currently at APT IV, and is presently enhanced by efforts of the Sculptured Surface Project of CAM-I, Inc., (Computer-aided Manufacturing International, Inc.) in Arlington, Texas.

Generally, APT deals only with problems of profile geometry definition, and it is quite efficient in this area. However, it also needs computers with large working memories, or more than 256 k-bytes, in order to prepare geometric programs for simultaneous interpolation of up to 5-axis machining.

As NC machines penetrated the manufacturing market, parts with simpler profile geometry had to be machined and simpler geometry had to be programmed. APT proved to be too cumbersome and too costly. Therefore, many smaller programming languages were created for geometric calculation. Some are derivatives of APT, such as UNIAPT, AD-APT, UCC-APT, TC APT, MICRO APT, MDC APT, and CON APT, and others are unrelated to APT, such as COMPACT II, SPLIT, and NUFORM. Many organizations designed languages which were simpler to use or more limited in their scope and hence could operate on smaller computers. Some were limited to a specific machine tool, or to the tools of one machine tool manufacturer, or to one class of tools, such as lathes. The degree of automation of computer-assisted part programming increased as the efficiency of the smaller computers increased. Soon there was talk about semi-computer-assisted part programming.

The continuous development of software and new efficient hardware allowed for programming performance similar to that of large systems in the past. Interactive computer graphics systems, with NC modules generating cutter centerline data, as well as microprocessor-aided portable programming systems with graphic display screens and simple function/symbol keyboards, even on the workshop floor, have become important NC programming aids.

A further development has been the generation of a master computer language based on group technology principles. As opposed to APT, which describes geometric elements, the master language describes function elements. This has led to a significant reduction in input data. However, not all profile geometry can be described by this method, and even today, these programming languages are limited to simple geometry, like that of turning.

5.4.1.6 Main Features of Internationally Known Programming Systems

A great many programming systems have been developed and are in practical use in all industrial countries. As of 1984, 60%-70% of NC machines in use were programmed with the aid of computers. The performance and main features of selected NC part-programming systems are given in Table 24. Most of them emphasize geometric ability and have capacitiy for further developments in production technology and computation (optimization, modularity, integration, and flexibility). An important trend of the past few years is the transition to interactive, dialogue-oriented operating modes.

Some examples of NC programming work stations with their most important features are listed in Table 25. The application of NC technology by specialized NC programming stations covers a wide field. Most stations allow for the possibility of program editing, but in general they cannot be used universally with higher programming languages or as DNC computers.

Table 24. Performance of selected NC part-programming systems

		APT	APT JGS	OKI-SURF	TIPS	EXAPT 1	EXAPT 2	EXAPT 3	MITURN	SYMAP		COM-PACT II	ELAN
Control system	Continuous path	x(5)	x(5)	x(5)	x(3)		x	x	x(2)	x(2)		x(2)	x(2)
	Point-to-point	x	x			x					x	x	x
	Straight-line	x	x		x	x	x		x		x	x	x
Technological decisions	Chucks								x				
	Machining sequence				x				x				
	Tool selection					x	x		x	x	x		
	Cutting data					x	x	x	x		x		x
	Tool path				x		x	x	x		x		x
	Path decomposition				x		x	x	x				x
Geometric descriptions	Point	x	x	x	x	x	x			x	x	x	x
	Point pattern			x		x				x	x	x	x
	Line	x	x	x	x	x	x		x	x	x	x	x
	Circle	x	x	x	x	x	x		x	x	x	x	x
	Vector	x	x		x					x	x		
	Curve 2nd level	x	x		x				x				
	Interpolation curve	x	x	x				x		x		x	
	Contour			x				x		x		x	
	Plan	x	x	x	x			x					
	Plan 2nd level	x	x	x	x								
	Sculpture surface	x		x									
Programm-technique features	Solid body			x									
	Macrotechnique	x	x					x		x	x	x	
	Procedure technique												x
	Control statements	x	x					x				x	
	Indicated variables	x	x					x					
	Error service								x	x	x	x	x
	Files			x		x	x	x	x	x	x		
Kind of processing	Job processing	x		x	x	x	x	x		x	x	x	
	Dialogue		x	x	x	x	x	x	x	x	x	x	x

Table 25. Examples of NC programming stations

Designation (producer)	Process modification	Higher programming language	Program editing	DNC store
MDSJ (Compact II)	P, S, B PS 3-D, 4-D 2x2-D lathe	x	—	—
FANUC (FAPT) (Siemens)	Turning 2-D milling 3-D milling (DIE I DIE II) Wire erosion, Stamping, nibbling	x	—	—
Olivetti	PS Turning Wire erosion, Stamping, nibbling	x	FORTRAN 4	—
Hewlett- Packard	PS 2-D Turning	x	HP-Basic	—
Teledynamics	Manual Programming station and acquisition (text processing)	x	—	x (up to 32 NCM)

P, point-to-point; *S*, straight-line; *B*, continuous-path

5.4.1.7 Selection Criteria for NC Programming Languages

There is no one language which is superior to all others. The geometry which has to be programmed, the NC machine tool on which the part has to be machined, and the ability and experience of the part-programmer must all be considered. An NC programming system can be defined as the process of creating meaningful NC control data from a graphic representation of a part configuration, or in simpler language, the changing of print data to machine control data. In pursuit of the optimal system we must bear our objective in mind: to obtain a part that will be bought on inspection. Our goal is not merely to produce a tape, but to produce a good tape.

The question, then, is: Should you choose a specialized language or a universal language? Languages created for certain special machining tasks, such as turning, often offer advantages for that specific operation. By means of existing macros, for example, one may have a very efficient programming cycle focused on that particular operation. However, the disadvantage is that other operations such as milling would still have to be programmed manually or with other computer-aided systems. A universal language, on the other hand, offers one the assurance of being able to cover all NC machines in one's operation with a single language. The solution of specialized machining technique problems can be formulated, very easily, via macro technology present in all modern languages. Flawless parts programs are an essential prerequisite for the profitable operation of NC machines. Fast, economical and error-free program availability is the goal. Deciding in favor of one of the existing systems troubles most NC users, since each system has specific benefits and drawbacks. As a basis for the decision one should consider:

1. The parts spectrum — the complexity of geometry
2. The diversification of machine tools — different machine tool/controller combinations
3. The strategic planning data — new purchases, lot sizes, repeat frequency
4. The organizational infrastructure — personnel qualifications, systems availability

5.4.1.8 Future Trends

Of particular interest is a new CAM-I, Inc. development called "Advanced NC". Systems such as the Geometric Modeling Project aim to create and store geometric information at the design stage, in what is termed "bounded" form. NC programmers and process planners will not need to respecify the geometry as is presently done in the APT-like systems. While this permits the description of any geometry, it is not the description of the part. It is the programmer's definition of the tool motion statement defining the path of the tool going around the component that actually describes the geometry of the shape. In APT the bounded component geometry is not defined. Using bounded geometry for the shape definition, the computation of the tool path is less difficult. Advanced NC will capitalize on bounded geometry definition developments in the computer-aided design field and will provide an NC processing capability to the geometric modeling project.

The current CAM-I specifications for the design and development of an advanced numerical control system are an outgrowth of approximately 2 years of advanced numerical control project effort. They detail many goals, capabilities, and constraints which are desired by numerical control processors. It is certainly specified that the new system possess capability equivalent to that of the APT numerical control processor. In addition, the specification indicates that a new approach should be based on evolving concepts such as "complete" part description, computerized process planning, and "generative" numerical control programming with user control via graphics interface. "Complete" part description refers to the fact that parts to be processed are completely defined mathematically — either in terms of the bounded faces and edges which describe them, or in terms of the primitive solids which are united, intersected, and differenced to form them.

"Generative" numerical control operation sequences are determined automatically by the system. This is as opposed to current APT programming, where all tool positions and commands are normally generated by explicit user input.

The task of the Advanced NC system is to prepare sequences of instructions for the control of numerical machines in specific production processes. A number of production processes are included in these requirements, but machining and inspection of discrete parts are most relevant to the current specifications.

From a geometric modeling viewpoint, the advanced numerical control machining problem is characterized as follows. Consider the geometric model of the current workpiece (a blank) and the model of the final part. The set difference of these two solid objects is the stock object. The numerical control process is then one of practical and economical determination of tool motion to convert stock into chips so that the part is formed according to tolerance and other specifications.

When all of the variables related to this process have been established, the basic simplicity disappears. For example, suppose a tool and tool path have already been specified. The necessary feed and speed calculations then depend on cutter geometry, part geometry, stock geometry, part material, tool material, part rigidity, tool rigidity, fixture rigidity, machine rigidity, coolant, and tool condition, to name a few factors. The problem of determining tool paths and selecting tools and machines for efficient production is correspondingly more complex, but nevertheless relevant to the Advanced NC system.

The inspection problem is also one of determining tool (or probe) motion. In this case, however, the objective is not to convert stock into chips, but to determine a relatively efficient pattern of motion to verify with certainty whether a part meets tolerance specifications, and coincidentally, to provide an analysis of any out-of-tolerance aspects of deviant features.

The system will optionally accept input from various data bases — especially a part specification from a geometric modeling data base and variable levels of process specifications from a process planning system. It will be modular so that it can be configured to operate in subsets of the total processing required (e.g., point-to-point, 2-D contouring and will have alphanumerical APT-like language and graphics input. The design of the language and graphics will be carefully coordinated so that graphics input may result in automatic generation of equivalent alphanumerical language.

The integration between graphics and language will be facilitated by the use of bounded geomeotry. Numerical control commands will be oriented toward numerical control processes as opposed to the determination of individual tool positions. The language commands will also be designed for compatibility with corresponding graphics requests. The output of this system will consist of both CL file (cutter location file), with optionally expanded CL information, and source file. Additional outputs such as operator instructions and management information will also be included.

The Advanced NC specifications have been devised by a statement of goals and objectives and by the subsequent effort of developing related requirements. The list of goals (33 in all) is available in the project long-range plan, but that list is by no means non-redundant or consistent.

In order to provide a simpler and clearer direction on project purpose, a user survey was conducted based on the set of goals. The survey results are quite detailed, but the following goals for an advanced numerical control stood out as the most desired:

— Generative programming
— Correction reporting
— Reliability/accuracy
— Automatic processing across part discontinuities
— String processing

In short, sponsors desire above all a system which performs numerical control processing more automatically and with high-level user graphics interface for verification and correction.

5.4.2 Programming Systems for Industrial Robots

5.4.2.1 Steps in Development

The development of the programming of manipulators followed that of industrial control engineering. The basic principle was the permanent circuit-technical wiring of specific sequences of functions, designed for specific application. Changes or adaptations could be made, but at great cost.

Permanently wired logic components were then supplemented by additional external switches or control units (program drums, crossbar distributors, diode matrices as program carriers). As with the analogue program controls of machine tools, this provided some flexibility, including the possibility of adaptation to different control sequences.

Next came the use of free-programmable (CNC) controls. In contrast to the program generation for NC machine tools, the aim with manipulators from the start was to program them almost exclusively at the workplace. For each given operation (setting, programming, test operation, automatic operation) relatively fast program generation, program correction, or the repeated input of already filed programs is possible.

In accordance with these stages of development, industrial robots are classified into three generations, as shown in Table 26. The complexity of the control functions to be described and the integration of material handling into highly developed solutions of automation (such as FMS) require a high level of programming.

Table 26. Sequence of generations of industrial robots

	Control features	
	Controlled axes	Type of control
1st generation	Two to four	Permanently wired logic
2nd generation	Up to six (point-to-point control)	Principle of analogue control (crossbar distributor)
3rd generation	Convenient point-to-point control; path control up to six axes	Free programmable; additional distribution of sensor information

5.4.2.2 Programming Systems

Following the development of industrial robot controls, the programming methods shown in Fig. 83 have been devised. The known methods of manual programming, play-back, and teach-in will not be dealt with in more detail here (see [183]).

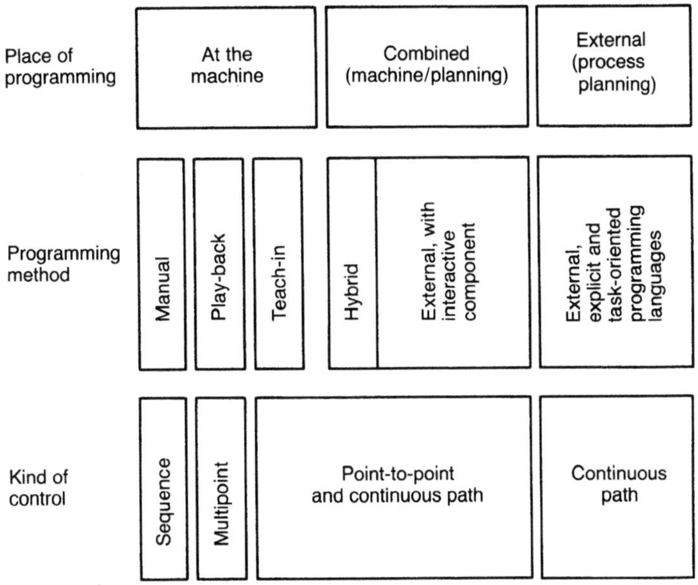

Fig. 83. Programming methods for industrial robots [243]

Just as with the NC programming of machine tools, programming of industrial robots attains a higher level with the aid of an external computer. The essential advantages of external programming methods are:

1. Rational program generation through utilization of the possibilities provided by high-capacity computer techniques
2. Reduction of unproductive idle time of industrial robots
3. Possibility of integration into efficient CAD/CAM systems, interconnection with suitable data banks

The basic idea of programming industrial robots is that few statements are given in advance for the external programming, while all detailing of the sequences of motions is carried out with computer assistance according to the specific character of the industrial robot and the problem of manipulation. A general classification of programming languages for industrial robots is given in Table 27.

Table 27. Programming languages for industrial robots

Explicit, movement-oriented, program-sequence-oriented	World-model-oriented, task-oriented, problem-oriented
— Direct movement orders — Elementary program sequence functions	— Description of the movement task — Simple and combined description of program sequences — Description of IR geometry — Description of working space

IR, industrial robot

5.4.2.3 Classification and Features of Languages for Industrial Robots

For *movement-oriented* languages, all movements of the robot which are required for the execution of a task have to be described individually. The exact statement of positions and orientations in three dimensions requires a programmer with good imaginative faculties. Typical motion-oriented languages are VAL (Unimation) and SIGLA (Olivetti).

VAL (Variable Assembly Language) is both the name of a programming language and the designation of a control-internal programming system which is widely used in industry with the PUMA 500 industrial robot. VAL includes movement statements for point-to-point controlled behavior and linear interpolation, as well as the possibility for relative programming. Conditional and unconditional program jumps, subprogram techniques, and statements for the interrogating and activating of peripheral devices or states are available. The opening and closing of the gripper jaws to any specified extent is possible by means of special gripper commands.

SIGLA (Sistema Integrato Genèrico per la Manipolazione Automàtica) is a programming language for the industrial robot SIGMA produced by Olivetti, Italy [197]. The simple programming system makes possible parallel control of up to four arms, with three to eight degrees of freedom each. The instruction set allows, among other things, conditional branching, gripper commands with sensor feedback, and the processing of external signals. Some other programming languages for industrial robots are listed in Table 28.
The difference of *task-oriented languages* from movement-oriented languages consists in the clearly arranged description of complex processes of motion and logical connections. Such languages are therefore especially suited to interlinked manufacturing plants within flexible manufacturing systems.

ROBEX [243], AUTOPASS [150], and RAPT [182] are the best known task-oriented programming languages; however, they are still being developed or tested. This is due to the complexity and the extreme dimensions of the systems. A large number of interface problems still must be solved, including their integration into existing CAD/CAM systems.

Table 28. List of several programming languages for industrial robots

Name	Features	Developer
AL (Assembly Language)	ALGOL-like, block-structured programming language for assembly robots	Stanford University, California [86]
RPL (Robot Programming Language)	Extended Backus-Nauer form for execution of — Input-output statements — Process statements — Check statements — Macro-statements	Berlin University of Technology [216]
ML (Manipulator Language)	Syntactically simple structure, similar to assembler language; developed for enterprise-internal simple manipulators	IBM [245]
LAMA (Language of Automatic Mechanical Assembly)	World-model-oriented programming language for the employment of manipulation systems in assembly	MIT [152]
LM (Language of Manipulation)	PASCAL-like language structure, processing of arithmetic, Boolean and symbolic variables; simultaneous operation of several IR possible	Grenoble University [147]

IR, industrial robot(s)

5.5 Problem-Restricted Solutions

Independently or as components of systems for process design, there are a number of computer-based decision aids which can be classified under (a) aspects of the organization of manufacture or (b) aspects of manufacturing engineering. Among these are:

1. Computer-aided comparison of variants
2. Methods for process optimization
3. Computer-aided decisions for efficient multiple-machine operation
4. Methods for selection of blanks
5. Optimization of production-technological working values
6. Design-technological programs of dimensioning (e.g., for design of broaches and other tools)

More details can be found in the literature [133, 135].

6 Geometric Modeling

The present stage of scientific-technological progress is characterized by rapid changes in production processes [93]. There has been rapid development and implementation of the man-production interface — CAD/CAM systems. The annual market growth of hardware and software for CAD/CAM is about 40%. It is expected that in the near future this rate will remain stable [85, 158].
At present there are companies producing turnkey systems as well as computer-independent application packages and systems. This allows enterprises not only to implement new systems, but also to expand the functions of already existing ones by integrating new programs into them.
One of the current trends in CAD/CAM systems development is an integration of subsystems with different structures and various design and technological functions around a product definition data base [15]. The main part of such a data base is the information which characterizes form, structure, and position of an object in space. The process of creating and changing this information is called geometric modeling. This process is realized by a software-hardware facility usually called a geometric modeler.
Evidence of the importance of geometric modeling for creating integrated CAD/CAM systems is the large number of papers on the subject published during the past few years [13, 59, 111, 172, 192].
As examples of integrated CAD/CAM or CIM (computer-integrated manufacturing) systems we may consider the CAM-I system of the CAM research organization, the ICEM system of CDC [110], and the system SDRC I-DEAS system of CAE-International [113]. Most of the leading companies selling hardware and software for CAD provide systems with various levels of integration (Fig. 84). The basic system is a data base containing a 3-D object geometry model and information about other characteristics of the object. We shall call this basis a "project model". All subprocesses of integrated CAD/CAM, the analysis by the finite element method, optimization, documenting, program preparation for NC tools, manufacturing, assembling, and robot programming, have access to the project model. The process of geometric modeling is normally initiated by the user at the design stage, but there are also cases when it results from application programs.
Forming the model picture for the graphic device and inputting from it are performed by special application programs, basic graphic systems such as Core System [2], GKS [52], or PMIG [221]. The project model storage and data exchange between different systems can be realized in IGES standard [7].

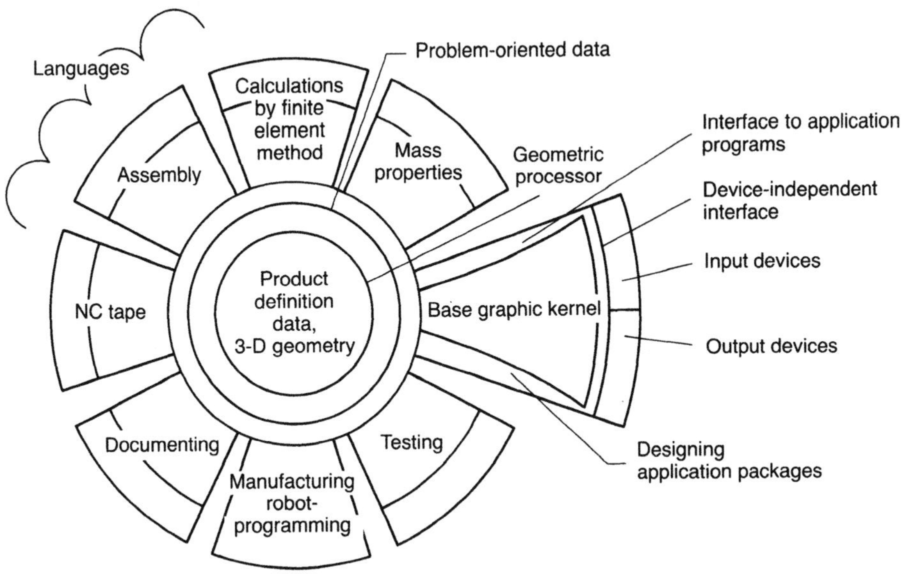

Fig. 84. Connection of integral CAD system subprocesses with data base

Geometric modeling accomplishes two important things:

1. It synthesizes the analytical definition of the geometric object with the predetermined properties in the computer.
2. It uses all numerical methods in mathematical experiments to determine and study the various properties of the designed object in the computer before the object is manufactured.

There are two approaches to creating the model of an object to be designed (Fig. 85). With the analogue modeling methods the generalized complex structure scheme is developed. Its variable parameters are dimensions. This scheme is put

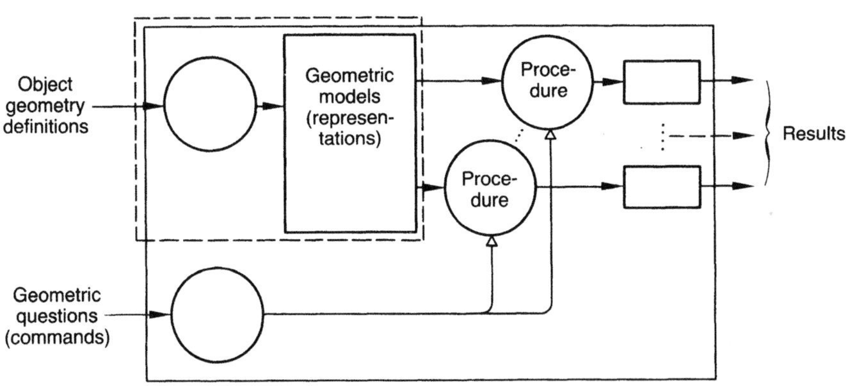

Fig. 85. Geometric modeling system

into the system storage. The actual dimension values are given by the user or determined through solving the given application problem. In future it will be possible to computerize the process of preparing design-technological documentation in accordance with actual structure parameters.

The analogue methods have the advantages of automatic problem solving, minimum requirements for computer power, and easy programming. Their disadvantages are a restricted subject field, low flexibility, program dependence on subject type, and the need for preparation work in developing the generalized drawing. The systems using these methods are SPLINK, FAP-KF, BUILD, and COMVAR.

The preferred systems are those which use the methods of variant and search modeling. The essence of these methods is that they give initially some properties and restrictions (geometric, physical, mechanical, technological, etc.) which are typical for the object considered. They also define the class of mathematical models, describing designed objects as a system of functions of many variables depending on unknown parameters.

The search for these parameters is carried out by mathematical programming methods. The advantages of these methods are the possibility of obtaining new, nontrivial solutions, flexibility, quick change of subject field, and the possibility of multicriteria problem solving.

The disadvantages are their high requirements for computer power and the complexity of mathematical programming in problem solving. Some of the systems that use these methods are GIPSY (France), PROREN (Federal Republic of Germany), CADD (United States), and APPLICON (United States).

For modeling the behavior of objects designed in order to determine their properties and to analyze structural strength, the finite element method, kinematic analysis methods, and visualization by computer graphic facilities are more frequently used.

6.1 Classification of Geometric Modeling Systems

The methods used in modern CAD/CAM systems allow the definition of designed objects in three dimensions. However, there are systems with 2-D objects that are used mainly for producing detail drawings and for designing schemes, printed circuit boards, etc.

Three-dimensional representation is more widely used and is applicable for describing large objects. It is very informative, but for in-computer representation and generation of drawings it requires more complicated algorithms.

According to Requicha and Voelcker [192], 3-D objects modeled in CAD can be divided by idealization level into 4 classes:

1. Wireframes
2. Polyhedral surfaces
3. Sculptured surfaces
4. Solid modeling

There are connections between these classes. Thus, from definitions of sculptured surfaces it is possible to obtain the wireframe definitions or to approximate them by polyhedrons. Solids, if logic functions are used, can be represented as linear inequalities, i.e., we can interpret them as polyhedrons, for example.
The various mathematical models are used to define the objects in these classes. The choice of a given model is determined by those design problems that should be effectively solved by it. However, from the point of view of a designer of a geometric modeling system in CAD, the following three forms of geometric object models are essential:

1. External form (A-model)

This is the explicit analytical writing of equations and their accompanying conditions or restrictions that allows the composition of a well-defined computational algorithm for determining the arbitrary point of the object or the element of mathematical or physical modeling.

2. Language form (L-model)

With the help of a special language oriented to the model or to a group of similar models, the designer may input graphic and semantic data which can be interpreted by the computer as a definition of the geometric object related to its model. With the help of this language the designer can usually manipulate both the form and the structure of an object modeled.

3. Internal form (D-model)

This form is an internal representation in the computer memory of a geometric object, which is represented in its environment by its own A-model and is described by L-language for input. The form is a structure of properly organized data that fully define the given class of geometric objects. This type of model is very essential for CAD, because all the application programs that use the CAD processes and geometric representation of objects "recognize" these objects as related data structures.
Thus, we shall consider that a geometric object is defined in CAD if the following triad is given for it:
$GO = \langle A, D, L \rangle$
The existing geometric modeling systems differ according to the variety of analytical model being used, to the semantics and syntax of the languages for describing and manipulating geometric objects, and to the data structure types that represent the objects in computer memory.
A-models of wireframe objects are equations of line, satisfying definite conditions, among them conditions for frame forming. D-models are data structures containing the following information: (a) line identifiers, (b) parameters of the mathematical line model in some canonic form, and (c) the sequence of its connection with other lines.
The advantages of object representation in the form of a wireframe are easy realization and the use of small computers and inexpensive graphic terminals. But such

representation does not provide unambiguity and completeness of mapping while displaying or plotting the frame image. This causes some difficulties when the image is interpreted by the user. In CAD/CAM systems, objects so represented often contain insufficient data for calculating some of their properties and performances.

The methods of representing objects in the form of polyhedral surfaces are more widely used. In this case, an A-model is an approximation of the object surface with the accuracy degree given by flat faces. L-models for such representation contain descriptions of these faces in terms of coefficients of linear equations and the matrix which indicates the sequence of their connection. D-model structure can be presented in the form of a logical function or a graph.

Object representation in the form of a polyhedral surface is useful for solving a number of geometric problems. In particular, it is widely used to solve problems of visibility and real visual effects. Surface representation by a set of flat faces makes the problem regular, i.e., reduces it to the finite element method.

A large number of examples of the algorithms used in such systems are given by Newman and Sproull [170]. The disadvantage of the systems for application to CAD/CAM is the need of allocating the large storage capacity for D-models. EUCLID [60], CDP (IBM) [244], and MEDUSA (AGS, Cambridge Interactive Systems) [160] belong to this class of system. Figure 86 depicts two objects designed in the EUCLID system.

The class of so-called sculptured surfaces is applied mainly in designing aerodynamic and hydrodynamic surfaces and in designing surfaces to meet certain aesthetic requirements.

A-models of objects in this class contain a number of descriptions of geometric characteristics of the object modeled, specified, for instance, by a point or a line frame and/or a number of approximating functions, the parameters of which are defined from the given description and the criterion for estimating the modeling quality.

D-models are list structures ordering the values of A-model parameters. The survey by Böhm and Kahman [20] gives the most complete presentation of mathematical solutions in this field.

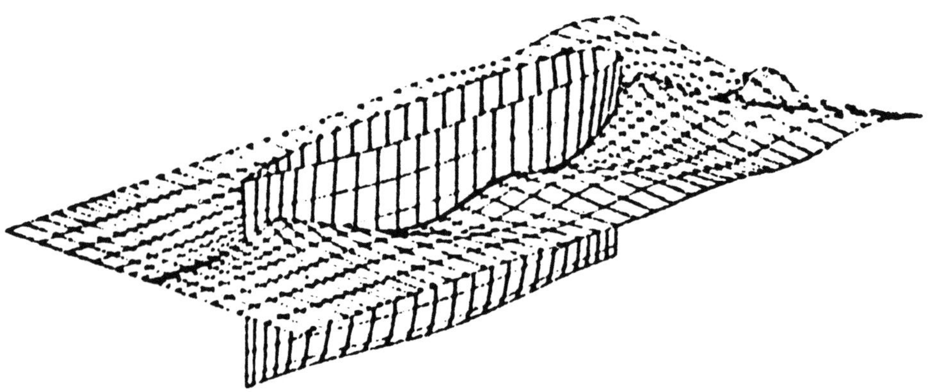

Fig. 86a. Object design with the EUCLID system; sculpture surface

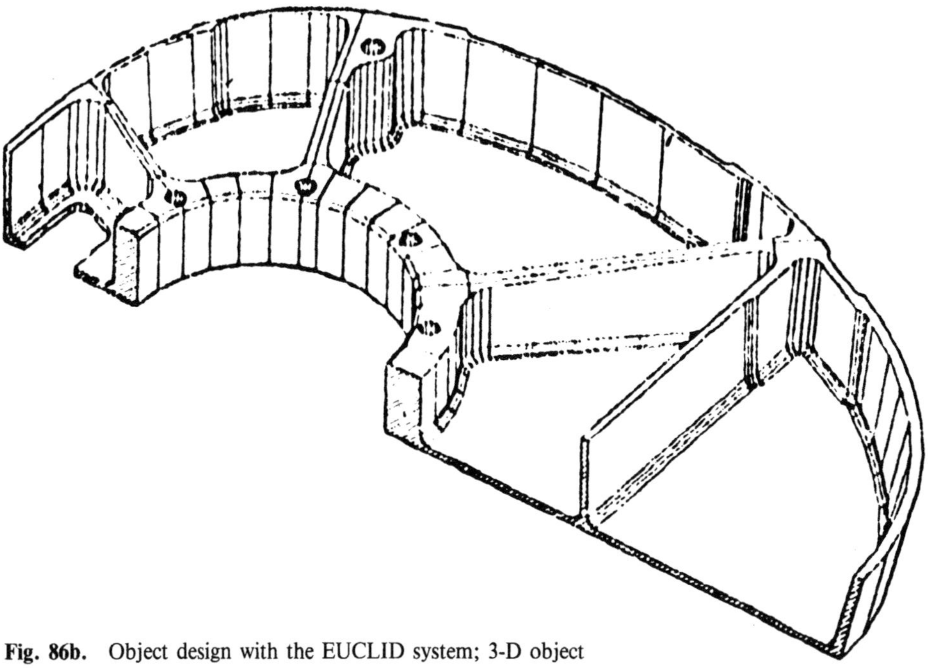

Fig. 86b. Object design with the EUCLID system; 3-D object

Systems based on solid modeling approaches are widely used in CAD/CAM. The two main representations of A-models, boundary representation geometry (BRG) and constructive solid geometry (CSG), are covered here (see Fig. 87). BRG requires the listing of all the solid faces bounded by edges and specifying halfspaces (the surface on which the face lies). The main idea of CSG is that the complicated solid is represented by a set of simple volume primitives from which the modified set-theoretical operations of "subtraction", "union", and "intersection" are realized. These operations are the basis of A-models. Figure 88 shows the results of the indicated operations. The volume primitives can be specified as topologically defined solids and also can be obtained from the intersection of halfspaces. For example, Fig. 87 b shows that the object can be designed in the form of the intersection of six halfspaces. However, the indicated procedure has the essential disadvantage that the correctness of the model obtained cannot be determined automatically and the task thus falls to the user. The representation of A-models by so-called translational sweeping and rotational sweeping is widely used in solid modeling. A large number of points of a solid result from the movement of some contour along the specified trajectory. Translational sweepings and rotational sweepings (Fig. 89) can be realized in different combinations according to user indications. Rotational sweepings are usually applied as input procedures, and the models obtained are transformed into boundary or space representation.

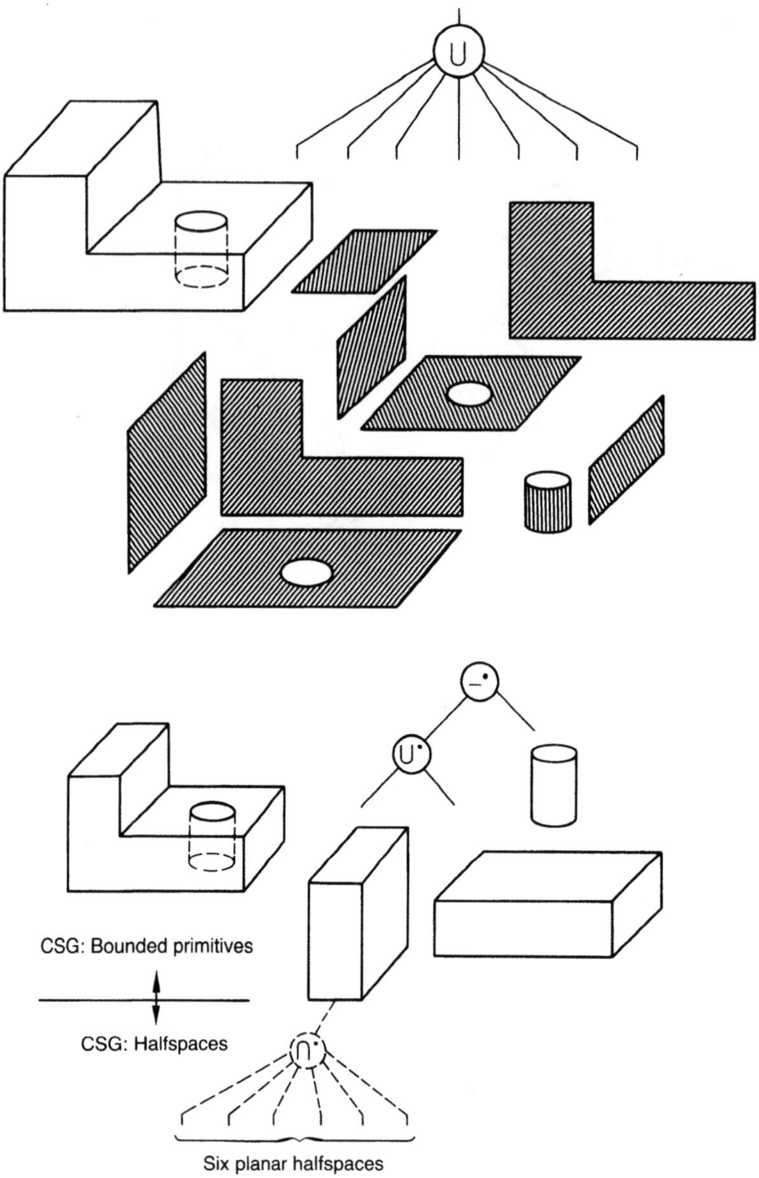

Fig. 87a, b. Two methods of representing A-models: **a** boundary representation geometry, and **b** constructive solid geometry **(CSG)**

The systems of geometric modeling using the idea of solid modeling can have boundary representation and apply the input by means of volume primitives and rotational sweepings. Such systems can also have mixed boundary and space representation, as do PADL 1/2 [26] and GM SOLID [24].
The theoretical foundations of CSG are described in the literature [26, 27, 28, 29], and the concept of regularized operator is the basis of the theory. The approach based on the use of Euler Operators is also widely represented.

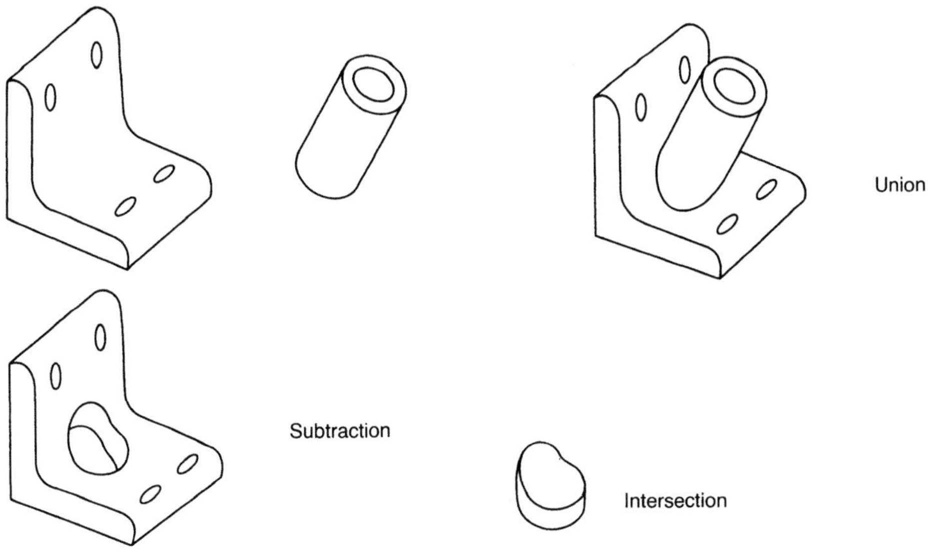

Fig. 88. Results of specified operations in constructive solid geometry

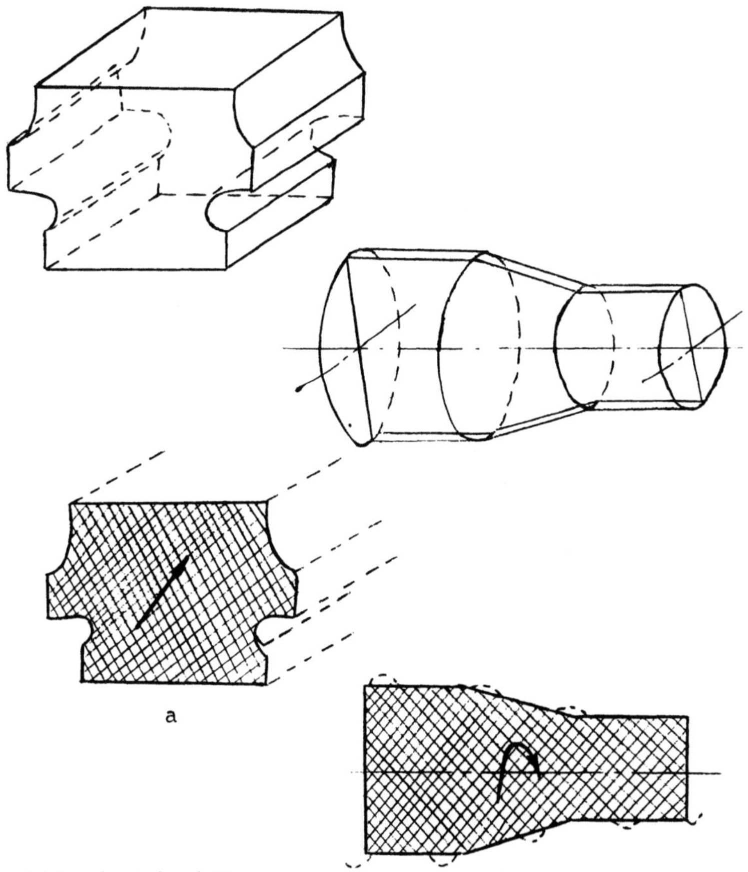

Fig. 89. Translational (**a**) and rotational (**b**) sweeping

Table 29. Characteristics of geometric modeling systems, and CAD/CAM systems based on them

Model	Organization	Host computer	Terminal capacity	Language (conversation, batch)
1. ANVIL-4000	MCS (USA)	MSK, VAX Family	Device indep. raster, 3-D	FORTRAN conversation
2. BDS/GLIDE	Carnegie-Mellon Univ. (USA)	DEC PDP 10, 11	Raster, 3-D	BLISS, PASCAL conversation
3. BUILD-1 BUILD-2	Cambridge U. (UK)	TITAN, PDP 11	Device indep. shaded	ALGOL-68, SAL conversation
4. CADD	MCAUTO (USA) McDonnel-Douglass	DEC PDP 11, VAX Family DG Eclipse	DDS 100 UNIGRAPHICS 1280x1024	FORTRAN, BAL, GRIP conversation
5. CAM-X	Ferranti (UK)	DEC VAX 11/750, 780	DEC VT 100, GMA·125.25 4096x3120	FORTRAN conversation
6. CDC SYNTHA-VISION	CDC (USA) Control Data	Sybev 6000, 8000 series	Device indep., 3-D, shaded	FORTRAN conversation
7. CDM-300	Kongsberg (Norway)	VAX 11/750, 11/780	DEC VT-100, 4096x3120 raster	FORTRAN conversation
8. COMPAC	TU BERLIN (West) ICOSS	SIEMENS 7.541, UNIVAC 1160		FORTRAN batch
9. DDM Solids	CALMA (USA)	Data General Eclipse	LEXIDATA/ 1280x1042, 3-D	FORTRAN, DAL conversation
10. DESIGN	MDSI (USA)	N/A		PASCAL conversation
11. DIKAD	University of Karlsruhe (FRG)	DEC 11/34		FORTRAN conversation
12. DUST	COMPEDA (USA)	DEC 11, VAX, IBM	Device independent	FORTRAN conversation
13. EUCLID (F)	MATRA DATA-VISION (USA, France)	VAX Family	Tektronix 4014	FORTRAN conversation
14. EUCLID (S)	FIDES Co. (Switz.)	CDC 6500		SUMBAL batch
15. FAP-KF	ITK AN BSSR	EC Family		FORTRAN batch

Modeling approaches	Analysis possibilities	FEM	CAM functions	Other CAE functions	Input basis
Boolean operations, scale, refl., translation	Mass properties, section			Tolerance analysis, dimensioning	CSG, Sweep
CSG, translation, EULER OP.					CSG, Sweep EOP
CSG, EULER OP CYLINDER SP.					CSG, EOP
SWEEP, sculptured surfaces	Mass properties	UNI-GRAPHIC GREM	2, 2.5, 3 Axes NC GIM, GMM, CMAX		Sweep
Romulus approach	Mass properties	FEMGEN	2, 2.5 NC TAPE GNC		CSG, Sweep EOP
CSG, boundary file, Boolean operations, sweep	Mass properties, section	UNISTRUC II MSC NASTRAN	5-Axes, EXAPT, APTIV	ICEM solution, IGES	CSG
AD-2000 approach	Mass properties, section	YES	5 Axes, EXAPT, APT-CLDATA		CSG, Sweep
CSG, Sweep, Quadric, SURFACES, B-SPLINE	Mass properties	COMFEM	EXAPT CAPSU	Dimensioning, GPM-APC APC-solution	CSG, Sweep
Boundary file, COPY, MOVE, extrude, B-SPLINE	Mass properties		3, 5 Axes milling APT, CLDATA	IGES	Sweep
CSG, quadric surface					CSG
Boundary representation, sweep, halfspaces		N/A	TC-APT CLDATA	Dimensioning, tolerance analysis	CSG
Bezier patches, sweep	Mass properties	YES	NC DATA, fa Compact II, APT PROMO SONC		Sweep
Boundary file, Boolean operations, sweep	Mass properties, section				CSG, Sweep
Bezier patches, Boolean operations, CSG					CSG
Quadric objects, boundary file, Boolean operations	Mass properties, section			Tolerance and topology analysis	CSG

Table 29. Continuation

Model	Organization	Host computer	Terminal capacity	Language (conversation, batch)
16. CDP/GRIM	IBM	IBM Family	Device indep.	PL/1 conversation
17. GEOMED	Stanford University (USA)	DEC PDP/10		ASSEMBLY conversation
18. GEO-METRICAL Modeling System	ISM (USA)	VAX 750/780, Apollo 1	Tektronix 4010 or 1024x1024	FORTRAN conversation
19. GEOMOD	SDRS (USA)	VAX Family, IBM 4341	Device independent	FORTRAN conversation
20. GEOMAP	University of Tokyo	TOSBAC 5600 (GE 635)		FORTRAN batch
21. CM SOLID	GENERAL MOTORS (USA)			P/1 conversation
22. HOSAKA	University of Tokyo	TOSBAC 5600 (GE 635)	Tektronix 4096x3120	G/L batch
23. IDS-80	GERBER (USA)	Hewlett-Packard. HP. 1000	Tektronix 4096x3120	FORTRAN conversation
24. INTERGRAF	INTERGRAF (USA)	DEC PDP 11/40 VAX 11/780	Raster 1280x1024	conversation
25. MEDUSA	AGS, PRIME (USA)	PRIME	1280x1024	FORTRAN conversation
26. PADL-1/ PADL-2	University of Rochester (USA)	PDP 11/40, IBM 300, 360, 370	Device independent	FLECS/FORTRAN conversation
27. PATRAN-G	PDA (USA)	VAX Family, PRIME, Harris	Device indep., 3-D, shaded	FORTRAN conversation
28. PROREN-2	University of Bochum (FRG)	INDEPENDENT	Device independent	FORTRAN conversation
29. ROMULUS	Shape Data (UK) E. Sutherland (USA)	INDEPENDENT	Device independent	FORTRAN conversation
30. SHAPES	DRAPER LAB (USA)	N/A		BAL-360 conversation
31. SIMAK	Moscow Power Inst., (USSR)	DEK 11/34	VT-100, VS-62	FORTRAN conversation

Modeling approaches	Analysis possibilities	FEM	CAM functions	Other CAE functions	Input basis
CSG, quadrics, boundary primitives					CSG
EULER operations, Boolean operations					CSG, EOP
Boundary representation, CSG, boundary primitives, sweep	Mass properties	N/A	N/A	N/A	CSG, Sweep
Boundary file, sweep, surface deformation, bending	Mass properties, section	SDRS SUPERTAB		Kinematics dimensioning	N/A
Boolean operations, CSG	Mass properties	YES	YES		OSG
HYBRID, CSG, boundary representation, QUADRIC SURF	Mass properties	YES	YES	Assembly	CSG, Sweep
Boolean operations		YES			CSG
Sweep, B-splines, Bezier patches		NASTRAN SDRS SUPERTAB	2, 2.5, 3.5 Axes NC APT-CLDATA	IGES, dimensioning	N/A
IGDS approach		NASTRAN STRUDI	2, 2.5 Axes APT-CLDATA	OMRS, OBMS	N/A
POLYGONAL schemes, boundary repr., QUADRIC, SCULPTURED SURF.	Mass properties, section		APT, GNC	Dimensioning, IGES	CSG, Sweep
CSG, bounded primitives	Mass properties, section	NONE	NONE	Dimensioning	CSG
Parametric objects, sweep, deformations	Mass properties	YES	YES	Kinematics, predesign	Sweep
Boundary representation, sweep, QUADRIC	Mass properties				CSG, Sweep
Boundary representation, quadric surfaces, Boolean operations	Mass properties, section	FEMGEN	$6M_cE$ Paction GNC		CSG, Sweep EOP
CSG, quadric surfaces		N/A	N/A		CSG
CSG, boundary representation	Mass properties	NONE	NONE	Assembly	CSG

Table 29. Continuation

Model	Organization	Host computer	Terminal capacity	Language (conversation, batch)
32. SOLIDVIEW	LEXIDATA (USA)	VAX Family, PDP 11, PRIME, Perkin Eimer, IBM series 1, DG	640/512	
33. SOLID MODELING	APPLICON (USA)	VAX 11/780, 11/750, PDP 11	Applicon / 768 x 576	FORTRAN conversation
34. SOLIDESIGN	COMPUTER-VISION (USA)	32-bit, CV	Wireframe 1280 x 1024 shaded 640 x 512	FORTRAN conversation
35. SYNTHA-VISION	MAGI (USA)	Any 32-bit, on mainframe	Chromatics 7400, IBM 3250, 3-D	FORTRAN conversation
36. TIPS-1	CAM-1 Hokkaido University (Japan)	IBM-360, 370, Syber, Hitachi VAX 11/780, Prime	Tektronix-4014, 3-D	FORTRAN batch

One of the recent publications in this field is the paper by Mäntyla and Sulonen [155]. A solution to the problem of object geometry model synthesis from base volume primitives on the basis of a formal axiomatic approach is suggested by Klimov and Klishin [131].

6.2 Solid Modeling Systems

Table 29 contains the characteristics of 36 systems of geometric modeling and CAD/CAM systems based on them. The table is composed from published data [13, 39, 56, 192] and from documents of firms and organizations received by the authors. It should be noted that, in practice, all the systems interface with the earlier listed CAD/CAM subsystems.

6.3 Structure of Systems

The PADL-1 system and its further development — PADL-2, designed at the University of Rochester (USA) [26] are the systems best represented in the literature.

Modeling approaches	Analysis possibilities	FEM	CAM functions	Other CAE functions	Input basis
display processor for objects, independent approach	Section	N/A	N/A		CSG
MAGI-SYNTHAVISION approach	Mass properties	SDRS-SUPERTAB	APT, SLDATA COMPACT II	Dimensioning	CSG, Sweep
Boundary file, sweep, Boolean operations	Mass properties, section	YES	5 Axes NC, APT CL DATA, COMPACT II	CADDS-IV IGES	CSG, Sweep
CSG, file representation, sweep, Boolean operations	Mass properties, section	YES	YES	IGES	CSG
Halfspaces, Boolean operations, rotation, sweep	Mass properties	YES	3 Axes NC	Dimensioning	CSG

PADL-2 is a program pack used for solid modeling. It uses the mixed representation of a geometric model: the main representation of the solid is introduced as the combination of primitives connected by set-theoretical operations. The boundary representation can be received by the process called "boundary evaluation". Such representation permits the solid model to be in the form of a set of related faces and edges. The D-model is represented as an oriented graph, where the vertices are base primitives and the arcs are the operations on them.

Along with representation in the form of a graph, a text representation of a program is formed with the input language of the system. It is possible to get the text representation from the graph representation, and vice versa, without losing the information. An example of data structure for generalized representations of the faces is given in Fig. 90. To perform the geometric calculations, the computer geometric subsystem CGPAK is used. CGPAK operates with the following objects: volume primitives, face primitives, halfspaces, curve segment/curve. The input language of the system permits the introduction of a hierarchy in the object description. PADL-2 supports a wide set of applied problems.

In this section we have described A-models and D-models of the geometric objects, realized in CAD/CAM systems. The transformation of the models is performed by L-models, realized in the user interface.

As shown in Requicha and Voelcker [192], geometric modeling systems should interact both with man and with other subsystems. The application programs (display, preprocessors, calculation according to the finite element method, programs of geometric and mechanical calculations) refer to the geometric modeling

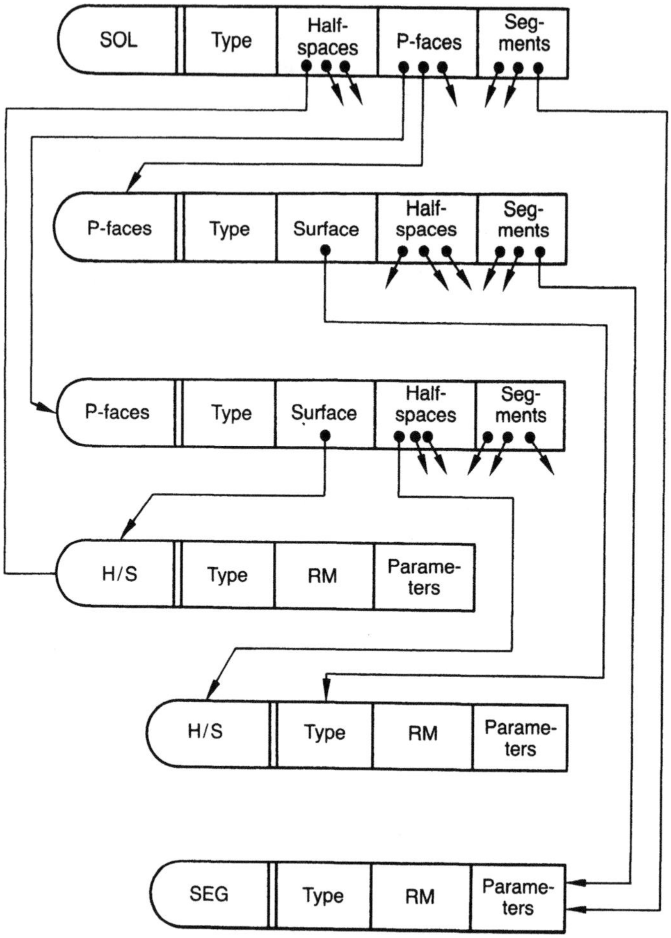

Fig. 90. Data structure for generalized representation of faces

system data through the corresponding program facilities, typical for modern management systems with data bases.

The method of building and realizing the user interface with the geometric modeling system is of special interest, since the convenience and effectiveness of applying the system as a whole mainly depends on this. In modern systems a number of peripheral devices are used to input geometric data into the system: alpha-numerical keyboards and functional buttons, various cursor control units, coding tablets, employing the menu technology through the light buttons [71]. The languages may differ in dialect according to which input devices are employed.

In the process of data input from the alpha-numerical keyboard the line recording language operators is used. In the process of data input from the coding tablets, or with the technique of light buttons, the languages are realized as the procedures of selecting the appropriate parameter, symbol, or indication from the menu. Here the functions of "prompt" and "echo" are widely used.

The input methods can be classified according to the nature of the operation mode. We distinguish batch mode, alpha-numerical conversation, and graphic conversation. In a batch mode, the data are prepared before the program execution and are then read into the data processing unit. The input devices are punched-tape readers, punched-card readers, and magnetic tape.

The typical input formats for batch processing are fixed format tables and input language with fixed dictionary. The fixed format tables contain all the information necessary for describing the input elements. The information can be only in the alpha-numerical form. In terms of processing, the table is one of the simplest forms of input.

In the interpretive-type languages the instructions for object description consist of sequences of symbols interpreted by the system. The object descriptions are read by the program system in the form of data. In conversational mode the data are entered only interactively, or a portion of the data is entered as in a batch mode and the rest are entered interactively.

The geometry description languages can include the operators for describing the models of different levels, e.g., specifying vectors, surfaces, base volume, primitives.

The nominal descriptive ability of a language depends on the set of available primitives, the type of operators suitable for combination of geometric solids, and the types of affine transformations (parallel transfer, rotation) useful for locating the geometric solids. As a rule, the systems using boundary representation as elements of the input language apply the operators of specifying volume primitives (PADL, ROMULUS), or specifying halfspaces and volume primitives (TIPS-1) [174] and operations on them. Application of primitives as the elements defined in the input language leads to reduction of the input time and the number of errors, and to simplification of the procedure of building a solid model. Frequently, in the systems oriented to input using volume primitives, it is suitable to apply translational sweeping and rotational sweeping, particularly in designing 2.5-D details and details of "rotation solid" type. The sequence of describing and forming the object image in the ROMULUS system is shown in Fig. 91. This system uses the sweeping procedures mentioned, as well as the volume primitives with Boolean operators.

A number of systems use input which allow the designer to remain with the idea of traditional designing technology, applying the drafting technique. In such systems the user forms two-dimensional images, projections, sections, and cross-sections using the "menu" technology. The information is then entered into the computer using translational sweepings or registration of images, while the model forming is performed automatically in the system. MEDUSA is one of the typical systems having such an interface with the user (L-model). A fragment of the "menu" applied in the MEDUSA system [160] is shown in Fig. 92.

Sweeping is suitable when the user assigns only the points of the object designed between which the characteristic of the edge (straight line, arc, spline) can be assigned, or when a number of points can be approximated by a certain spline. Having assigned the required graphical element and interpolation function by a set of points, the user can enter the surfaces by this procedure, and by translating the surface the user can assign the volume solid. Assigning of the solid can be achieved by translating the closed loop along some trajectory and closing it with appropriate

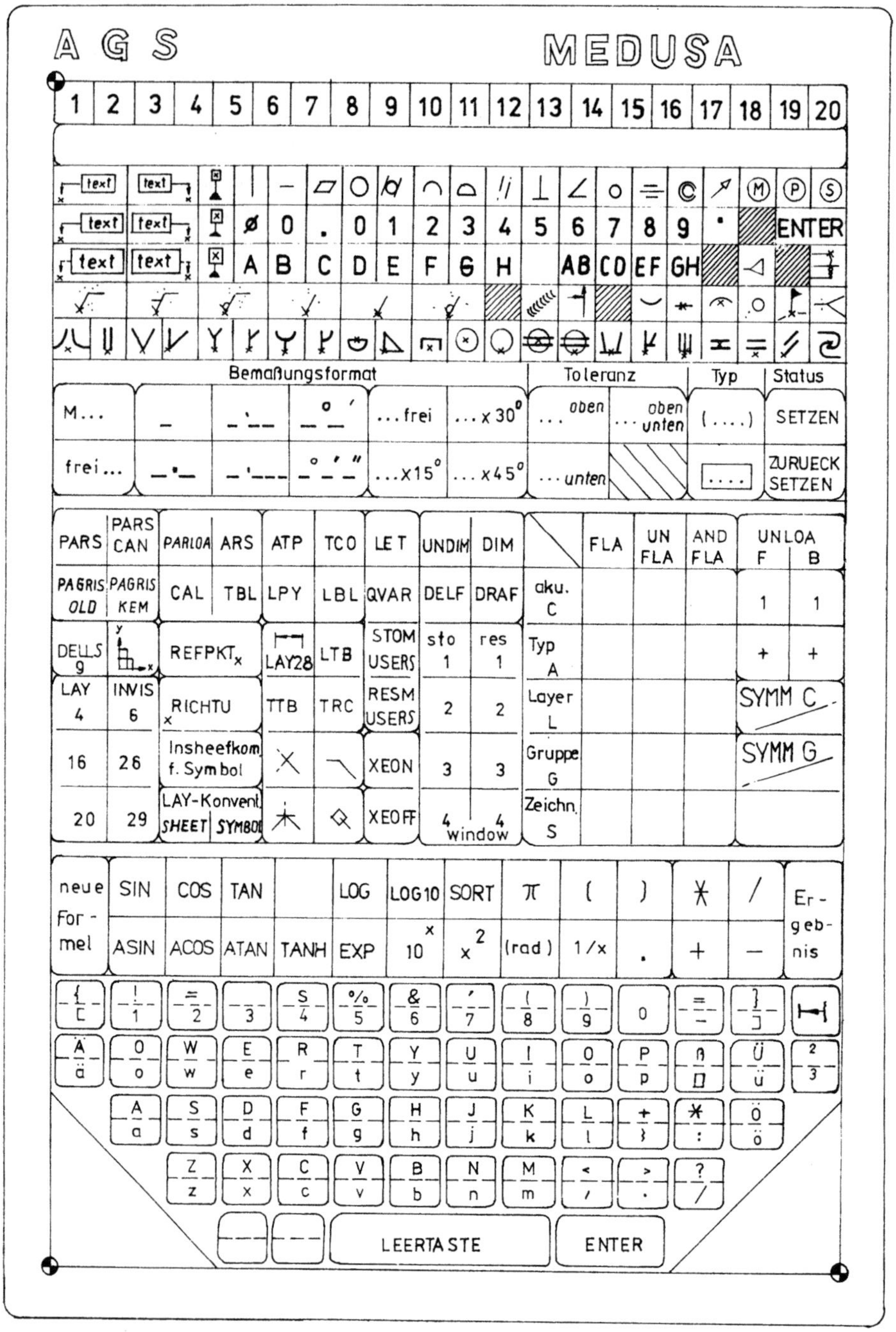

Fig. 91. Description and formation of the object image with the ROMULUS system

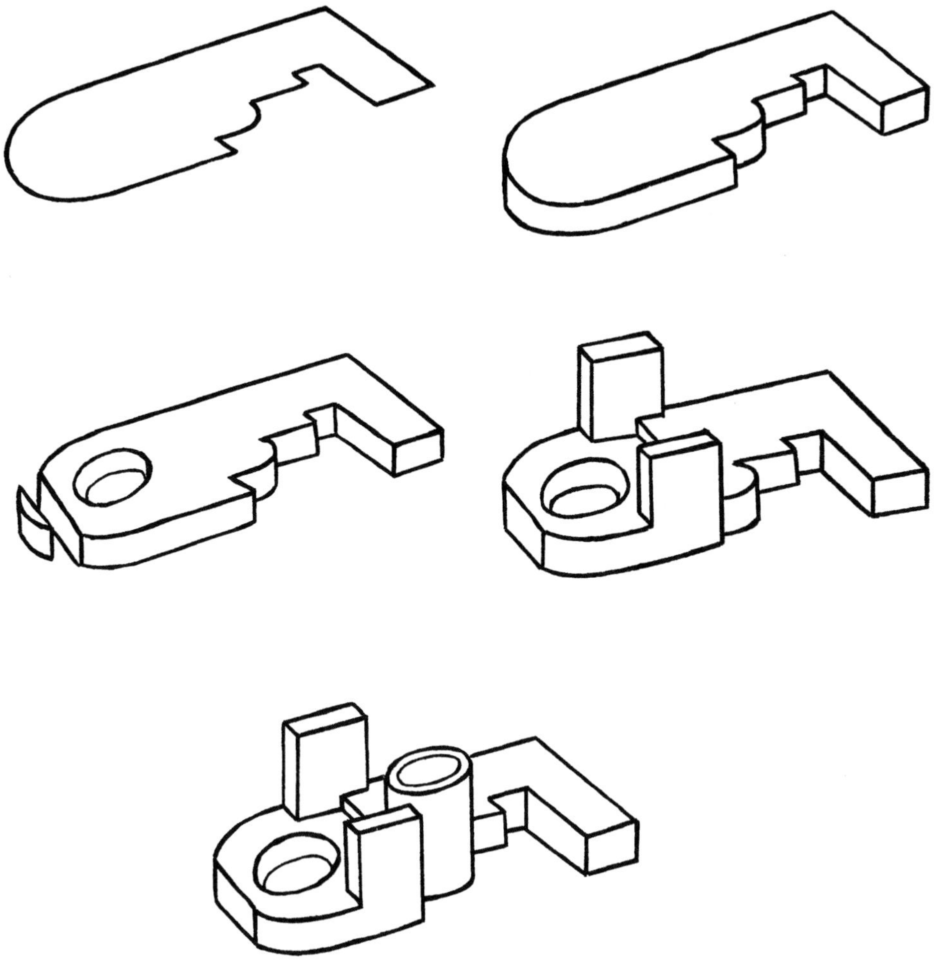

Fig. 92. A fragment of the MEDUSA menu

surfaces. The object of variable section can be obtained by varying the configuration of the loop being translated. This principle is used in the DUCT system.
Much attention has been paid lately to the methods of input based on the traditional drafting technique, when the user makes a drawing for the project draft of the object designed and the system performs the synthesis of the three-dimensional model [67, 119, 181].
The inputting into the computer of the complex forms of geometric information, objects formed from less complicated ones and geometric constructions of identical topology but of different metrics, requires the application of structuring and nesting mechanisms in input languages, as well as entering the object with implicit parameters and subsequently substituting them with the real ones. Such mechanisms are described in connection with the PADL-2 system by Brown [26] and with GDP by Wesley [244].

The convenience of using the interface depends on the level of the input language used. In the systems of designing and production preparation, the user's language, with which the geometric properties of the object are assigned, can have the designer's vocabulary. In this case the user inputs the geometric description by applying syntactic constructions close to the constructions of natural language. An example is the DICAD system [90]. The user specifies the geometry with the following sentence: "Drill the hole of 70 mm diameter with the center located on plane I at 5 mm distance from the edge (line 2) and 10 mm from line 3." Use of this language demands certain computer resources, but the user is released from the need to carry out the complicated calculations in specifying the geometry of parts.

Input of group and unit descriptions require specification of object names which form the group, coordinates of their base points, and relations between the objects in the group. Examples of such systems are AUTOPASS [244], GM SOLID [230], and SIMAK [131, 132].

The publications concerning the automated generation of dimension sequences, tolerances, and fits [97, 98] and automatic generation of tapes for NC tools [32, 239] are of great interest.

The CADD system (McDonnel-Douglass) can be used for:

1. Investigation of part configuration
2. Representation of parts, sheet elements and their modifications, assembly drawings, pipelines
3. Modeling of unit translations

The EUCLID system (France) is of interest from the point of view of language resources. Objects of the system are described by a special language and approximated as polyhedrons. Cylindrical, conical, and other surfaces can be described. The EUCLID system allows definition of dimensions of complex parts on the basis of assembly drawings and representation of new parts by sectioning the existing ones. The system is realized in FORTRAN. It offers the possibility of analyzing part repositioning and the conversational development of technology for NC tools. With EUCLID it is possible to represent parts whose forms are approximated as polyhedrons with plane faces.

The input language provides specification of solid elements (tetrahedrons, prisms, pyramids) as as well as of two-dimensional elements (points and planes). Two ways of specifying object form are provided for in the system. In the first the space position of some surfaces is specified, the boundaries of the surfaces joining each other. In the second the defined surface is used as the main element of the specified profile body. The operations of set-theoretical joining and addition and subtraction of some solids are possible, giving representations of complex drawings. The system reproduces representations obtained from various aspects.

The GEOLAN system (Messerschmitt, FRG) uses an input language analogous to APT. The language commands are interpreted. The system allows definition of any geometric objects (points, straight lines, surfaces, sections of bodies, lines of body intersections, etc.) and their representation. Parts are described by parametric curves and surfaces. With GEOLAN, part and assembly drawings are ob-

tained automatically. The specified geometric contours are loaded into the memory and can be fully or partially reused.
The GEOMAP system (Japan) is intended for three-dimensional graphical operations. Part forms are specified by simple solid elements which have contact through certain surfaces or can intersect with each other. The system allows representation of parts and units in various positions in space and geometric modeling of mechanism kinematics.
The CLIDE system (BDC-10) is also intended for three-dimensional graphical operations. The input language consists of commands analogous to ALGOL statements and is used to specify not only commands but also geometric definitions and programs. The system is written in BLISS. It is also used for data base management where geometric descriptions of parts are stored. By geometric definition in the input language geometric objects can be specified. The introduction of new geometric elements with arbitrary topology or structure is possible. By addition or subtraction of simple geometric elements the form of complex parts is described. The language has a command for checking the collision of construction elements. Parts are represented in perspective projection or in orthogonal coordinates, hidden lines being marked for choice.
The MOREAL (Wents) system is written in FORTRAN and intended for three-dimensional graphical operations. Part forms are specified by surfaces of the second order as results of the intersection of these surfaces. The semiautomatic dimensioning of drawings is possible. The system performs all calculations of geometric characteristics of parts (surfaces, centers of gravity, etc.). Geometric data in the TIPS system (Japan) are stored in two sections. The first includes two tables of general geometric data with description of elements and a table where interrelations of elements are described. The second contains the data of the actual user, for example, a computation scale for the finite element method.
The systems here described have many common characteristics. All are designed for 3-D graphical operations. For part description they use languages, tables, or interactive modes of operation using a graphical display. Simple geometric elements are combined through contact surfaces or by intersectioning. Hidden lines can be specified automatically or in a conversational mode for visual representation.
The distinctions between these systems lie mainly in their differing design descriptions of part geometry and in their use of different geometric elements and ways of interrelating.
The next stage in the development of the systems described is integrated CAD systems. The perfection of systems depends on broadening their hardware capabilities. The application of complex description elements is presumed. As the complexity of elements increases, their field of application narrows. For this reason, in describing parts it is advisable to use a set of elements having a different complexity. Complex parts have free formal parameters by which the required part parameters are specified. In addition to type elements, a designer can specify original elements which are included in the system without being changed. Regarding elements, lines, shaft elements, holes, parts, etc. can be chosen.
The COMVAR system allows operation with the description elements regardless of their complexity. Therefore, elements of any complexity are considered

complex elements (parts) which represent a part of a drawing and can have different dimensions. The complex parts are described using a generation principle.

6.4 Future Trends

The scope of this section does not permit consideration of a wide range of problems, ideas, and decisions in this intensively developed area of application of computing science methods and facilities, programming theory, artificial intelligence, and applied mathematics. In conclusion, however, we outline the structure of a hypothetical system of geometric modeling which would embody the most interesting decisions of this area.

6.4.1 Models

The problem of model selection is undoubtedly a key one for any system. It is determined by the user environment in which the given model should function. Thus, the environment for A-models is the variety of material objects in the space and of man's mental images. Here, not only object form models but also models of their location in space are important. Therefore, when defining a set of models necessary for a system, we have to look for a compromise between a large number of highly specialized, object-oriented — i.e., more effective — models and a comparatively small number of sufficiently universal — i.e., on average less effective — ones in each particular application. Obviously, the development of the system should proceed in both directions, enlarging a library of specialized models and at the same time generalizing them, aggregating their strongest features.

L-models define the user's language. The main requirement here is that the system communicate with a designer in his professional language. The language should allow description of an object model, formulation of design problems, control of their solution, and operations of data editing and of design information dumping.

D-models should function in the environment of multiple application systems represented by a family of CAD application packages. The main requirements here are completeness of design object data representation and simplicity of their manipulation in providing volume- and time-effective access to these data from the application package. The specific character of such representation is the possibility of separate storage in the computer memory of topological and geometric information, supplemented by necessary attributes of a technological and physico-mechanical nature and providing for their separate processing. In contrast to A-models, the range of D-models should not be large.

6.4.2 Integration with Application Systems

Geometric modeling systems themselves in CAD conditions usually interact closely with a number of functional subsystems: machine graphics and documenting, strength analysis, program development for NC tools. As a set they should form a single, integrated system on the basis of unity of technical and program-informational bases.

6.4.3 Accumulation of Geometric Data in the System

There should be a bank of the geometric objects previously designed by the system. Having the parameterized object descriptions in a continuously supplemented data base of the system, a designer can save a considerable amount of time in designing analogous objects. The effect becomes more appreciable if the data base stores, along with geometric object-analogues, all create design information, including, for example, manufacturing processes, equipment, and programs for NC tools. The CAD designer should have the possibility of creating a language and program information system which is oriented to him or her personally.

6.4.4 Transfer of Characteristic Properties of Artifical Intelligence Systems to Geometric Modeling Systems

In practice, this means that the system itself should be able to perform new and more complicated design operations without requiring instructions from a designer on how to do so. The means of realizing such system properties are banks of geometric knowledge and a methodology of semantic models. The application of the artificial intelligence technique is possible in several directions: first, development of languages which are as close as possible to natural ones; second, development of expert systems that allow planning and evaluation of the solution process on the basis of a given general problem description, and that choose automatically the optimal way of achieving the solution.

The geometric modeling system should be considered a complex one, including a number of its functional subsystems with their specific informational resources:

1. Object formation
2. Mass properties and geometric analysis of objects
3. Modification and editing of forms and structures of geometric objects
4. Location of geometric objects on the surface and in the space
5. Modeling kinematics of geometric constructions
6. Optimization of formulation and solution of geometric problems

All informational system resources should be accessible to a designer directly from the workplace at any time and in the required quantities. In other words, there should be easy and natural access to the system from remote graphic terminals and terminal stations through communication channels.

The requirements and properties enumerated here do not cover all the problems. Nevertheless, they give an idea of what features a prospective geometric modeling system as an important component of CAD should possess.

7 Integration of CAD/CAM

7.1 Introduction

The term "CAD/CAM" has been used in the language of engineering since the beginning of the 1970s, CAD meaning computer-aided design, CAM meaning computer-aided manufacturing.

According to the CAM Guidebook of 1983, a simplified interpretation of the content of the term reads as follows: "CAD/CAM is simply a general name or umbrella term for the many ways the computer can be used as a tool to help design, make and sell all kinds of products. More specifically, CAD/CAM implies the automation of this process."

Independent of this very general interpretation, in various branches of industry and language regions different explanations of the terms have been accepted (see Table 30). Other essential, generally accepted equivalent terms are CAE, computer-aided engineering (originally introduced for computer application in the fields of research and development; increasingly used for all computer-aided engineering), and ICEM, integrated computer-aided engineering and management.

In Sect. 10, special explanations are given for the even more comprehensive term CIM — computer-integrated manufacturing.

Table 30. Various interpretations of the term "CAD/CAM"

Design	Preparation of manufacture	Manufacture, control of manufacture	Main area of use
CAD	CAP CAPP	CAM	USA, Japan, FRG, CMEA
CAD			West Berlin, FRG
CAD		CAM	
	CAM		USA (standards committee)
CAD	CAM		UK, partly FRG

CAP, computer-aided planning; *CAPP,* computer-aided process and production planning; *CMEA,* Council for Mutual Economic Aid

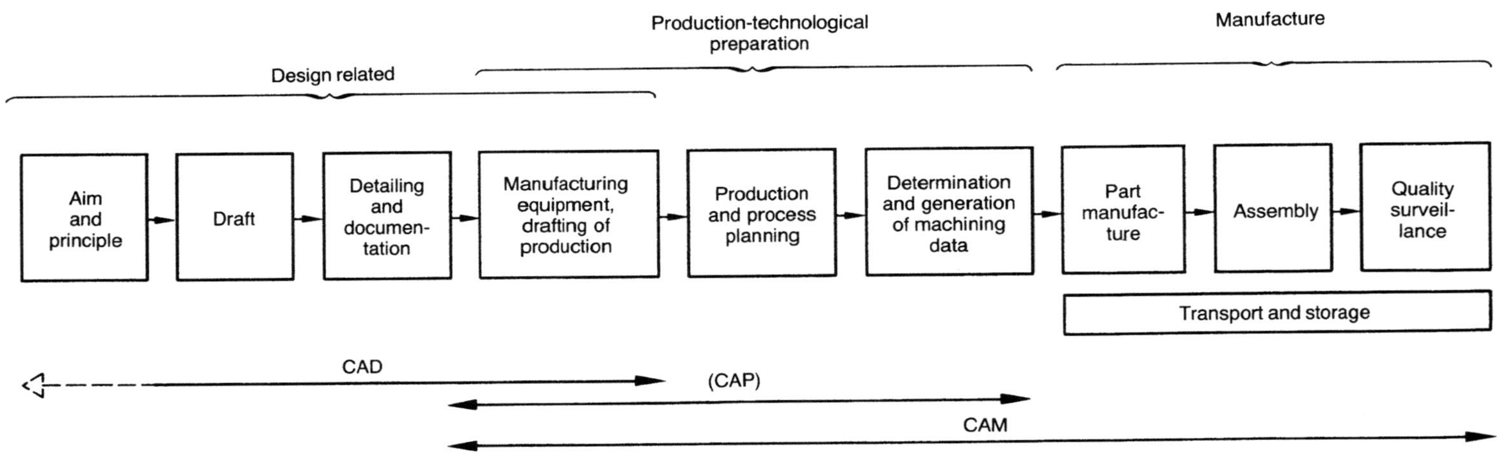

Fig. 93. Fields of application for CAD/CAM solutions in various branches of industry (main field: metalworking)

On principle, the fields of application of CAD/CAM solutions can be represented by the main field, metalworking, as shown in Fig. 93. The following features are generally valid for all fields of application:

— Preferential use of interactive working methods for different classes of tasks on the basis of efficient workstations for designers and production engineers
— Availability and utilization of comprehensive program and data bank systems in cases of different possibilities of access, and the application of systems of support guaranteeing portability
— Interfaces designed according to uniform principles for the computer-internal representation between design-related and production-technological preparation
— Transfer of information according to uniform rules from the preparation of manufacture to the manufacturing process by means of conventional or automated machines

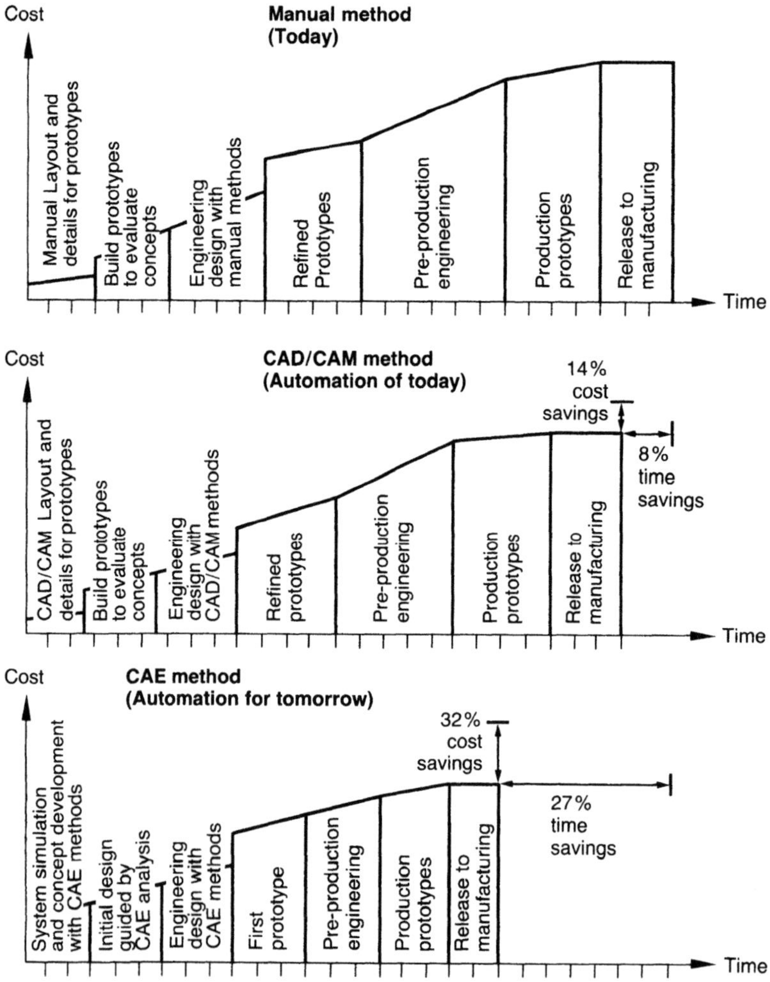

Fig. 94. Comparison of the development costs and times for conventional, computer-aided, and integrated problem processing [149]

In the fields of design and technological planning, the application of programming systems for computer-aided job processing (CAD systems) — following measures of systemization and classification — is a preferred means for increasing productivity [211]. For this purpose a multitude of systems are available, by means of which mainly partial jobs are processed in an automatic way [138]. To a certain extent, the single computer-aided job processing activity only slightly increases productivity. For special tasks, however, single systems can result in considerable increases in productivity, i.e., improvements regarding door-to-door time, reproducibility, and the avoidance of routine activities. With the interlinking of individual systems for continuous computer-aided process planning, input effort is reduced and productivity can be increased.
In the USA there are statements showing — for the entire range of product development — a cost reduction of 14% and a time reduction of 8% for CAD/CAM application compared with conventional methods. One can expect a cost reduction of 32% and a time reduction of 27% with the use of integrated computer-aided processing for all ranges of product development, as against conventional processing [149]. (Figure 94 shows that the greatest reductions in time and cost are to be expected in the areas of prototype development and production preparation as the consequence of comprehensive product analysis and simulation.)
Figure 95 shows the changes in a Japanese automobile company when an integrated CAD/CAM system is applied for car body manufacture instead of the conventional approach. With conventional processing there are essential intermediate steps in the overall process from car body styling to actual production: the development of a master model as a result of body design and the development of a die model as a result of die design. If production of the car body design is prepared with the use of CAD/CAM systems, the product model acts as a data base which in turn serves as the basis for the die design and manufacture. One can expect a

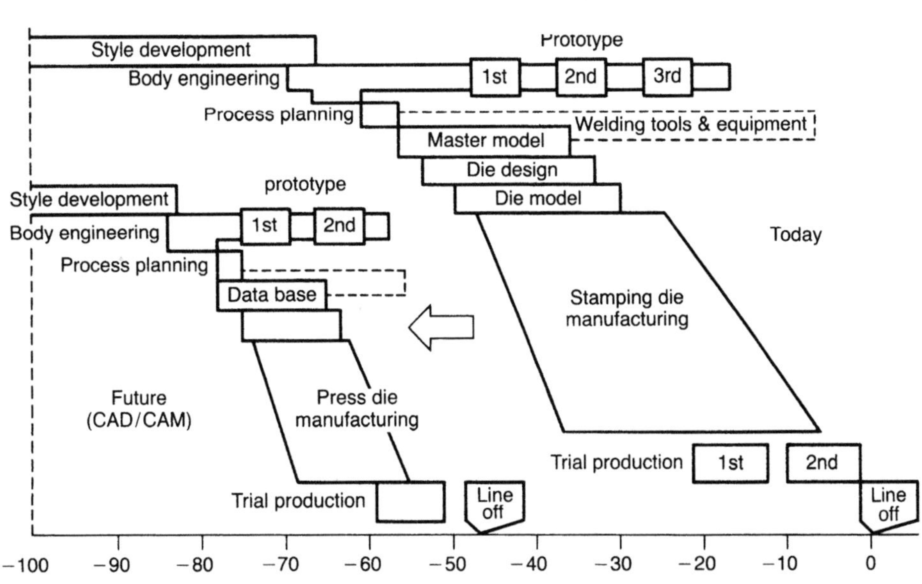

Fig. 95. Future production of auto bodies using CAD/CAM system from Toyota Motors

reduction of the necessary time by approximately 45 % in the field of car body production with CAD/CAM systems, which use integrated task processing, as compared with the conventional approach.
These increases in productivity with computer-aided integrated task processing are based on multiple factors. Figure 96 shows the advantages which result in increased productivity [238].

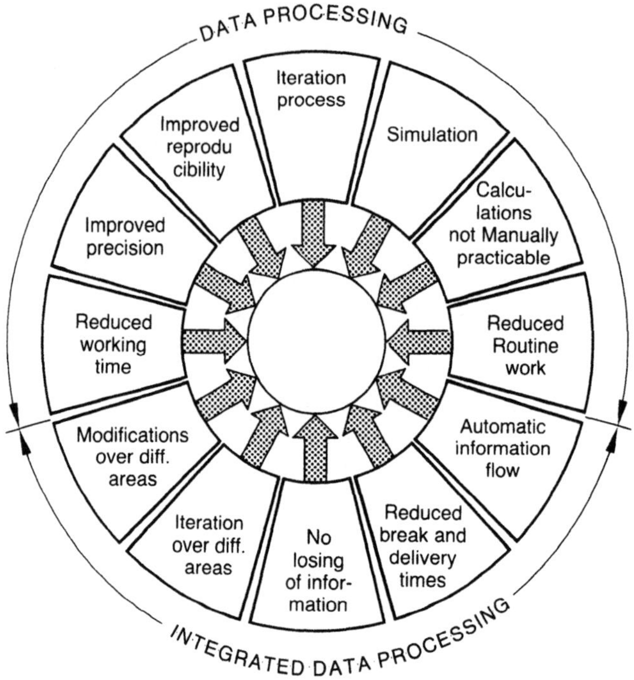

Fig. 96. Advantages of computer-aided task processing [238]

7.2 Meaning of CAD/CAM Integration

The following terms are used [238] for application of computers in various technical fields:
— CAD, computer-aided design
— CAP, computer-aided planning
— CAM, computer-aided manufacturing
— CAQ, computer-aided quality assurance
— CAE, computer-aided engineering
— CIM, computer-integrated manufacturing

These terms are currently in use but are always subject to change.
The computer was initially used in the technical field for program systems such as the method of finite elements or for the programming of NC machines. These forerunners of CAD technology were used even prior to 1955. The starting point for defining CAD terms for computer-aided task processing was the NC programming system APT, developed at MIT from 1955 to 1959 [10]. Parallel to CAD/CAM development, and important to integrated CAD/CAM processing, were the development of the programming language FORTRAN (1954) [12], the introduction of device-independent graphic output routines (1966) [248], the conception of complete three-dimensional computer-internal object representation (1971) [143], and standardization in the field of technical data processing. Other important developments were the updated program language FORTRAN (1966) [46], the APT language syntax (1974/1983) [48], the CLDATA for NC machine instructions (1974) [47], the GKS for standardization of graphic output (1983) [49], and the IGES as standard format for system-independent data transfer (1981) [25].
For more integrated CAD/CAM software systems, the system ICES for structural engineering has been known since 1965 [194]. Some integrated CAD/CAM systems under development are APS [9], CAM-I [210], IPAD [114], and ICAM [109]. Figure 97 delineates the development of computer application, from its beginning in various parts of production to the planned automatic performance of the entire production process.
Depending on the approach, the term "integration" has different meanings [99]:
— The composite of parts to a whole as a process, i.e., an active or process-referred approach

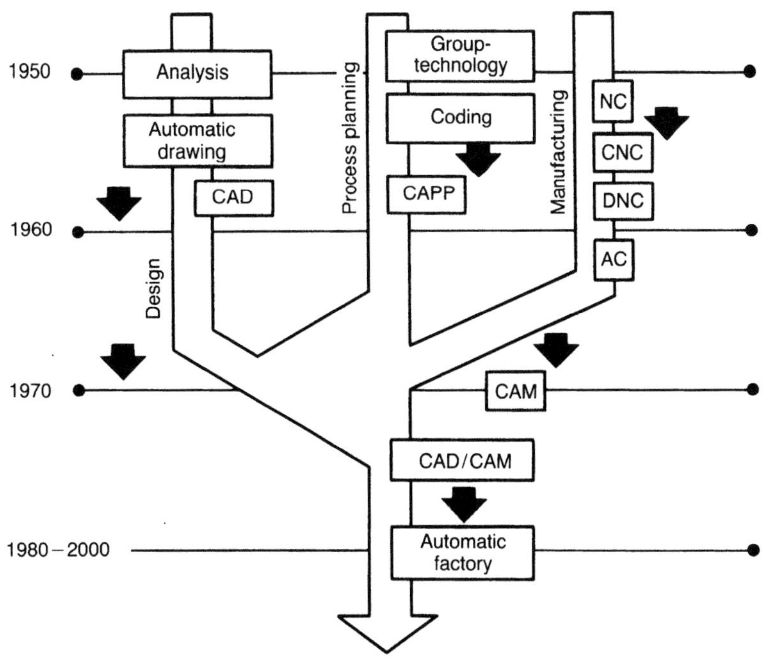

Fig. 97. Historical development of CAD/CAM [161]

— The result of the process "composite" as a stage, i.e., a passive or state-referred approach

For the process of integration to a composite of a greater whole various objects can be defined in the field of information processing (see Fig. 98). The process or state of integration can be described in different ways. Figure 99 shows various definitions for this integration.

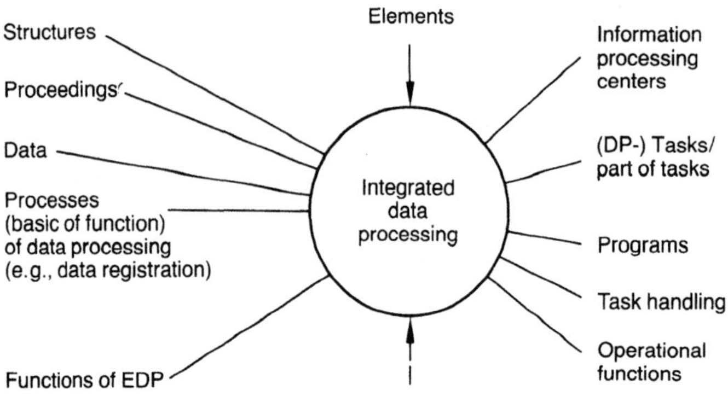

Fig. 98. Objects of information processing [99]

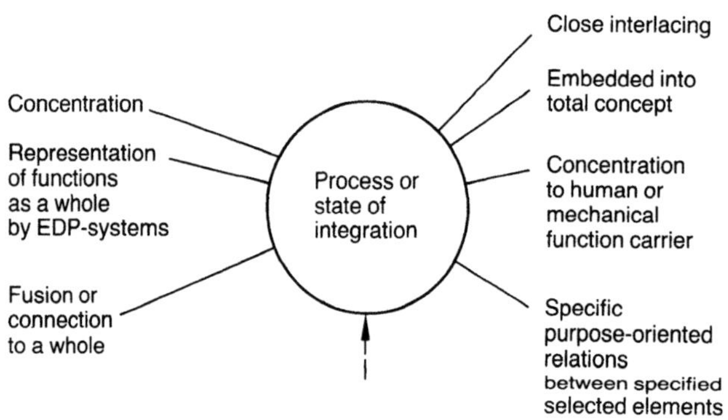

Fig. 99. Definitions of process or state of integration [99]

In connection with integration there are two problem definitions to be mentioned for computer-aided design and technological planning [76]:

1. Layout of an integrated CAD system with definition and selection of the planning functions and the information-referred interlinking of partial functions to an executable overall system
2. Inclusion of an integrated CAD system in the operational procedures in consideration of the organizational requirements, the training of staff members, and the utilization of the generated planning results

The interlinking of tasks in design and technological planning can be performed with differing degrees of automation (Fig. 100). With the application of isolated systems, the definition of the partial task to be performed with the help of a com-

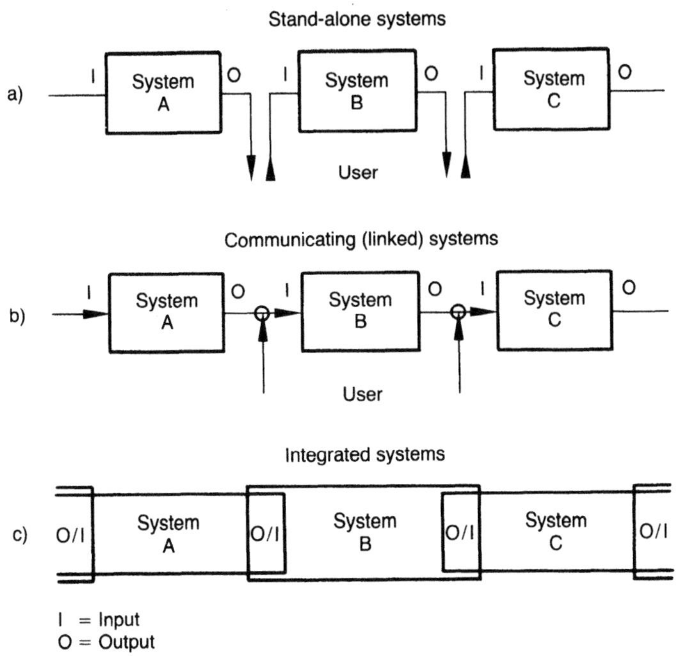

Fig. 100. Forms of integrated technological planning

puter is formulated by the originator, at whose disposal the obtained result is placed. This result serves the originator in describing the task for the next partial system. Integration is possible here only with regard to the number of staff members, in that several CAD tasks are performed by one employee.

With communicating systems, data transfer ensues between the individual systems, i.e., the computer-aided information flow is integrated. The results of the individual systems are used as a problem definition for the relevant partial system that follows. Modification of the information can be performed between the individual systems, either automatically by coupling modules or by operator intervention.

With integrated systems the interlinking of the partial systems is not done by staff members but by common utilization of programs or parts of systems.

The fact that the interlinked stand-alone systems can still be used as isolated systems distinguishes them from communicating and integrating systems. Partial utilization of a CAD system designed as an overall complex is no longer possible.

With regard to the extent of the partial tasks considered, various integration steps can be shown [238]:

— Computer-aided processing of individual tasks
— Connected computer-aided processing of several individual tasks, i.e., partially integrated processing
— Connected computer-aided processing of all partial tasks as fully integrated processing

In addition to the requirements resulting from the aims of integration, the layouts of the CAD systems to be used are important for the execution of integration.

Task description	Task execution	Software design	Hardware design

Information search
 	
Designing
– Calculation
– Evaluation
– Drawing
– Generation of list of parts
 	
Process planning
– Manufacturing
– Assembling
– Inspecting
– NC programming
	Integration range
	– Stand-alone
	– Linked
	– Integrated
	
	Processing principles
	– Retrieval
	– Adaption
	– Variant creation
	– Generation
	
	Mode of operation
	– Batch
	– Dialogue
	  – Alphanumerical
	  – Graphic
		Area of application
		– Product-related
		– Task-related
		– Company-related
		– Function-related
		
		Organization
		– Central
		– Stand-alone/integrated
		– Counteroperating
		– Direct operating
		
		Hardware relation
		– Hardware-related
		– Hardware-flexible
		
		Software structure
		– Single/multiple components
		– Data storage
		– Methods storage
		– Modularity
		– Single/multiple user
			Computer
			– Performance
			– Linking
			
			Periphery
			– Input
			– Output
			– Input-output
			– Storage

Fig. 101. Forms of implementation for CAD systems

Figure 101 shows implementation forms of CAD systems having an influence on integration measures [238]. The layouts can be subdivided according to task description, task performance, software design and hardware design. The multitude of differing CAD system layouts results in obstacles to the integration measures:
— Differing tasks
— Differing task performance
— Missing or hidden system interfaces
— Voluminous datasets
— Differing information contents of the CAD systems
— Computer dependence on CAD systems
— Differing peripheral requirements
— Application exceeding departmental boundaries

These obstacles can be overcome by suitable measures in the fields of operational organization, man-computer communication, software-system design and hardware-system design.

7.3 Information for Integrated Processing

For the manufacture of products there is geometry-oriented and administration-oriented information flow (Fig. 102). This includes product data, and the geometric (topology, dimension), and administrative (numbering, classification) data are supplemented by technological (processing, properties) and commercial (cost,

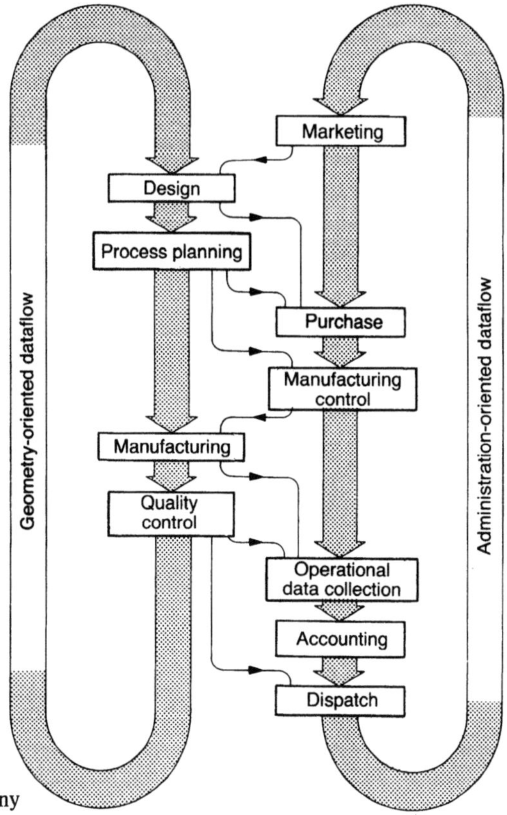

Fig. 102. Information flow within a company

suppliers) data. These product data are generated in the individual divisions of the companies and are included in the information flow. The generation of information ensues on the basis of the data existing in the information flow. The functional areas of a company are related to each other by means of the information flow. These relations apply to task processing by conventional means as well as to data processing [238].
The exchange of data between the functional areas takes place with conventional processing by transfer of the design and manufacturing documents; with integrated computer-aided processing this is accomplished via automated information flow between the partial CAD systems.
The integration of planning tasks is directed mainly at the information occurring most frequently in the manufacturing documents. An analysis of manufacturing documents regarding the content volumes of the individual kinds of information is represented in Fig. 103. It turns out that the great majority of information is of a geometric nature, with information of an organizational and nonpersonal nature, i.e., administrative data, ranking second. This corresponds with the essential information flow in a company, shown in Fig. 102.

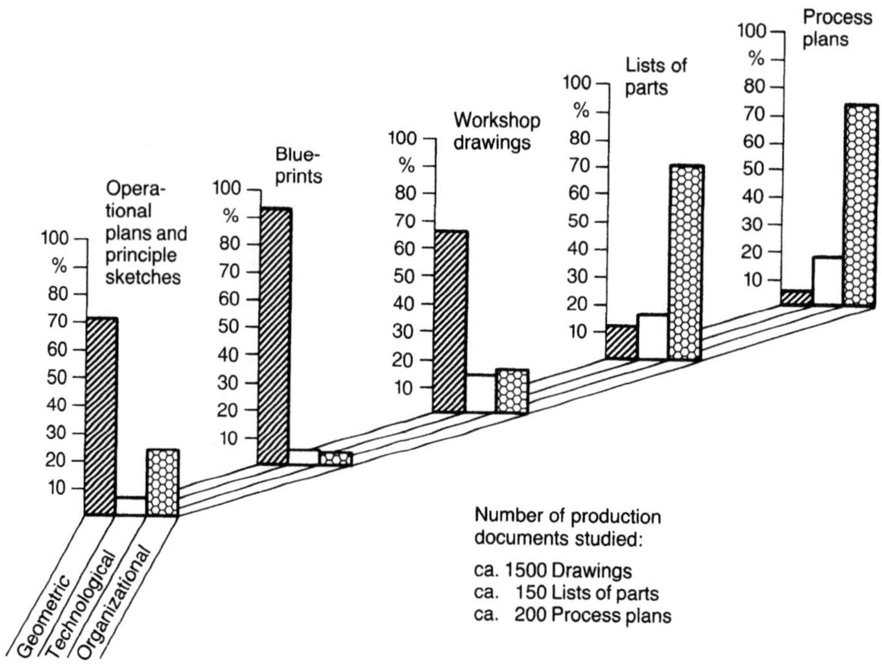

Fig. 103. Distribution of types of information from manufacturing documents [224]

	Range of application																			
Production document	Identifying functions	Identifying principles	Design	Detailing	Generating part lists	Def. of blank parts	Def. of operation sequences	Selection of machine tools	Selection of auxiliaries	Time calculation	Generation of work documents	Capacity planning	Disposal of material and external parts	Accounting/salary	Purchase	Manufacturing	Assembling	Long-term company planning	Product planning	Marketing
Operational plan	●	○	○															○	○	○
Principle sketch		●	○															○	○	○
Blueprint		●	○	○							○	○	○	○	○			○	○	
Workshop drawing			●	●	○	○	○	○	○	○	○	○	○	○	○	○		○		
General drawing/ Assembly drawing			●	●	○		○	○	○	○	○	○	○	○	○		○	○	○	○
List of parts					●								○	○	○		○	○		○
Process plan						●	●	●	●	●		○	○	○	○	○		○	○	
Assembly plan						●	●	●	●	●		○	○	○			○	○	○	

○ Use of information ● Use of information for similar tasks

Fig. 104. Utilization of information from manufacturing documents [224]

The utilization of information from manufacturing documents within production planning is evidence that multiple use of much information is possible. Figure 104 shows the ranges of application for the information from particular manufacturing documents.

7.4 Software Structures for Integrated CAD Systems

The formal consideration of CAD systems reduces the functional components to communication modules and method modules and the data file to a storage module (Fig. 105). With a heterogeneous system structure, coupling modules may also be necessary, which, however, correspond to the method modules. The individual system components are interlinked via system interfaces, the user and the system via communication interfaces.

The product data are processed as central information of the design and technological planning area. They are provided as computer-internal object representations in all system components, or transferred at the system and communication interfaces.

Communication modules are used to implement the communication interface between the users and the CAD system. The information input by the user at the user terminals is thereby transformed to be suitable for data processing and is provided in the method modules for processing or stored in data files. The information generated or processed in the method modules or stored in the data files is transformed to be understandable for the user and displayed on the user terminals.

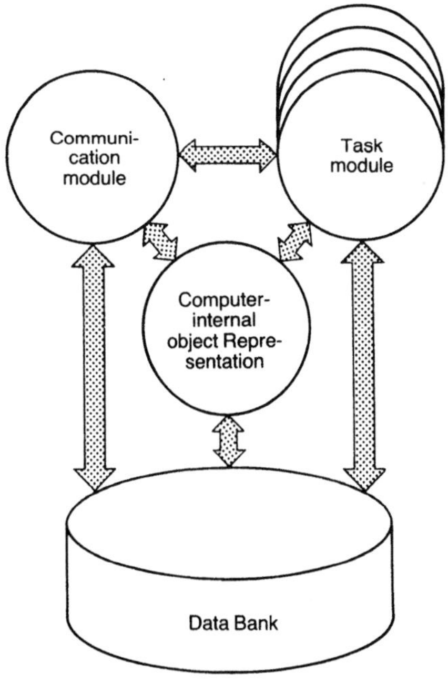

Fig. 105. Principle structure of CAD systems

The calculations and determinations for solving the tasks allocated to every method module are performed in the method modules. These tasks include data generation, data transformation, and data management. Mathematical correlations in the form of algorithms or logic relations are used, permitting nonalgorithmic determinations.

The data bank contains the functions of long-term data recording and data management. Provided are the computer-internal representations of design and technological planning tasks and results as well as the planning data required for these tasks. Here division is made into general information, design information, drawing information, technological planning information, and NC data. The processing programs for data recording and data management are provided by the data file system. If the information is provided in separate individual data files, the data file management ensues mostly via the operating system of the computer. Apart from data recording and management by means of a central data bank, data holding can be performed by a data file distributed to several logically interlinked data stores.

Computer-internal representation (CIR) includes all current object data and all relations between object data which are necessary for an object image and object processing in the functional modules [238]. Features for computer-internal representations are the data structure, the interface to the computer-internal representation, and the model algorithms used. "Data structure" means the logically set up correlation of data by which the object is described. The range of application depends on the layout of the computer-internal representation. The more precise and complete the object image developed, the wider the application spectrum. The layout and design of the CIR is determined by process-technological influences and user-specific requirements.

Data transfer between the functional modules of the CAD systems is performed via the system interfaces. The result is the logic interlinking of communication modules and the data file. Consequently, these system interfaces are of particular importance for integration measures.

Using the above-described functional modules and interfaces, CAD systems can be designed according to two different methods. The central component of CAD systems can be a connecting system core with standardized techniques of access to the common data file or a connecting data file system for common filing of various kinds of system information.

The designing of CAD systems with system cores is based on the systematization of interface with uniform data and program management [1]. Here a subdivision is made into the problem-neutral system core and the problem-specific user modules (Fig. 106). The system core performs functions of system management, sequence control, and data storage, as well as the provision of data complexes for user modules; it can also contain higher processing levels with communication modules or method modules for job-neutral basic functions. The user modules contain the job-specific processing algorithms, and are set up on the system core. The entire system is not based on a uniform data model; a separation of data and programs is not given. Consequently, the coupling of several user models has to be considered with their provision. A subsequent interlinking is possible only via the specific CIR of the user modules.

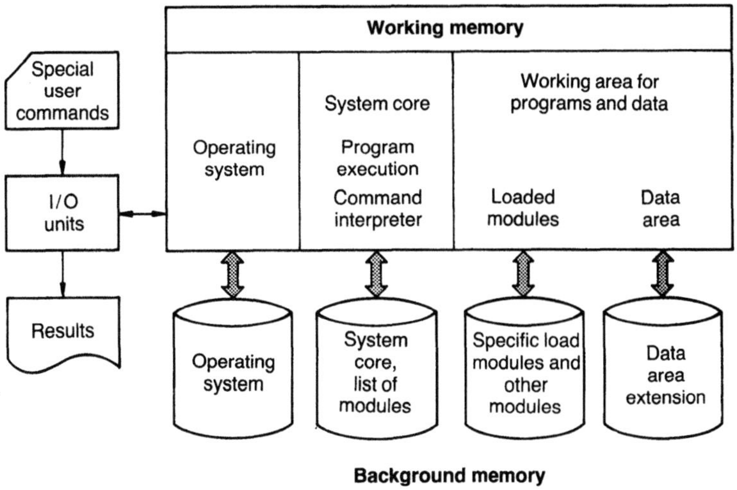

Fig. 106. CAD system design with system core [1]

If the CAD system is designed on the basis of commonly used data, the data bank forms the center of the system. All partial systems included in the CAD system hold their data in the common data bank and make them available, directly or via the coupling modules, for partial ranges (see Fig. 107) [1].
The data bank assumes here the tasks of data management and storage, and most possibilities of data handling such as data privacy, protection, and data security are available to the user. Due to the separation of method modules and data, it is relatively simple to tie up further partial ranges.

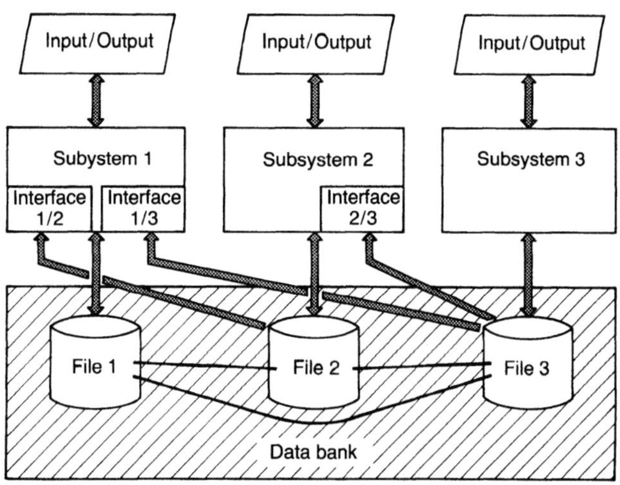

Fig. 107. CAD system design with data bank [238]

Integration via a data bank is available for the interlinking of existing homogeneous and heterogeneous partial systems. It may be accomplished by means of a common CIR for all partial systems, by separate CIRs for all partial systems, or by a separate CIR for every individual system (see Fig. 108) [214]. If the integration of the individual systems to a heterogeneous overall system is performed without common CIRs, each of the individual systems is always connected with all others to which information transfer is necessary via one special coupling module (Fig. 108 a). This is the simplest way of achieving integration of individual systems. With a greater number of individual systems the quantity of required coupling modules increases quickly; likewise, greater quantities of data have to be stored and managed, since every individual system holds the required information completely ready.

With the interlinking of the individual systems to a homogeneous overall system, all individual systems operate directly with a common CIR and a common interface to it (Fig. 108 b). Because of the homogeneity with interface and CIR this form of integration is applicable only for new overall concepts. An adaptation of existing individual systems regarding interface and CIR would entail considerable program-technological interventions into existing individual systems. The avoidance of multiple storage of information, the unnecessary conversion of information, and the avoidance of information losses by data adaptation are advantageous.

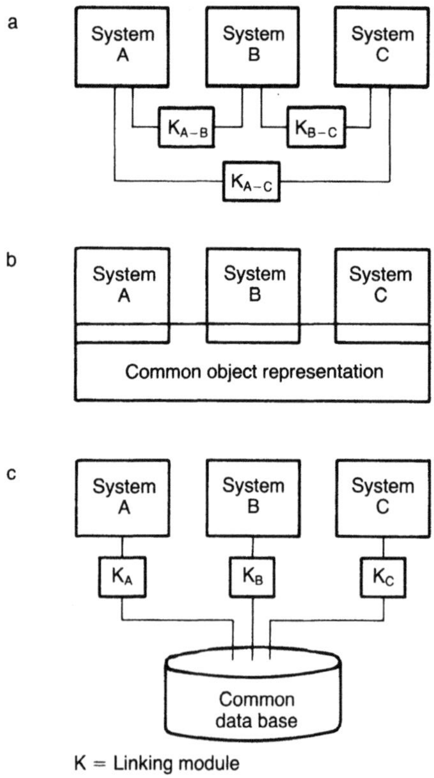

Fig. 108. Coupling possibilities of CAD systems **a** with linking modules, **b** with common data base, **c** with linking modules and common data base

Likewise, standard modules with access to the computer-internal representation can be used multiply.

If a heterogeneous overall system with a common data base is the aim of the integration of the individual systems, every individual system is connected with the common data base via a coupling module (Fig. 108 c). A program-technological intervention into the individual systems is not necessary here; however, the information has to be converted between the individual systems and the common data base. This may entail information losses.

Access of all individual systems to the overall information is possible via a common interface to the common data base. But, as with homogeneous integration, the common data base has to meet particular requirements, since it must include all information of the individual systems.

The structure of the integrated overall system can be designed in a linear, hierarchical, or cross-linked way. With a linear system structure a defined sequence has to be observed within the overall system; branches and returns are not possible. A hierarchical structure permits returns and branches, but here, too, a defined hierarchical sequence has to be observed. Only with a cross-linked structure is any processing sequence of the individual system possible.

The problematical nature of the interface definition with computer-aided task processing was recognized early. The first standardizations were made for the application of NC machines by the definition of CLDATA (DIN 66215) [47], for the machine-independent description of NC information for processing machines, and for the definition of input information for numerical controls (DIN 66025). Farther-reaching standardization efforts aim at the definition of IRDATA (industrial robot data) for the machine-independent description of NC data for freely programmable handling devices (VDI 2863), and of GMDATA (general measuring data) for the machine-independent description of NC data for measuring machines [17].

In the field of computer-aided provision of drawings, a standard for processing graphics information (DIN 66252) [49] was developed with the standards proposal GKS (graphical kernel system), which emphasizes the possibilities of the various hardwares. GKS is based on the definition of a so-called meta file, containing the graphics information in a hardware-neutral form as a standard interface. Similar tasks are performed by the standard graphics software GPGS-F, likewise aiming at an instrument-neutral processing of graphics data [68]. The independence of the hardware used — be it machine tools (CLDATA, IRDATA, GMDATA) or the user terminals (GKS, GPGS-F) — is the basic aim of these standardizations.

Only the latest developments in the field of interface definitions deal with the problem of the standardized transfer of object data between various CAD systems. In the USA the concept IGES (initial graphics exchange specification) was developed for the exchange of graphics data and computer-internal object models (ANSI Y14.26M) [25]. The aim is to secure the interchangeability of model data between differing CAD systems. A number of IGES elements and their representation in data format are standardized.

Figure 109 shows the structure of an IGES data file for a simple component. The concept satisfies the original target of IGES, i.e., the transfer of graphics information. Limitations with data transfer occur if the contents of the converted informa-

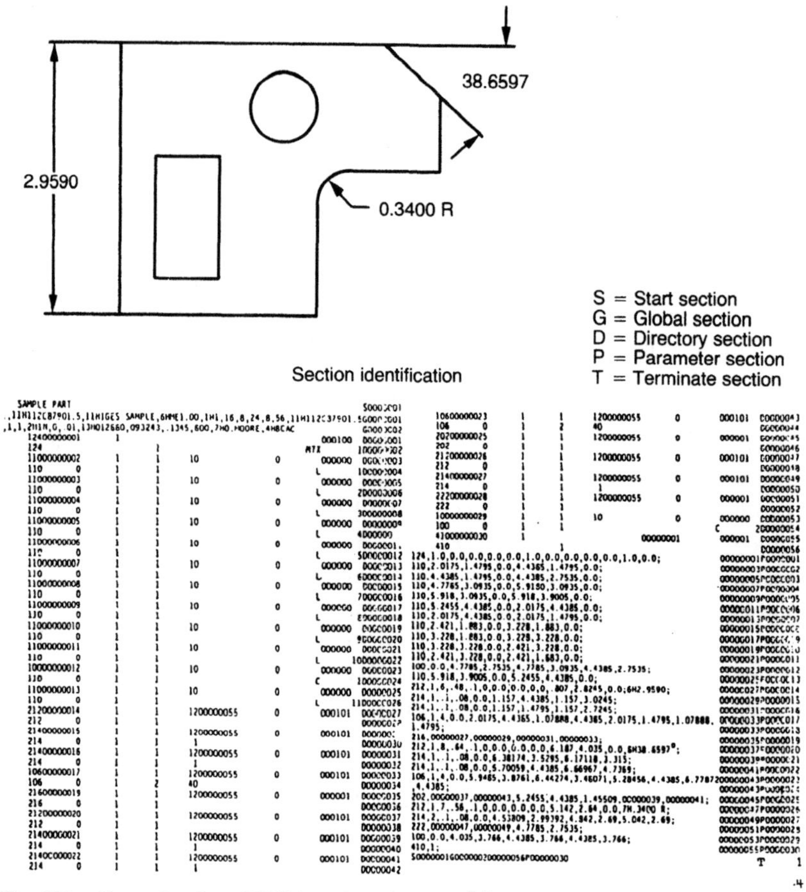

Fig. 109. Example of an IGES interface data file [8]

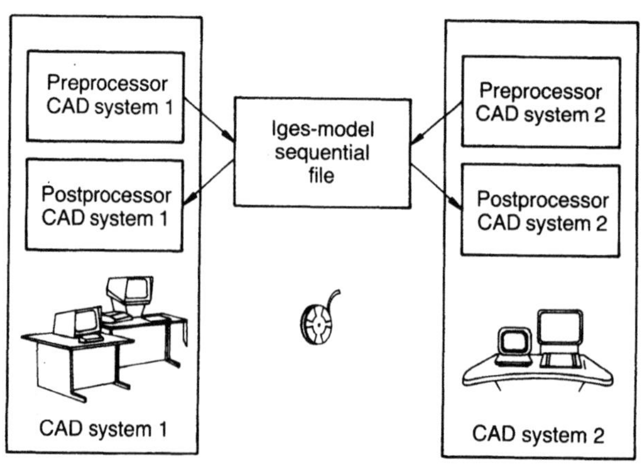

Iges = initial graphics exchange specifitation

Fig. 110. Model exchange with IGES [8]

tion have to be further processed. All graphic elements are transferred; however, the meaning of the elements is partly lost (surface signs, tolerance details, etc.). Likewise, the topology of geometric objects is not clearly mapped. By the introduction of user-specific elements these limitations can be removed, but in the process the standardization effect is sometimes lost. In addition, the elements stock becomes so comprehensive that application becomes rather slow.

With IGES, a postprocessor — generating the data file in IGES format — is required for every individual component according to the interlinking and/or a preprocessor which transfers the IGES format data file into the system-internal form (Fig. 110). The definitions of the IGES-standard interface have not yet been finished. Future developments must consider the geometric, topological, and technological requirements in a more far-reaching way than is the case with the present interface.

Figure 111 shows various possibilities of integration between CAD systems and the generation of control data for numerically controlled machine tools. The linkages shown can be made with the application of theoretically realizable interfaces.

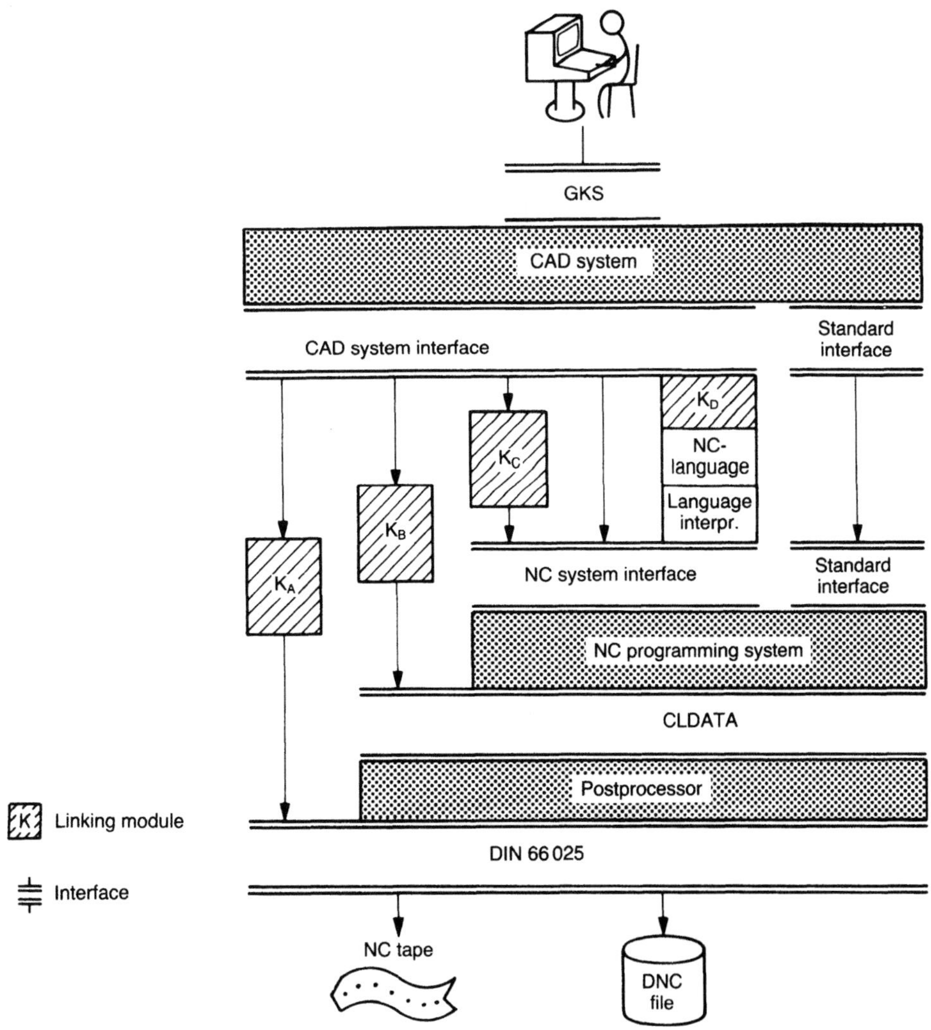

Fig. 111. Possible interfaces for coupling CAD systems with NC programming systems

7.5 Hardware Structures for Integrated CAD Systems

In connection with the attempts to decentralize data processing plants and the increasing volume of programs and data with CAD systems — in particular with integrated systems — the distribution of functions or of data banks to several computers has to be considered. In addition to centralized individual computers, decentralized hardware configurations are used to a greater extent. In essence, hierarchical and network structures are realized (Fig. 112).

With a hierarchical configuration the hardware group may comprise the central computer, satellite computers, and workstation computers. The distribution of the system components to the relevant computers is effected with reference to the function, in order that the available computer abilities may be utilized optimally. The central computer performs tasks of product data management, time-consuming calculations (FEM calculations), and commercial data processing tasks. The satellite computer is concerned with tasks of geometric data processing, process planning, and NC programming, while the workstation computer undertakes the generation of graphics and interactive communication [149]. The organization of the data bank must therefore be designed in a correspondingly hierarchical way.

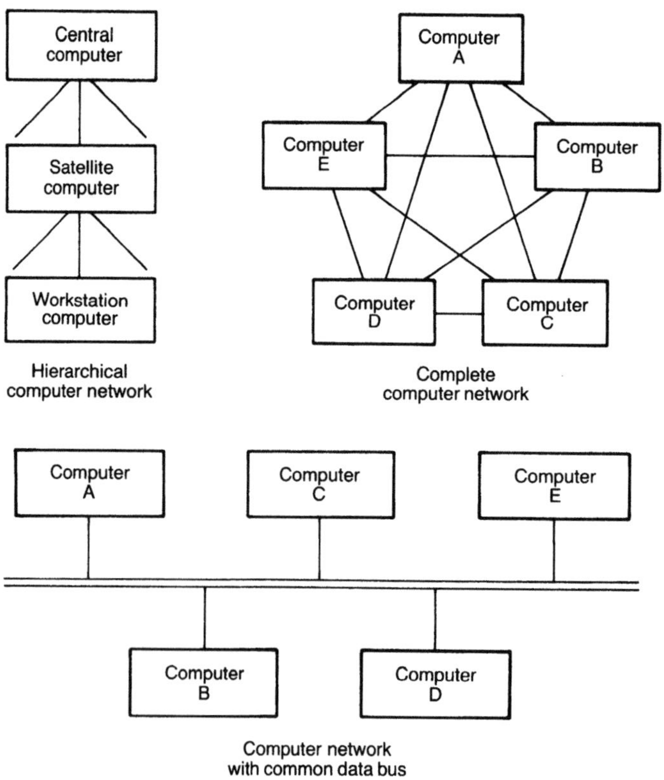

Fig. 112. Distribution of hardware for CAD systems

The administration of material, manufacturing, and commercial information is allocated to the central computer. The satellite computer operates with product information and planning documents while only those active product and planning data necessary for the momentary performance of the tasks are allocated to the workstation computer.
Due to the functional distribution there are different requirements for data transfer between the individual hierarchical steps as regards transfer speed and data volume. The transfer of large volumes of data is required between the central computer and satellite computers, and the duration of the data transfer may be long. Data transfer between satellite computers and workstation computers is designed for medium data volume at high transfer speed, while data transfer between workstation computers and user terminals requires a small data volume at extremely high data transfer speed [149].
In case of computer interlinkage in a kind of network the distribution is made with reference to tasks (design, provision of drawings, process planning, NC planning). Contrary to the case with hierarchical distribution, similar hardware requirements are made here for each component. The distribution of the data bank — if provided for — will also be made with reference to tasks. Data transfer between the partial systems has in all places the same requirements regarding transfer speed and data volume. With this kind of distribution problems occur with the dissemination of information due to collisions of the various data flows. The transfer of information between the individual computers can be made via local networks or by means of remote data transmission systems.

7.6 Examples of Integration

As examples the possibilities of integrated task processing for design and process planning are presented. The aim with the development of the first example was the integrated generation of design and production planning documents for the fabrication of cam plates for machine tool controls. A cam design system is interlinked with a variant drawing system and an NC programming system (Fig. 113) [63]. The description of the manufacturing task ensues with the input for the cam calculation system KUKON [136], where coordinates tables as well as time, path, speed, and acceleration diagrams are generated.
For the provision of workshop drawings, KUKON was interlinked with the variant drawing system COMVAR. The KUKON system generates an output file containing the information for generating the workshop drawings. The COMVAR system operates on the basis of complex components from which the variant is obtained by the input of parameters [9]. By means of a coupling module, and from the results of KUKON, the required parameters are generated as input of COMVAR, which then provides the required workshop drawing.
For generating the NC information for the fabrication of the cam contour, the calculation results of the KUKON system are also used. The output file generated by KUKON is converted into the form required by the NC programming system,

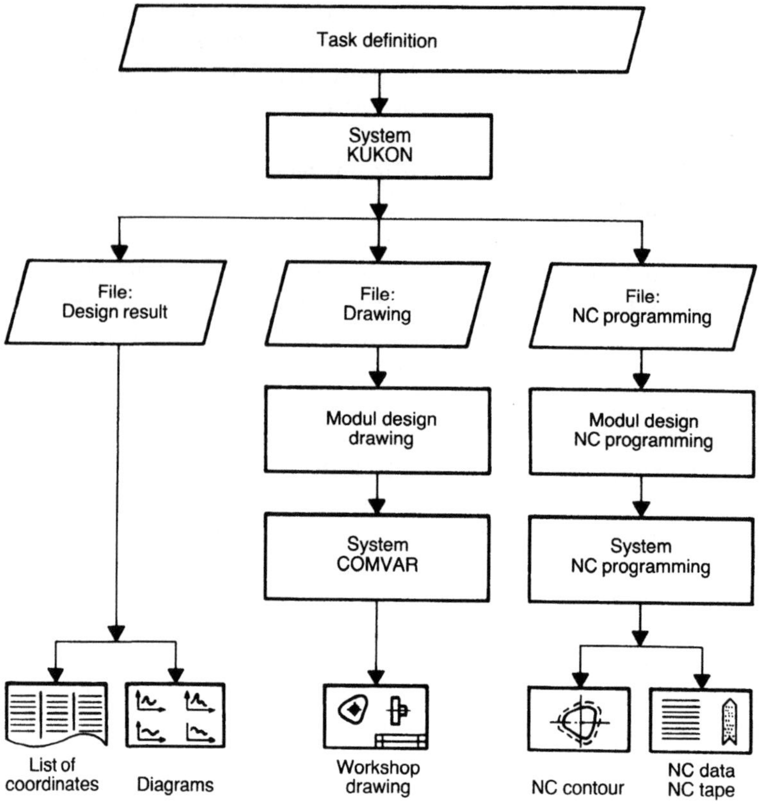

Fig. 113. CAD system for integrated task processing

which then generates the NC data, including the control tape and a graphic representation of NC contours and tool paths.

An estimate showed that with the provision of only three variants the costs are covered. The comparison of manual and automatic designing results in a time effort reduced by the factor of 20 [63].

At the *Rheinisch Westfälische Technische Hochschule* in Aachen an integrated system for the generation of manufacturing documents was developed which serves to process workpieces according to the variant principle [185]. The functions generation of design, generation of detail drawing, generation of process plans, and generation of NC punched tapes are performed in an integrated way (Fig. 114).

The individual modules are coupled with one another via the workpiece information basis. The information required for the generation of drawings, generation of process plans, and NC programming is fully covered from this data base. For the generation of the design the workpieces are included in the data base. If the machining is performed without the designing systems, the workpiece description is made by means of an input module. The description is geometrically and technologically fully generated by a shape-element-oriented language. With this con-

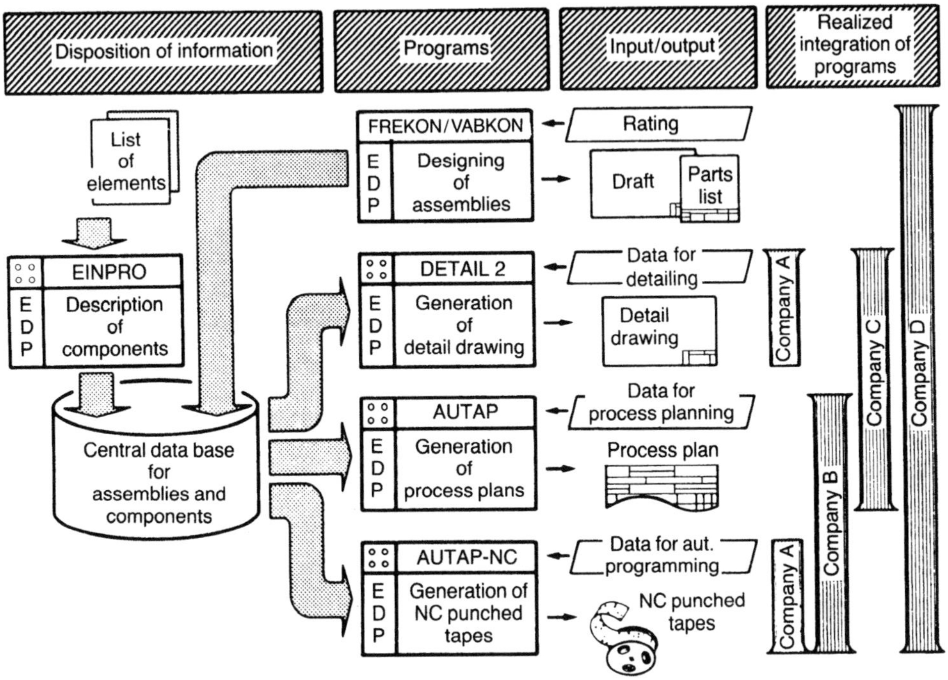

Fig. 114. Integrated system for generating manufacturing documents [185]

cept all partial systems can be applied in a singular and integrated way. With the interface definition in the form of a data file, new partial systems can be included without problems.

The generation of drawings is based on the available workpiece description of the integrated system. The drawings are automatically generated by adding drawing and system commands, e.g., for kind of dimensioning and title field [186].

With the generation of the process plans, and for fully automatic processing, other information such as lot sizes and order data is required to perform the planning task by means of the available workpiece description [76]. Figure 115 shows the description of the component and the process plan, both of which are generated by the third represented component — the workpiece description. The result of a shape-element description can be retrieved in the workshop drawing as well as in the process plan.

NC programming in the integrated system can be done according to two principles. The NC control information is generated in CLDATA form, or the part program is generated as input of an NC programming system. Figure 116 shows the linkage of drawing and NC partial program via the shape element defined in the workpiece description. The results of economic research into the integration of the tasks of drawing generation, process plan generation, and NC data generation for the system presented are shown in Fig. 117 [76]. It is apparent that the contiguous processing of just two partial tasks already achieves a reduction of the

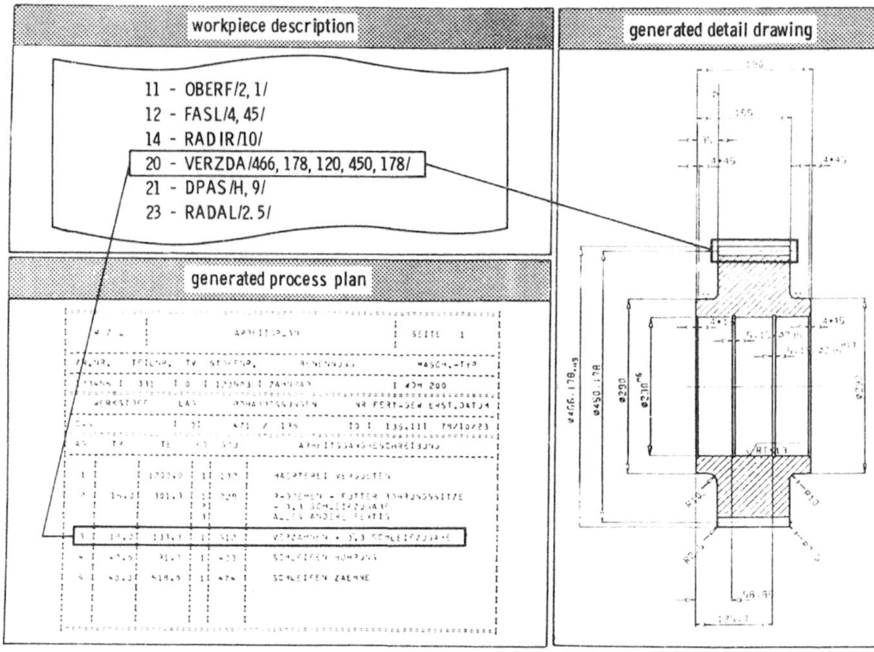

Fig. 115. Process plan generated by means of AUTAP [185]

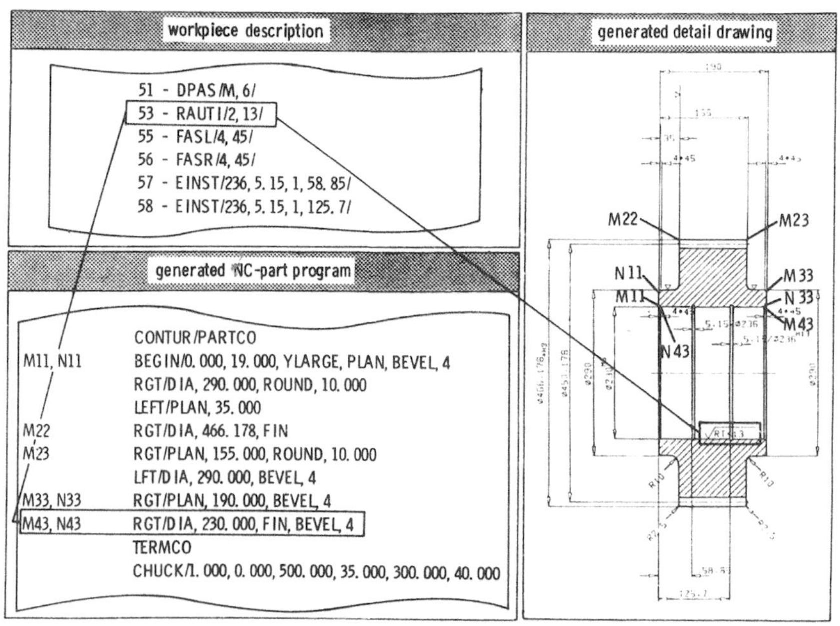

Fig. 116. Generation of an NC part program [185]

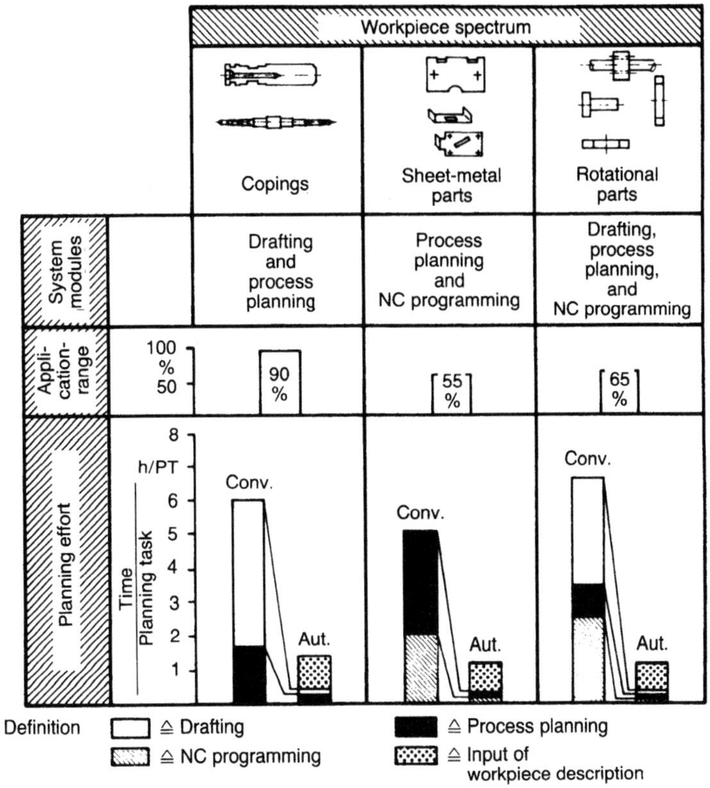

Fig. 117. Effects of integrated processing on the planning effort [238]

planning time per planning task of approximately 40%. With an integrated generation of workshop drawing, process plans, and NC part program, the reduction can amount to approximately 75%.

Within the framework of the German-Norwegian Research Project APS (Advanced Production System) an integrated CAD/CAM system is being developed. The overall system comprises three functional partial sytems: the design system, the technological planning system, and the geometric modeler (Fig. 118) [9]. Functions and cognitions of existing systems are to be used for the implementation of the APS system. Within the overall system the individual systems retain their single applicability.

The following are used: an interactive computer, ICUE [72]; a geometric processing system, COMPAC [78]; a sheet-metal processing system, GPM APC [79]; a drawing generation system, DETAIL 2 [186]; and a system for processing free-shape areas, GPM SS [79] for the geometric modeler (APS Geometric Modeler). The APS design system is based on the parts for the workpiece definition and the resulting representation of the systems used in the modeler, as well as on calculation programs [220]. In the field of process planning, cognitions and parts of the

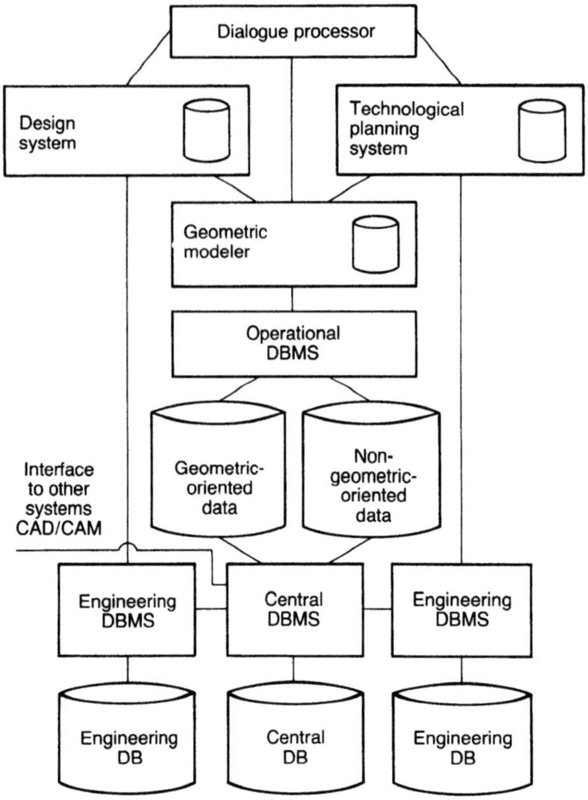

Fig. 118. Concept of the integrated APS-CAD/CAM system [9]

process planning and NC programming systems CAPSY [180], AUTAP [61], DISAP [62], CAPEX [29], and EXAPT [27] are used.

The connection of the partial systems is based on a common product data base which is generated during product definition and design. With this data base various tasks such as component calculation, drawing provision, technological planning, and NC programming are performed. The planning data required for performing the planning tasks have to be made available in job-oriented data files, and a central data file management system assumes the common management. The modeler by means of which the computer-internal object representation is generated holds a central position in the overall system. For a wide spectrum of application different kinds of geometry have to be processable by the modeler. Correspondingly, components with analytically recordable geometry (Fig. 119), with analytically non-recordable geometry (Fig. 120), or with problem-specific geometric definitions — e.g., for sheet-metal machining — (Fig. 121) can be processed. For these different geometries there are partial systems in the modeler, interlinked with a common interface via the common data base. The design system is based on the possibilities of the modeler. Functions such as workpiece definitions or application of the data base for design calculations and drawing provision are performed.

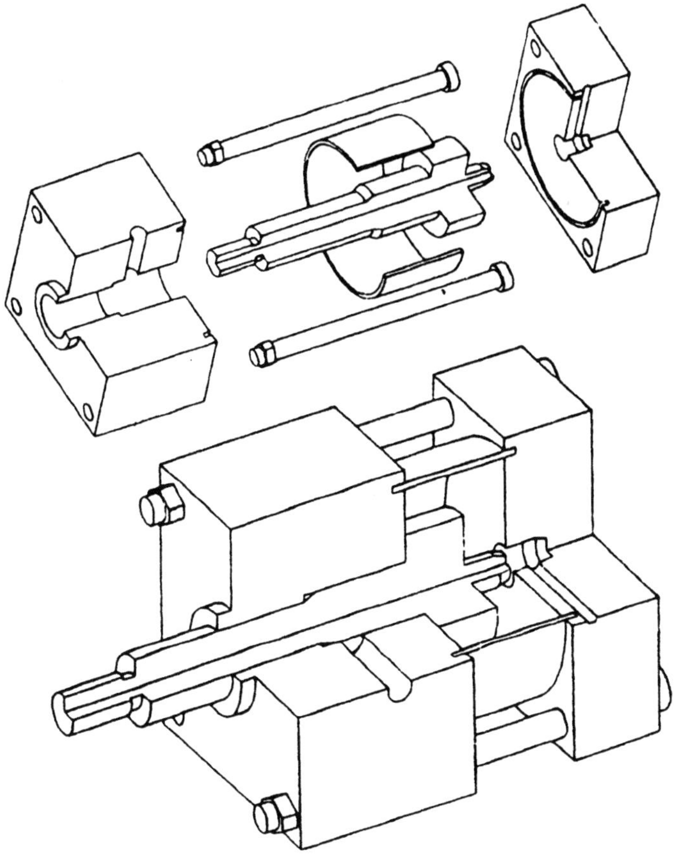

Fig. 119. Perspective representation of a subassembly [139]

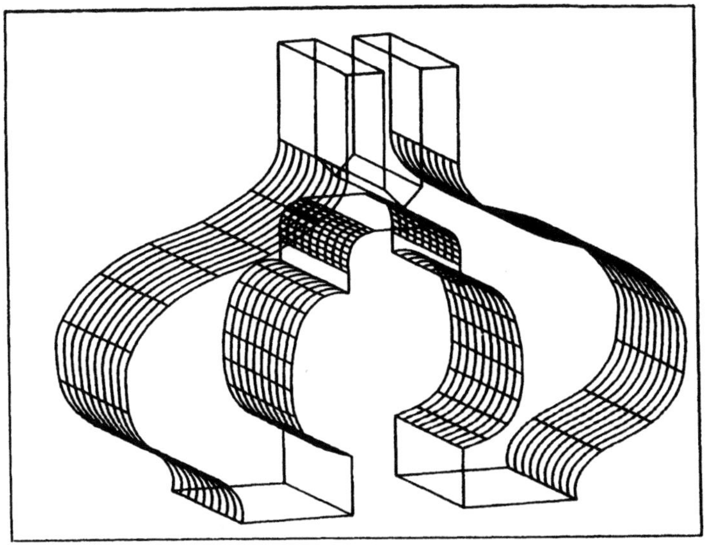

Fig. 120. Example of a product with analytically nonrecordable geometry [9]

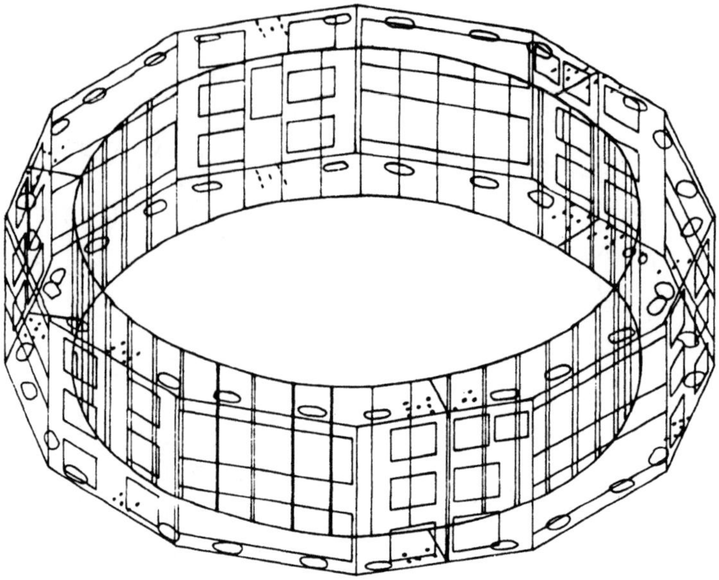

Fig. 121. Representation of a component of sheet-metal machining [79]

For the workpiece definition, a hand-sketch input with reconstruction technology for 3-D description of the component (Fig. 122) [119] is available in addition to a design language, the interactive input of geometric elements, and a shape-element-oriented description language. For the calculation component, finite-element networks can be generated, proceeding from the computer-internal object representation, and corresponding calculations can be performed (Fig. 123).

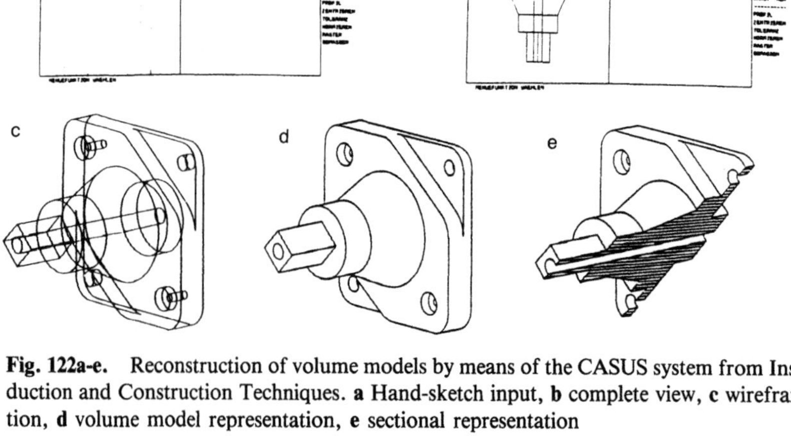

Fig. 122a-e. Reconstruction of volume models by means of the CASUS system from Institute for Production and Construction Techniques. **a** Hand-sketch input, **b** complete view, **c** wireframe representation, **d** volume model representation, **e** sectional representation

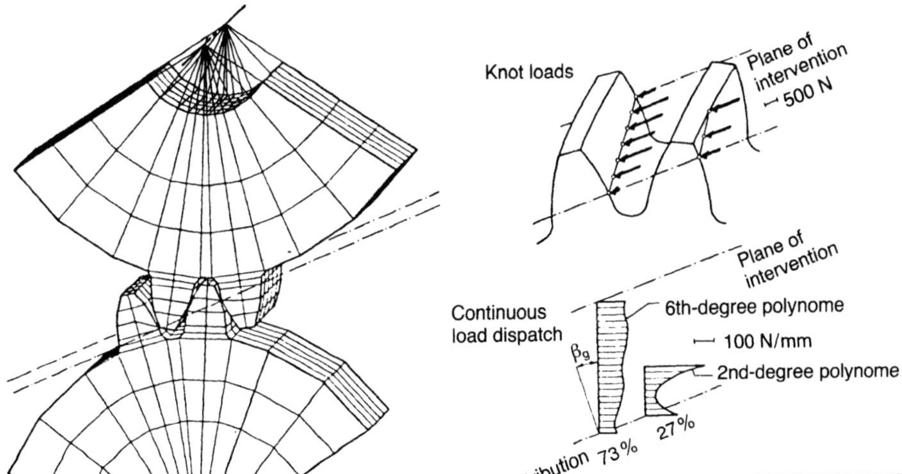

Gearing data	
$m_n = 5$ mm	b = 40 mm
z = 20	$\alpha_{no} = 20°$
x = 0.461	$\beta_o = 19°$

Fig. 123. Result of an FEM network generation [220]

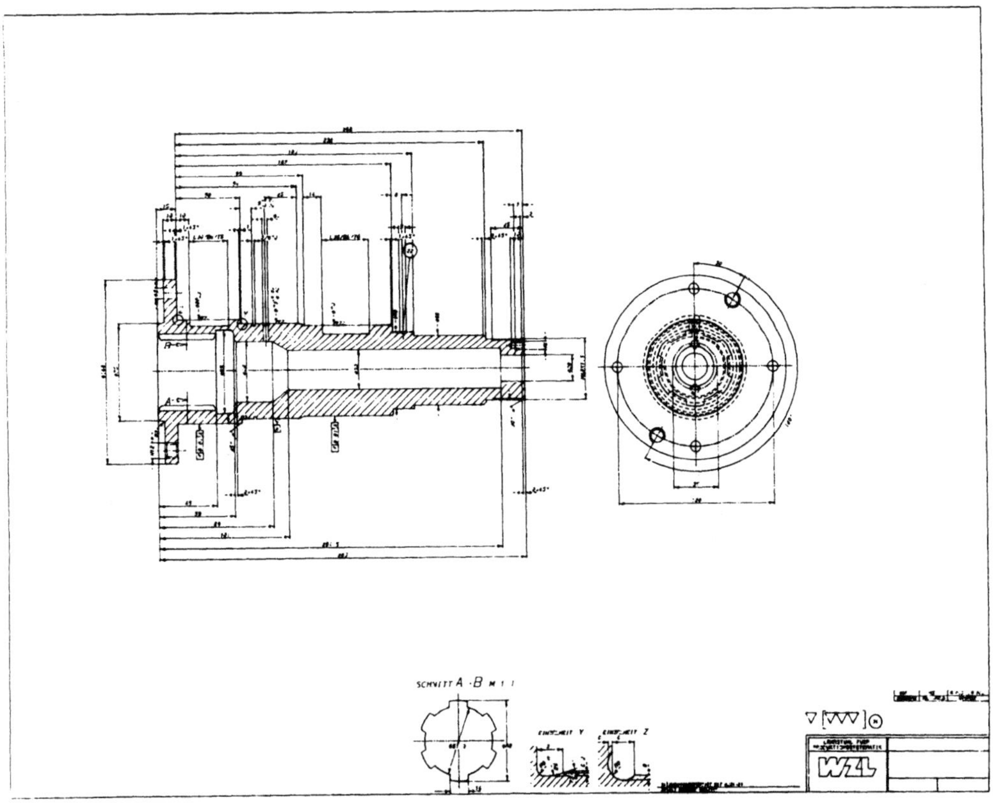

Fig. 124. Automatically provided workshop drawing (from APS [9])

The workshop drawing is provided either interactively on the screen, using views and sectional views of the computer-internal representation, or fully automatically on the basis of shape-element-oriented workpiece description (Fig. 124).
Technological planning and NC programming in the integrated overall system also work with the common computer-internal object representation. The global sequence of operations for a workpiece is fixed, and for individual manufacturing processes a detailed determination of the operational data is performed (Fig. 125).
With the determination of the operational data for turning, a graphic simulation of the operational progress appears on the screen (Fig. 126). Control data for manufacture on NC machine tools are generated in the NC programming system also included.

APS — MANUFACTURING PROCESS PLAN — VERSION 0 — PAGE 1

PART.....:
PART-NO..:
MATERIAL...:
REFEREE....:

SEQUENCE OF OPERATIONS
DATA OF OPERATION

SEC	DEP	STA	WAG	TEXT	D	N	S	F	TOOLS.JIGS.FIXT.	ST	OT	CT
10	16030	11000	1001	SAWING	0.00							9.00
20	16030	11000	1002	CENTRERING	0.00							10.00
30	16045	11000	1010	TURNING.KONV	0.00							
	6	12	7	STRAIGHT TURNING A N12 D 0.0	7.60	5	450.0	0.350	ABGES. ECKDREHM.		5.40	1.70
	6	12	7	CROSS TURNING A	5.00	2	1400.0	0.280	ABGES. STIRNDREHM.		7.80	0.10
	6	12	7	CENTER DRILLING D 6.8		1	900.0	0.063	STUBOHRER 12		10.00	0.10
	6	12	7	DRILLING D 5.0		1	1400.0	0.045	SPRBOHRER 31		12.60	0.40
	6	12	7	BORING D 12.0		1	560.0	0.100	SPRBOHRER 32		14.40	0.50
	6	12	7	RECLAMPING							14.40	
	6	12	7	STRAIGHT TURNING A N11 D 0.2	5.00	2	450.0	0.560	ABGES.SEITENDREHM.		19.50	0.50
	6	12	7	STRAIGHT TURNING A N11 D 0.2	6.25	4	560.0	0.350			25.10	1.00
	6	12	7	STRAIGHT TURNING A N11 D 0.2	6.46	2	1400.0	0.310			30.10	0.10
	6	12	7	STRAIGHT TURNING A	7.07	1	1400.0	0.560			34.60	0.10
	6	12	7	PRODUCTIVE TIME						3.90	34.60	4.50
	6	12	7	DELAY ALLOWANCE						0.50	73.40	0.60
40	12110	10406	1015	MILLING	0.00					1.10		7.00
50	12110	10406	1015	MILLING	0.00					0.85		7.09
60	12720	13535	1045	DEBURR	0.00					1.00		6.00
70	12110	10226	1017	DRILLING-RAD	0.00					0.10		5.20
80	12710	18112	1010	GRINDING	0.00					1.50		6.00

Fig. 125. Production plan with sequence of operations and data of sequence of operations (from APS [9])

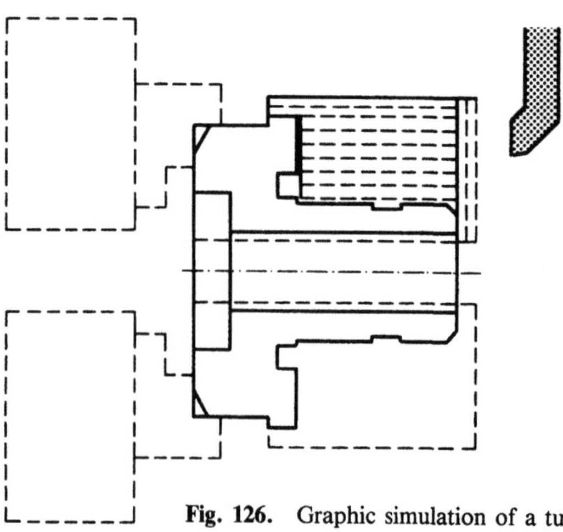

Plain turning
Diameter: 70 mm
Length of shoulder: 85 mm
Depth of cut: 7,20 mm
Number of cuts: 6
Speed: 355 U/min
Feed: 0,31 mm/U
Tool: tool no. 12
Cutting time: 4,6 min
Non-cutting time: 4,0 min

Fig. 126. Graphic simulation of a turning operation [180]

7.7 Working Techniques with Integrated Systems

Through the application of integrated systems, working situations are also modified for the designer and the process planner. The most important changes occur with information retrieval and information transfer and in the activities of the user.
With the information retrieval any information of the integrated partial systems is at the disposal of the user, increasing the volume of information that is quickly available. This information flood, however, can be coped with by a purposeful application of data processing. The use of conventional aids will steadily be reduced since the information is held ready in the integrated system.
Between the individual fields of tasks, i.e., design and process planning, information is transferred less and less in the form of drawings, piece lists, and work schedules; the application of computer-internal representations predominates with integrated processing. Consequently, the user has only the information available which he requires for his tasks, unless he generates intermediate output from the computer-internal representations. With automated information flow there are better forms of communication between the partial tasks, making possible the interactive processing of tasks — e.g., with the modification service.
The use of integrated systems is likely to further reduce the percentage of manual activities in design and process planning; the task description takes place only once. This, in turn, increases the volume of creative activities for the user. Depending on the layout of the CAD classification in the company organization, there may be a loss or a growth in job content. By concentrating several different partial tasks at one workstation the job content can be considerably extended.
Due to the multitude of tasks with integrated processing, certain things are required of the communication interface between the integrated CAD/CAM system and the user. The wide spectrum of tasks with integrated processing results in various forms of information being generated at the user interface. Alpha-numerical and graphic, complete, schematic, or symbolic forms of representation can be used. Fig. 127 shows differing forms of representation for the geometric object information to be processed mainly in the CAD/CAM field.
Other requirements from the user interface result from the tasks to be performed by means of an interactive symbol operation. Particular mention should be made of input support, decision finding, user guide, and user information. The system utilization by batch processing — common in the initial stages of CAD application — is used today only in fields requiring long computing times, resulting in long response times with an interactive processing, e.g., with the method of finite elements.
Man-machine communication in integrated CAD/CAM systems can be realized with various techniques. In use are programming languages with various degrees of abstraction, interactive input via interactive communication, menu or graphic reconstruction techniques including pattern recognition analogous to the conventional design techniques, and higher-level design and process planning languages for task description and sequence control. Fig. 128 shows a subdivision of the input techniques with a view to the convenience attainable with the system.

More far-reaching developments deal with processes for system utilization by means of human language for more convenience and speed; information is input with the voice rather than with the hand [118]. The aim of higher-level design languages is the definition of complex objects with as little input information as possible.

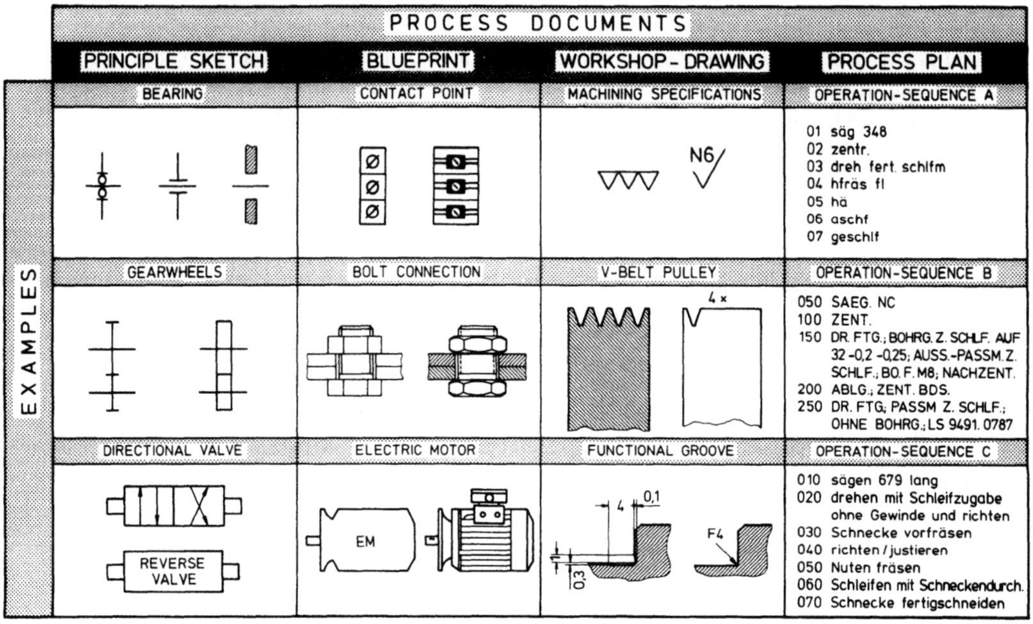

Fig. 127. Differing forms of representation for geometric information [224]

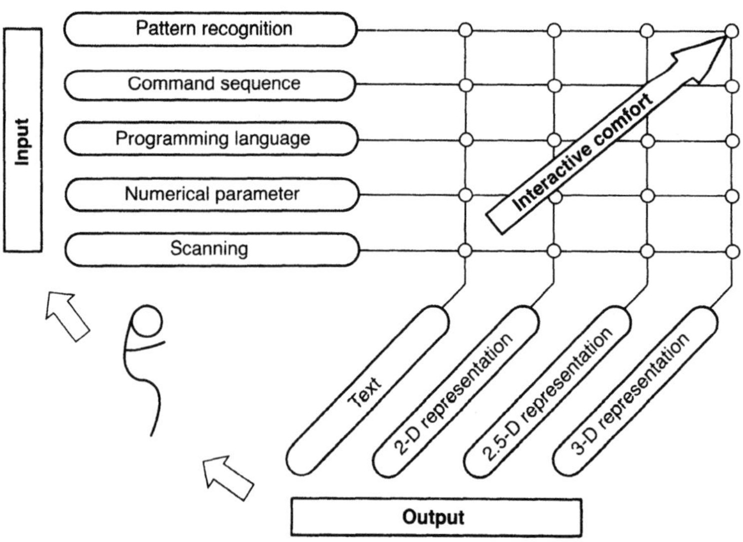

Fig. 128. Steps to convenience of interactive communication in CAD processes [118]

With integrated processing a different form of communication can be provided for every individual system; operation with one uniform communication interface for all partial tasks is also possible. If a special form of communication is selected for every partial task, this means a simplification of the system linkage. For task processing the users have to adapt themselves for every partial task to the modified form of communication; if necessary, they even have to change workstations.
For an integrated system with a uniform communication interface considerable program-technical interventions into the communication modules are often necessary to interlink individual systems. If the communication modules exist as separate modules, integration can be made easier by using a superset command module. The uniform communication interface eases the application of the integrated system through a like communication form for all partial tasks. Standardization of any required peripheral units is also possible.

7.8 Factory Aspects of Integrated Systems

There are many aspects to be considered with the operational introduction of an integrated system or the realization of an integration concept. If the integration concept is to show short-term success and is to be replaced by a higher-ranking concept in the long run, then the linkage of task-specific systems with special interlinking solutions is proposed, a solution which emphasizes the optimal utilization of the individual systems. In case of a long-term concept, the integration ability of the individual systems is the predominant aim, and limitations of the individual systems are allowed for with task processing.
With the integration concept questions of maintenance and advanced development must also be considered when using systems from various producers. The aim of integration may be a vertical data flow within the operational hierarchy, where problems of data compression and detailing predominate, or a horizontal data flow within the operational hierarchy, with which the task-adapted data transfer is essential to avoid information losses.
The practical application of integrated data processing presupposes its classification into the existing planning structure of a company. Figure 129 presents two different principles.
With a small percentage of automation, incorporation into existing planning fields is useful; since the additional load in the planning field is small, causing only a change in the sequence organization and a minor spatial modification, but requiring no new organization of the planning field itselfs with a high degree of automation (integrated processing) it is useful to centralize the processing. This eases production-technological cooperation between design and operations scheduling, required for the further processing of determined information. Such centralization also eases the solution of problems, the coordination of tasks, decisions on investments, training measures for the staff, and the maintenance of EDP equipment.

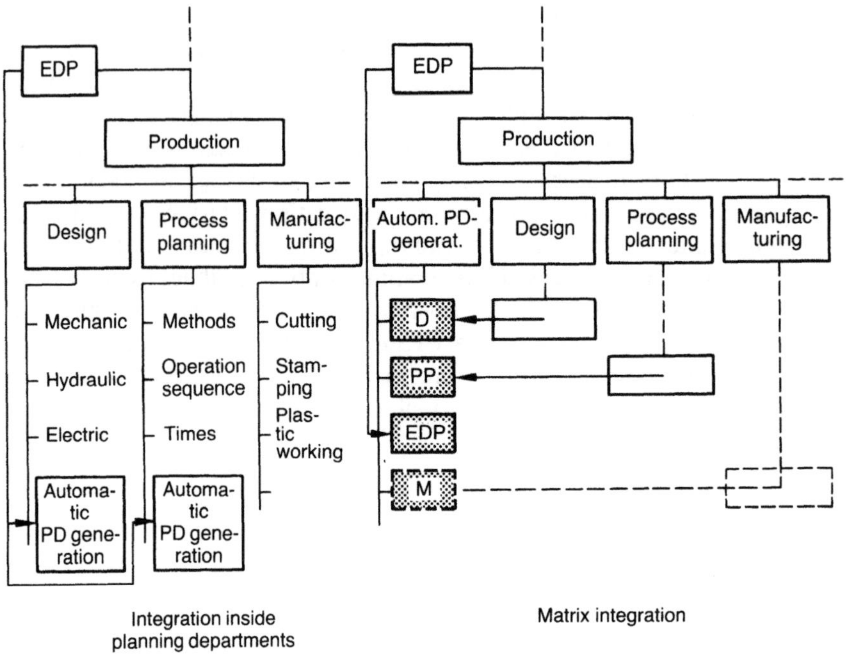

Fig. 129. Organizational incorporation of integrated systems [224]

Figure 130 shows a conventional company structure and the modified structure after CAD integration.

The result of this restructuring — performed in the American aircraft industry — was amortization of investments within 1 year due to avoidance of double operations, a quick-acting modification service, and facilitation of the inspection of drawings [21]. Improved drawing quality reduced modification efforts in provision of production equipment and manufacturing facilities, as well as in process planning and NC programming.

Compared with individual systems, integrated CAD systems require greater effort with system updating, data maintenance and updating, and modification of documentation and user documentation since the requirements of several functional areas or departments have to be considered simultaneously (Fig. 131).

Extensions and improvements with integrated systems also require more from the CAD-system service man, since partial systems with different task spectrums have to be serviced. Another crucial point of organization is the coordination of utilization time, calculating power, and peripheral offers.

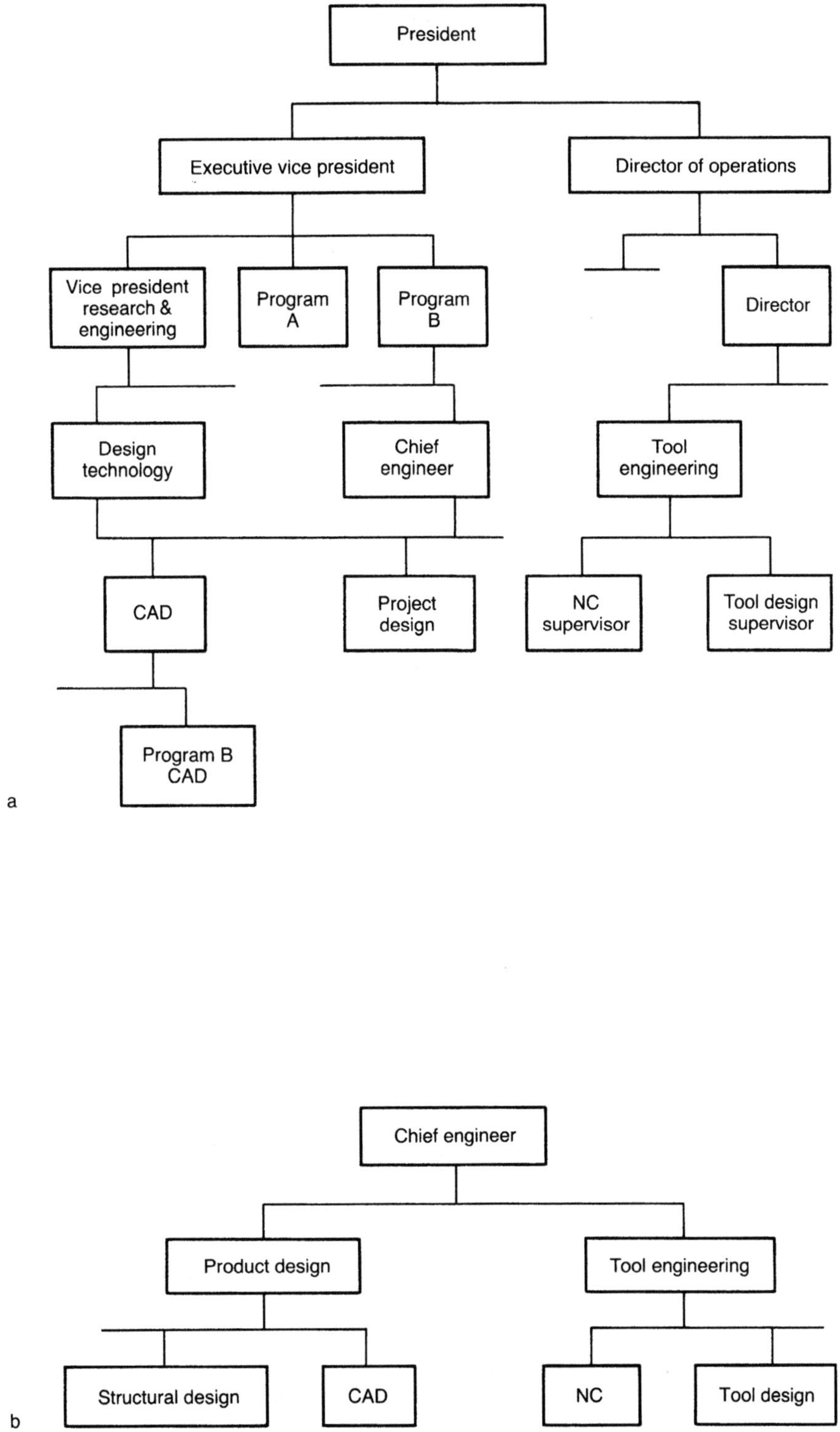

Fig. 130a, b. Conventional **(a)** and CAD-oriented **(b)** operational structure [21]

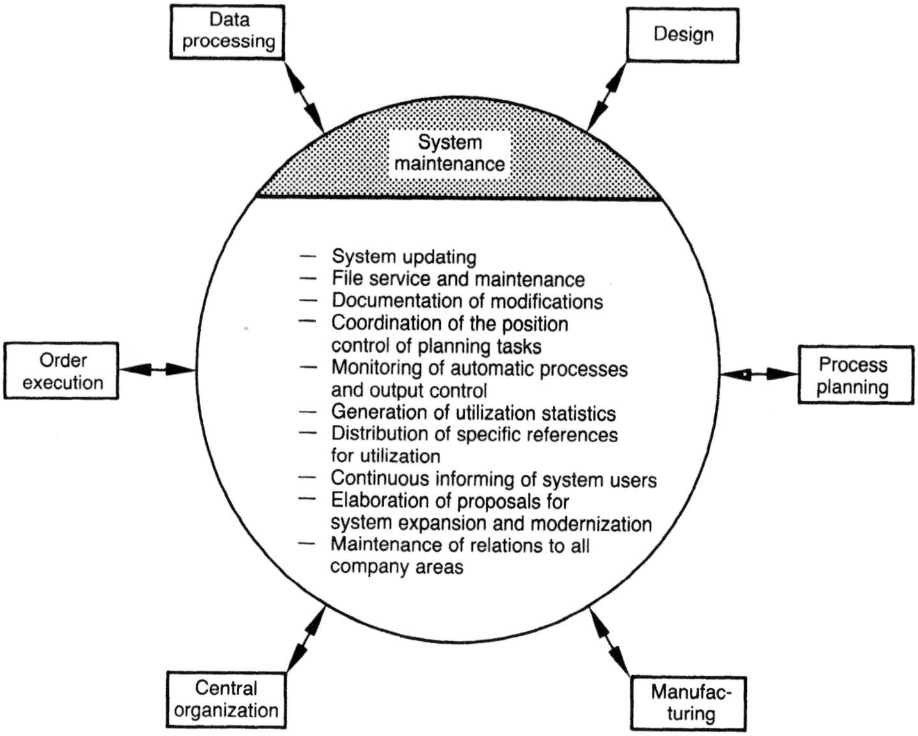

Fig. 131. Tasks of system maintenance [224]

7.9 Trends

Until now, developments in the CAD sector have been aimed predominantly at the provision of job-specific CAD systems for processing functions of design and technological planning. Only during the past few years has there been increased activity regarding standardization in the CAD field. By now these standardizations show a special character, whether they be for a special function (NC control, graphics generation) or for the solution of a limited problem (transfer of drawing information). To obtain the most from CAD application in the long run, generally valid standards for the transfer of computer-internal representations from any CAD system must be defined. To this end a volume-oriented, three-dimensional object model could form the basis of a standardization.

Greatest attention should be paid to the advanced development of data banks, which today are of central importance in integration measures. In this connection, problems of access time play the greatest role, after the organization of data distribution.

Regarding hardware, decentralization can be a means for reducing response times and for parallel processing of CAD tasks. The problems of a smooth exchange of data between the partial systems and of the distribution of data files are not yet solved.

8 Techniques of Software Design for CAD/CAM

Computer-aided design and manufacturing is a field where software engineering plays a predominant role. Programming CAD/CAM systems has proved to be a very expensive task, nearly on the same level as for programming computers integrated in large real-time or military weapons systems. Moreover, the reliability of results is rarely satisfactory, and the facilities offered by CAD/CAM systems frequently look awkward to the final user. The main reason for the high cost is that CAD/CAM systems imply the use of sophisticated and not always well-mastered techniques such as interactive programming, data base management, real-time programming and graphical input and output. CAD/CAM systems are very complex and thus not very reliable due to the lack of appropriate and efficient software engineering methods. They do not have the power the final user would like, due to their lack of some kind of "intelligence".

This is why CAD/CAM people have been the pioneers of several computer science topics: computer graphics, interactive programming, and relational data bases.

A partial solution to the deficiencies of present CAD/CAM software systems can be found through the use of more powerful languages [3], permitting a higher quality of programming in conjunction with an efficient programming environment and rigorous programming methodology. Moreover, specific software tools are being developed or are in use for graphics, interactive programming, and data base management. They are not yet integrated in operating or programming systems, but are becoming more and more the basic tools of the CAD/CAM man.

A further recent improvement has been the development of artifical intelligence (AI). Man-machine communication becomes easier through the use of quasi-natural language and of pattern recognition for text and diagrams input. Expert systems provide powerful data-retrieval techniques. Associated with problem-solving methods, they can provide a guide for the designer and propose possible solutions to his or her problems.

Artificial intelligence languages are the natural way of implementing AI techniques. Moreover, they provide a better and easier way of programming, at least in the development phase, because of their relative efficiency at execution time. Tremendous work is being done in industrial and academic research laboratories for their development. LISP [157] is on the verge of being used in industrial CAD/CAM applications. New AI languages are being more and more frequently experimented with in CAD/CAM research laboratories.

This section deals not only with the CAD/CAM systems designer's point of view about software production, but also with the user's point of view. It is a fact that in CAD/CAM the separation between the designer and the user is narrowing. Modern systems offer the possibility for the user to define language extensions and to add new methods and processing facilities. Three aspects of CAD/CAM software production are considered here: programming languages, programming aids and tools, and artificial intelligence tools and languages.

8.1 The Evolution of Programming Languages for CAD/CAM

It can be said that CAM really began in the mid 1950s with the development of numerical control machine tools. At the beginning, these machines were programmed through the manual punching of paper tapes, using a code directly understandable by the electromechanical control devices of the NC machines, which consisted mainly of binary coordinates and a few commands (spindle and feeding speed, for instance). This was very similar to machine languages for computers but with a much smaller set of instructions. It became clear that some symbolic language, permitting at least the use of mnemonic commands and decimal numbers, would make programming much easier, leading to some kind of assembly language. But that did not solve some basic and repetitive problems, mainly linear and circular interpolation, when not included in the control device, and the computation of the tool's path, given the surface to be obtained after machining.

Decisive progress was made in that direction with the development of the APT language [195] and of its processors and postprocessors. APT has a very simple structure of the type command/parameter 1, parameter 2, except for the use of parenthetic arithmetic expressions and of loops and macros. Its processors do all necessary interpolation and tool path computations in two and three dimensions.

APT is still used extensively and has been the father of nearly all NC programming languages: IFAPT, EXAPT [64], MINIFAPT [40], MINIAPT, which are usually called "APT-like". The extensions of the APT family have been in three basic directions:

1. Increased geometric power as in UNISURF [18], including general parametric surfaces processing, here BEZIER surfaces
2. Restrictions to small systems with limited geometric possibilities, but available on mini- and microcomputers (e.g., with MINIAPT, MINIFAPT, or ELAN [145]); geometry is limited to 2.5 dimensions (i.e., plane drawings with possible discrete variations in the third coordinate)
3. Interactive NC programming systems, the first being NCG (Numerical Control Graphics [184]), designed at the beginning of the 1960s and used mainly in the aircraft industry; NCG offered the possibility of defining 2-D geometries on a graphic display and then defining the tool's path through APT-like interactive commands

A graphic simulation of the machining process allows the interactive verification and correction of NC programs. Interactive NC program editors have been developed on small computers. VULCAIN [40] is an example of such a system where modifications of the geometry and the NC program can be done permanently (but without any possibility of backtracking to the previously defined NC program). The wide use of these languages in planning and design offices has led many CAD systems designers to use an APT-like syntax for the languages they define, such as COMPAC [140, 213] or drafting systems such as PRODES [165]. Moreover, most interactive CAD systems such as CADAM and CATIA [16] use similar structures for man-machine communication, where the main difference is that the command may be replaced by pointing to a menu, and some arguments by pointing to a drawing.

APT is a language for the final user in the planning office rather than for the system designer, although NC languages may be used totally or partly in CAD or drafting systems. It is a good illustration of the stability of the use of languages for industrial CAD/CAM systems. The main reasons for this stability are the high quality of the design of APT, the cost of CAD/CAM systems development, and the effort mechanical engineering people must make to learn programming languages.

FORTRAN [12] may be the best illustration of this stability, however. Most present CAD/CAM systems are programmed in FORTRAN, except for the man-machine communication and data-base modules, which unfortunately are too often still in assembly language. The reasons for the success of FORTRAN in most industrial applications are standardization which implies a good portability, ease of learning, and the existence on nearly all computers of efficient and frequently optimized compilers which produce a machine code nearly as efficient as the direct use of assembly language is: But along with its success, FORTRAN has many drawbacks (see Table 31):

— Lack of rigor
— No interactive facilities
— Elementary data structures
— Poor (or absent) programming environment
— Processing power nearly to arithmetic expressions

Table 31. Features of programming languages

	FORTRAN	Highly developed languages
Data structures	Arrays	Lists Relations
Interaction	Read Write	Commands Recovery from control points Graphic
Programming structures	None	Block structure avoiding side-effects
Programming power	Arithmetic expressions	Recursion Backtracking Associative retrieval

Regarding lack of rigor, it is well known that the structures of FORTRAN are ill-adapted for program verification with modern programming, an issue which is fundamental to users of large current CAD/CAM systems. From this point of view, PASCAL [246] is a much better language, and it is becoming the universal academic language. Its use in industry is increasing, and this will speed up when present students have executive positions in industry. It is clear that ALGOL [168] came too early, without general and industrial support. PASCAL will not fail as ALGOL did, from which it borrowed many features, such as recursion and block structure.

Neither FORTRAN nor PASCAL provides facilities for interaction and graphics, which are fundamental for CAD/CAM systems. Engineers constantly work with drawings, diagrams, and images. The basic task of design is to construct step by step, to modify a project, and sometimes to come back to a former state of it. Thus, in good CAD, interaction with the execution of the CAD programs is an absolute necessity.

Graphics programming has made much progress recently. GKS [82] has now been accepted by the ISO as an international standard, and enjoys a growing consensus. It allows for the production and modification of images on any type of graphics terminal. Portability toward different computers should be ensured as soon as the standards are fixed, and there will soon be GKS subroutine packages for each computer and GKS graphics processors for each graphics terminal. Whether there are problems in using GKS with high-level languages is not yet clear, as GKS can be shaped into either a set of functions (or subroutine package) or into a language extension.

In parallel with these standardization achievements, a special protocol, IGES (initial graphics exchange specification) [124], has been developed and is now in industrial use in many companies. IGES is a 2-D standard permitting the communication of 2-D definitions of products. It is designed mainly for communication of technical drawings and information between CAD systems.

Interactive languages have existed since the beginning of programming languages. BASIC was one of the first examples and is used more and more by engineers in the design office on mini- and microcomputers. It is more or less an interactive FORTRAN with a few added possibilities (recursion, free formats). It is also used as the supporting language of new mini- or microcomputer CAD systems which are becoming more and more popular, although the word CAD may not be considered appropriate here. Except that it allows for interactivity, it has the same drawbacks as FORTRAN.

APL [66] has led to very flexible CAD systems such as GRAFIXI (a product of GIXI, subsidiary of CISI, Paris). It is considered by its "fans" as the ultimate interactive language for CAD and many domains. It permits easy interactive definition of new functions, which is — for the designer as well as for the CAD system designer — a fundamental facility. It also provides simple data retrieval facilities for vectors and arrays. However, its limits on data types and on the possibility of dealing with large amounts of data are prohibitive for new CAD systems, which can currently deal with millions of data elements relating to complex surfaces or complex products such as automobiles or planes. At the same time,

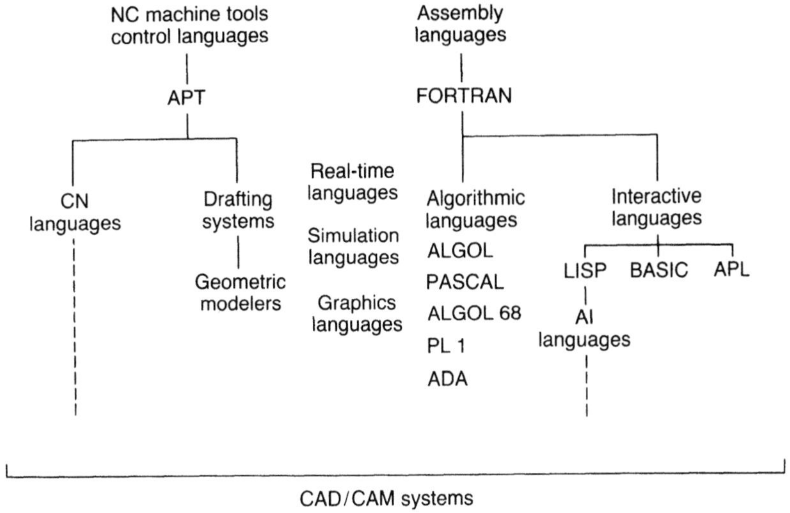

Fig. 132. Programming languages for CAD/CAM

the low price of memory allows even minicomputers to deal with such large amounts of data.

LISP is becoming more and more widely used in CAD/CAM laboratories. It is certainly a very good interactive language, offering good data structuring facilities; however, its syntax is rather awkward and its efficiency at execution time is rather poor. It is the father of AI languages, although it does not include any real AI facilities. It will be discussed in more detail in the section on AI.

To complete this survey of programming languages used for CAD/CAM (see Fig. 132) we should mention real-time languages and simulation languages (e.g., SIMULA [173]), which are used for CAD/CAM but constitute very specific topics.

The main debate at present is about whether we need universal languages which offer all kinds of needed facilities, such as ALGOL 68 [151, 237], PL/I [187], and ADA, or simpler languages such as PASCAL, APL, or LISP [157], which offer a great processing power. ALGOL 68 and PL/I have failed as far as CAD and many other applications are concerned. It looks as though ADA [3] will be — if provided with the large programming environment which is announced, and if available on most computers — a very successful language for system programming. But it will not be sufficient for the CAD/CAM designer who uses high-level computing facilities to build tools for solving industrial problems in his technical field.

8.2 CAD/CAM Software Engineering Tools

This section is concerned with tools and techniques specific to or in wide use in CAD/CAM. General CAD/CAM systems were first developed to encompass the deficiencies of former operating systems. They have been replaced by more powerful operating systems in conjunction with good programming environments. Still, some problems are not yet well solved by these means, namely data-base management and interactive programming. They are presently the subject of much research and development by computer scientists as well as by CAD/CAM specialists, since CAD/CAM is one of the fields where these problems arise with the most emphasis.

8.2.1 CAD/CAM Systems

Operating systems of the 1960s were designed mainly for batch processing. The development of CAD systems made evident the need for communication between different independent programs, for flexible linking between them, and for the possibility of sharing and transmitting data. Simultaneously, the use of graphics and interactive programs grew rapidly. Operating systems offered only poor, and frequently no help to CAD/CAM designers on these points. They had either to adapt the operating system and to develop it, or to implement directly in their programs the needed facilities. All of this had to be done in machine language, leaving no hope of portability, and as it is not the normal job of the CAD/CAM designer, the result was usually rather poor.
Thus, in the early 1970s came the idea of building specific systems for CAD/CAM, offering programming facilities for data handling, man-machine communication, and interlinking programs. There have been several attempts in that direction, such as GERMINAL [117] and RAINBOW [247], which were presented as general-purpose CAD systems [31] (Fig. 133).

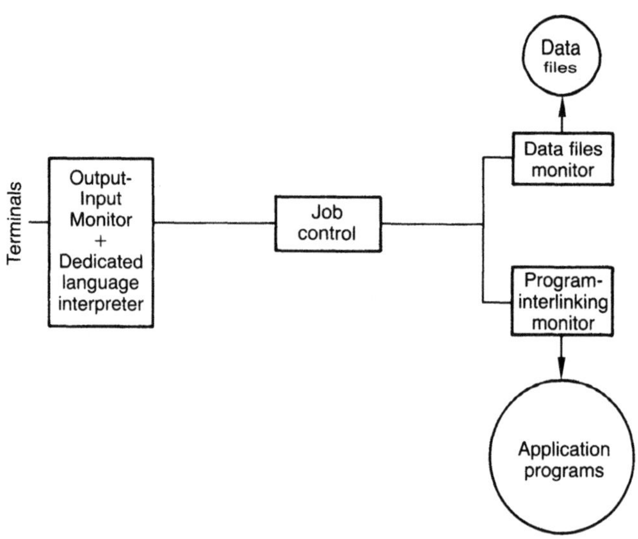

Fig. 133. Scheme of a general-purpose CAD/CAM system

In GERMINAL, for instance, a central data base can be accessed through a library of programs and through specific interfaces. Graphic and interactive facilities are offered. The data base consists mainly of FORTRAN arrays plus simple data structuring capabilities more or less similar to those offered by ALGOL 68. Interfaces can be programmed, matching the data structure of the data base with that of the application program. These interfaces are automatically invoked when data are input or output in an application program. Simple graphic facilities are provided, independent of the graphics display. Each display needs its own interpreter. Interactions can be described in a way analogous to state diagrams.

These systems have been experimented with successfully in various applications, but they are no longer in use. This is due mainly to their lack of portability and flexibility. General-purpose CAD systems were designed for a given operating system and a given programming language (FORTRAN, in most cases). Although some possibilities of extensibility could be found, the systems remained huge and rigid, and therefore not very adequate for the constant evolution of CAD/CAM systems. Moreover, they offered rather poor data base management and interaction capabilities. Therefore, these systems, which were supposed to become the core of all large CAD systems, can be considered failures. But they showed how advanced CAD/CAM systems should be organized: a library of programs around a data base plus sophisticated man-machine communication.

It appears that current modern, comprehensive operating systems offer most of the capabilities of general-purpose CAD systems, mainly program interlinking, data communication, and interactive use. This is the case with UNIX, which is available on a large set of different computers. Moreover, some new languages are accompanied by programming environments that offer advanced program design, test, and debugging facilities. This is so with ADA [3] which is expected to become *the* programming language for complex systems, and hence, perhaps, for large CAD/CAM systems.

Software engineering in the complex domain of CAD/CAM is more a problem of methodology than of programming itself. Many programming methods already exist and are in industrial use. The realization of large software is considered to be no longer possible without efficient methods, but proper methods for CAD/CAM do not really exist yet. Current methods are designed mainly for management or real-time purposes. A programming methodology cannot be considered efficient if it does not have a good programming environment that provides tools to translate the program scheme from the methodological approach to a real program, and to debug the last program used.

It thus becomes clear that CAD/CAM systems must be organized around high-level operating systems, flexible and dynamic data bases, sophisticated man-machine communication, and interaction, and will be realized using high-level languages with good programming environments.

8.2.2 Data Bases

Data bases constitute the core of most modern CAD/CAM systems, and the data structures used for CAD/CAM are mostly complex and dynamic. Several data

base management systems are presently proposed, none of which are fully satisfactory. Verification of consistency and coherence, which constitute specific problems in CAD/CAM, remains unsolved.
The immediate problem is how to handle huge amounts of data. Modern maxi-, mini-, or even microcomputers provide the hardware to do it. The question is how to store data in such a way that the user will be able to retrieve it in the usual manner.
The first task is to store the present state of a product during the design process, to modify it or to add some information, making sure that all information is known and that the consequences of modifications are taken into account by all those working on the product. Bolts need to have the same axis on different parts of an assembly. It is a problem of integrity and coherence of data in a data base (Fig. 134). The user and the system designer need tools to say which relations or constraints must be maintained in the data base when the design of the product is evolving. This is usually done with logical expressions whose validity will be checked every time a change is done. Propositional and predicate calculus are used in most AI systems, but are not practically suited for mechanical engineers.
A further step is to retrieve former data about previous steps of the design or about the design of former products. Filing the different versions of an industrial product is an extremely time-consuming job, which must be automatized in a CAD/CAM system. Moreover, it is important to have easy access to former drawings or designs. Some modern CAD/CAM systems allow the user to backtrack the history of the design of a given product.
The knowledge gained from designing many products must be available to the designer. This knowledge is stored in what is called a "knowledge base", which also includes current standards of the firm and current design traditions. Retrieving these data can be done only in an associative way. Those problems will be further

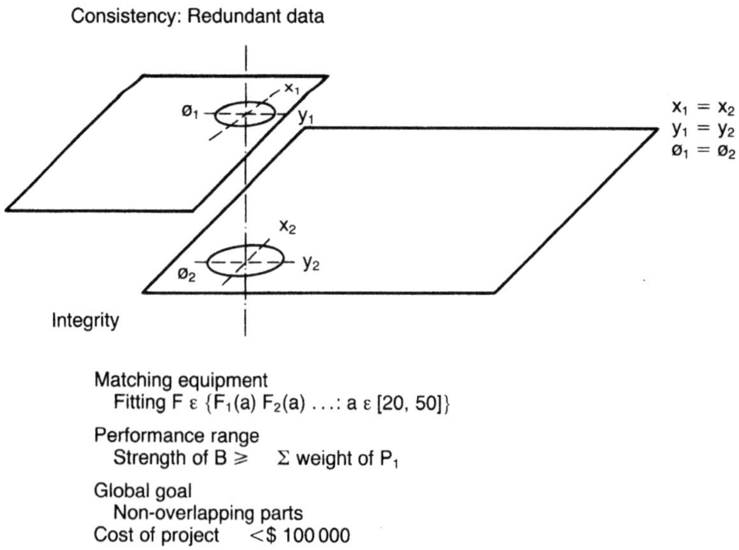

Fig. 134. Coherence and integrity of data in a data base

examined in the section on AI, as their solution may depend on AI techniques. Modern languages offer many possible data structures — arrays, trees, files, lists — which are necessary to represent the complex structures of objects to be designed or manufactured. Even those which are used in connection with virtual memory computer systems do not offer real data-base management facilities. In addition, data retrieval is done in a single deterministic way, which does not correspond to the numerous types of searches a designer or a design system has to do.
It appears that the requirements for CAD/CAM data bases are quite different and on a much higher level, except for security, than those for management or administrative data bases. In CAD/CAM data are dynamic (project data base), and their access is complex. For example, the representation of polyhedral volumes requires access to points, edges, faces, and volumes, through their connectivity relations. The ordered set of points of the contour of a face as well as which faces or edges contain a given point will be asked for. Moreover, for many geometric algorithms, such accessing must be done very rapidly if the algorithm is to maintain acceptable performance (a few hundred thousand accesses are usual for a hidden line or volume intersection algorithm).
The complextiy of accesses more and more frequently implies the use of *relational data bases,* or data bases including a relational model: data are accessed through relations which can be defined dynamically. These geometric relations, which are n:m, i.e., n input — m output, n and m varying between two occurrences of the relations, are a classical example of CAD/CAM relations. Moreover, in many cases, the validity of certain occurrences of the relations is uncertain, whence the use of fuzzy sets [249].
The problem of performance versus CAD/CAM data bases has not yet been satisfactorily solved; research is now being done on this point [23, 208]. Aside from the use of new theoretical results on data-base access techniques, two solutions now appear feasible:

1. To make copies of the part of the data base needed by some given algorithm, with a structure adapted to that algorithm
2. To use data-base machines such as LISP machines, which are beginning to be available

8.2.3 Design of Interactive Programs

Diverse techniques such as character recognition, scheme recognition, and quasi-natural language communication are being used more and more frequently. They facilitate the man-computer dialogue, especially in a field where drawing is a basic skill. They do not directly intervene in programming languages for CAD/CAM, as all these facilities have to be implemented with a given language.
The use of interactive languages is only a small part of interactive systems design. Aside from programming methods, the main task remains *specification.* Much work has been done in the past decade on the specification of languages, processors, and programs. Specifying an interactive program is not far removed from specifying a compiler. More recently, several publications have been devoted to the specification of man-machine dialogues [41, 141, 206].

The basic idea is to build an algebra of the sequences of couples of elementary input and output (Table 32). These algebras use concepts of formal languages and automata theory. However, it seems that none of these is powerful enough for large interactive CAD/CAM systems, or is sufficiently simple to be used by non-informaticians, who are those at the workshop level in charge of specifying CAD/CAM dialogues. Thus, if the underlying theories are now becoming known, the main problem remains: how will a normal user or CAD/CAM system designer describe the man-machine dialogue using his own professional language?
Although some academic applications of formal specification methods have already been realized in CAD/CAM, they remain far from being current industrial practice.

Table 32. An example of dialogue specification

Input	Output
Mode segment drawing... Pointing Point 1. Point 2	Draw segment Point 1 Point 2
Mode segment modification Pointing Point 1. Point 2	Erase segment (x) Point x. Point 1 Create segment (x) Point x. Point 2
Mode modification Bezier surfaces Pointing Point 1. Point 2	Erase surface of Point 1 Create new surface with Point 1 replacing Point 2.

The only method widely in use is that of *state diagrams* [169], which cover only a small part of the interaction process. The principle is to represent the dialogue possibilities with a finite automation. The present state of the process is a state of the automation (there is a finite number of them); from each state are issued a finite number of transitions, each one corresponding to a subset of possible inputs, and to a procedure realizing the consequent actions and output. It is clear that the finiteness constraint limits the use of finite state diagrams to only approximate models of the dialogues. Nevertheless, it is a very important first step in the direction of specification.
Interpreters and editors of finite state diagrams have recently been developed [95]. They permit an interactive construction of the diagram, and generate the interactive program once the given procedures have been written.
A more precise approach is to regard the dialogue as part of a language. This means that the technical language used by the CAD/CAM user during his or her work in connection with the CAD/CAM system will be defined as a programming language and processed in the same way. Thus, the technical problem here is one of compilation or interpretation. The use of *compiler-compilers* is growing as a means of easy definition of specific technical languages and their automatic processing. Several experiments have already been done [70, 154].
With regard to *real-time* problems and *concurrency*, there are now several specification systems with corresponding interpreters based on Petri nets [178, 236]; GRAFCET [73] is now in industrial use.

8.3 AI and AI Languages for CAD/CAM

LISP may be considered the predecessor of AI languages. The main reasons for its success among AI specialists are:

1. Interactivity: LISP is an interpreted language. Data and programs can be modified interactively. In fact, there is no basic difference in LISP between functions (or programs) and data. Both are considered LISP objects. This makes it possible to design self-modifying functions, or programs which create new functions — a very important facility for future CAD/CAM systems, where functions and procedures will have to be defined dynamically (integrity test functions, for instance).
2. Data structures: The *list structure* of LISP offers many possibilities for organizing data. Trees, sequences, arrays, or graphs can be easily organized in lists. Moreover, the possible recursivity of lists leads to very powerful structures. The storage of knowledge in lists is rather easy, although sometimes awkward and not very efficient where data retrieval is concerned.
3. LISP is a *functional* language. Sets of functions can be easily defined and used just by typing their names, so that specialized languages are easily created and used. Moreover, recursivity allows the direct use of formal definitions of functions and of backtracking procedures.

These explain why a great many AI languages and programs were first implemented in LISP, although they were later, for efficiency reasons, implemented in algorithmic languages. LISP is very widely used in AI laboratories and now more and more frequently in CAD/CAM laboratories. It is not used in industry mainly because of its poor processing power where computation time is concerned, or for the manipulation of large amounts of data. The development of LISP processors should change this radically and open the field of industrial CAD/CAM programming to all LISP-like or LISP-based languages.
All the facilities provided by LISP are found in the more recently created AI languages. Compared with LISP, the main features which have been added are data bases with possible associative data retrieval and integrity and consistency checking, and logics processing power, including filtering, pattern-directed invocation, and theorem proving. These features will be briefly reviewed here, with respect to CAD/CAM use.

8.3.1 Data Bases

The first idea was to retrieve data through contents, and not through a fixed, predefined structure. Questions have to be answered such as "Which parts already manufactured in the firm have three holes and five edges?" or "Which mechanism that has already been designed in the current project realizes a given movement transformation?". *Relational data bases* answer these questions satisfactorily. Such associative data searches are implemented in most AI languages, e.g., in PROLOG [38, 242]. An important use of such possibilities is in group technology, where for efficiency and ease of design and process planning, parts are grouped depending on shape and size similarities [45, 148].

A further step is *pattern-accessible data,* where data are accessed not only through simple relations but also through complex patterns which they must match. This means that a question like: "Which mechanism made with flat parts and pins is fixed with three holes, has a linear motion input, and has simultaneously linear and rotational motion outputs?" should be answerable, even if the data have not been structured in the same way.

An important use of these data base facilities is the construction of *knowledge bases.* In manufacturing it is necessary to have available, if possible immediately, all information concerning the technical field, current industrial products, the components of present projects or products, and the former products of the firm and the processing and design methods used in producing them. Even if some of these data can be easily stored and searched for through a fixed data structure, such as files of standardized parts (and even here it is never easy to find one's way through the jungle of standards), it is clear that there is no way to know in advance how and for what purpose the engineer will use all this knowledge about methods and products. The search will be essentially associative, with the use of more or less complex patterns, and in some cases will have to use heuristic and/or logical reasoning to answer questions such as: "Where can I find ideas about how to solve my vibration problem, i.e., which similar vibration problems have already been solved in different contexts?". Good AI languages should provide, with their data handling facilities and ultimately their logic processing facilities, efficient tools for this kind of search and easy ways of describing and specifying the knowledge base.

8.3.2 Logic Processing

The next step is *expert systems.* The problem is not only to make all information available to the user, but also to use the relevant information automatically to guide him or her in search of good solutions or of good decisions in design and planning, and in some cases to find these solutions automatically as well. An expert system consists of a knowledge base plus a set of heuristics and rules which it can use to achieve some goal set by the user. The main tools here are filtering, backtracking, and logical calculus. They are implemented in nearly all AI languages [50]. *Filtering,* or *pattern matching* (Fig. 135), is already used for pattern-accessible data. It is used here for pattern-directed invocation of functions, that is, applying automatically selected functions to data whenever possible. When no solution has been found in one direction, *backtracking* permits looking in another

Pattern	All parts P, P has a hole H, center (x, y), radius equal R
Data	Part P′, hole H′, center (x, y) → center (x′, y′), radius R → radius R′
Result	For all parts P, hole H″, if center (x, y), radius R″, (x, y) → (x′, y′), R″ → R′

Fig. 135. Pattern matching in CAD/CAM used for integrity maintenance

direction. *Logical calculus* is not only the basic tool for deriving consequences from previous results or assertions, but also nearly the only way to express assertions or relations.

8.3.3 Automatic Decision

Automatic decision is rarely possible in CAD/CAM, due to the complexity of the products considered and the diversity of methods and techniques which can be used to solve a given problem. It is thus quite unrealistic for the designer to give a command such as: "Design a product with these specifications." This is the main difference, for instance, from theorem proving, where the domain is well defined and the basic techniques belong to only a single theory. In CAD/CAM, two approaches are possible and complementary:

1. The user defines subgoals which the computer will be able to deal with efficiently.
2. The task of the user is oriented by proposals of possible actions in a given context; in turn, he or she chooses among these actions or indicates where the system is to search. In CONNIVER [222], a user-controlled procedure-invocation facility has been implemented which gives some control to the user — after filtering — over the choice of procedure invocations.

   Another very important problem for CAD, which can be partly solved by AI and the use of AI languages is maintenance of *integrity* and *consistency* [55] of data bases. Basically, it consists of maintaining the validity of given relationships between data during the life of a data base (see Fig. 135). It is clear that pattern-directed invocation of procedures which test the validity of logical relations fulfills this task. Implementing tests is thus made very easy for the systems designer, and for the CAD/CAM user who will very often have to introduce dynamically new relations or constraints. Nevertheless, in this latter case, which is discussed below, the syntax offered by AI languages is not exactly what a mechanical engineer would like. At execution time, of course, the efficiency problems encountered in the use of AI languages remain.

   The problem of maintaining integrity of data bases was first encountered in commercial or management data bases, where it is important, for instance, that one person is not invoiced twice for the same product or service. In CAD/CAM this problem is more complex, as the relations or constraints to be maintained may take a large variety of forms — logical, technical geometric — and are defined dynamically. A typical example is that of two parts designed by different design offices which have to be assembled. There are many local relations to be maintained between the computer representations of both parts during the design period.

8.3.4 Syntax of Logic

Propositional, predicate calculus, is not the language the industrial designer needs. It is ill fitted to express the type of assertions to be found for CAD/CAM problem

Table 33. Some AI languages

Language	Developer/Year	Advantages offered
Q LISP Q A4	MIT 1969	
PLANNER	MIT 1971	Pattern matching
CONNIVER	MIT 1972	User-controlled invocation
SMALLTALK		Logic Actors Ease of use
PROLOG	Colmerauer 1973	Relational Unification Ease of use
PLASMA	Hewitt 1975	Pattern matching Actors Semi-unification

solving. Some efforts have been devoted to expressing them in a more practical way, such as in SMALLTALK [122]. Though defining simple relations can be done easily by using, for instance, sets of tuples, it is not clear how to efficiently define the complex relations, assertions, and goals — numerical, technical, geometric — found in CAD/CAM. Moreover, the problems usually dealt with are often poorly defined, that is, a large part of the knowledge and the goals is implicit in the designer's mind. Expressing this implicit environment is very difficult, and will perhaps need some real "intelligence" from CAD/CAM systems. It is not clear whether there is, other than logical expressions, any universal tool for it. Specialized input languages plus some intelligent learning facilities may be the solution. Nothing has yet been investigated in depth on this point.

This section does not presume to give an exhaustive and comparative review of AI languages. That has been well done elsewhere, as early as 1974 [19]. As far as CAD/CAM is concerned, some material can be found in Latombe [146] or in Crestin [42]. From the creation of LISP to the present, many languages have been developed for AI, from the MIT series QLISP [188] to PLANNER, CONNIVER, SMALLTALK, PROLOG, PLASMA [96] (Table 33), and many others. Many of them have been used for experimental robot planning systems, very few for other CAD/CAM areas, but their applications here are rapidly increasing.

8.4 Conclusion

The great efforts being made in software engineering are most significant in CAD/CAM and in real-time operational systems. ADA, developed for defense purposes, will probably be used largely for CAD/CAM. Graphics and part description standards have been intensively developed by CAD/CAM people, and are at the point of being internationally recognized. Work on interactive programming is being pursued and should soon find some practical realization in CAD/CAM.

AI will certainly take a fundamental place in the development of the next generation of CAD/CAM systems, and AI languages are the best tools for it. Present AI languages have been developed for research in more or less theoretical fields of AI, and for a few experimental applications, mainly robot planning. It follows that the present AI languages do not offer the syntax or the semantics needed for CAD/CAM: the syntax, because the largely implicit contents of CAD/CAM are hard to express through the presently used logic; the semantics, because the general facilities offered by AI languages are more or less awkward when used to solve the technical and geometric problems of design and manufacture. Much is now being done to develop AI languages. Interest in them is growing in the CAD/CAM field, and we can reasonably hope that many of their defiencies will soon be eliminated, at least where experimental systems are concerned. PROLOG is already being used more and more frequently for experimental CAD/CAM programs.

There is still much work to be done to improve the efficiency of the new, sophisticated CAD/CAM systems which are presently arriving in the field. The great amount of research being done on the various tools needed will certainly bring greater improvements. A decisive step forward should be the use of dedicated machines, data-base, or LISP machines. Several LISP machines are now available or in development [177]. Their success will open up possibilities for the intensive use in CAD/CAM of all the facilities offered by AI languages — list processing, filtering, and backtracking. In connection with the improvement of general computing performance, it can be reasonably thought that in the near future desk-top computers will be used for industrial-size AI applications.

It is clear that progress in CAD/CAM systems development will result from significant progress in all these fields. Based on present efforts, it can be expected that the CAD/CAM system on a microcomputer in the design or planning office, connected with the factory or firm's network, will soon be a reality and will allow CAD/CAM users to introduce and use their own methods of thought in their work.

9 Artificial Intelligence

9.1 General Trends and Basic Concepts

Intelligent functions are becoming more and more effective in all components of fully automated systems for the technical preparation and execution of production.

The concept of artificial intelligence is very important, but there is no uniform definition of it. According to Turing, a machine behaves in an intelligent way when the questioner is not able to determine whether he or she is confronted with a human being or an automaton [129]. Although this definition is plausible and gets to the core, it covers only the field of imitation of human intelligence. With reference to technical relations, therefore, the following formulation is obviously more adequate: "On principle, artificial intelligence is a category of more highly automated information processes in technical systems" [196].

The field of artificial intelligence was pointed out for the first time by Steinbuch [219]. A more general characterization of the possible fields of application is given in Fig. 136.

In connection with technical and production-technological problems, one preferably speaks of mechanical or machine intelligence [212]. In general, the same tasks are outlined here as those in Fig. 136, but under the aspect of production-technological systems. The so-called knowledge basis or expert system is very significant here. "Expert systems are for automated problem solving in restricted fields of tasks. In the role of an expert in automaton-aided consultations, they assist and advise the user in working out problem solutions. As an essential characteristic they have a basis of knowledge which contains the relevant facts and rules" [22]. Such systems are already being applied, e.g., in the field of medicine.

In the field of technical and especially production-technological problems there are prototype solutions. "The optimization of integrated manufacturing systems can be aided by expert systems establishing special knowledge cells for the individual system range of the factory, which are used by intelligent operation systems for decision finding. Thus, knowledge cells can be created for development, design, preparation of manufacture, manufacture, and assembly, as well as for quality surveillance and for sales" [212].

With such local knowledge cells the system is able to react to different disturbing factors or influencing parameters (location-related, time-related, and geometry-

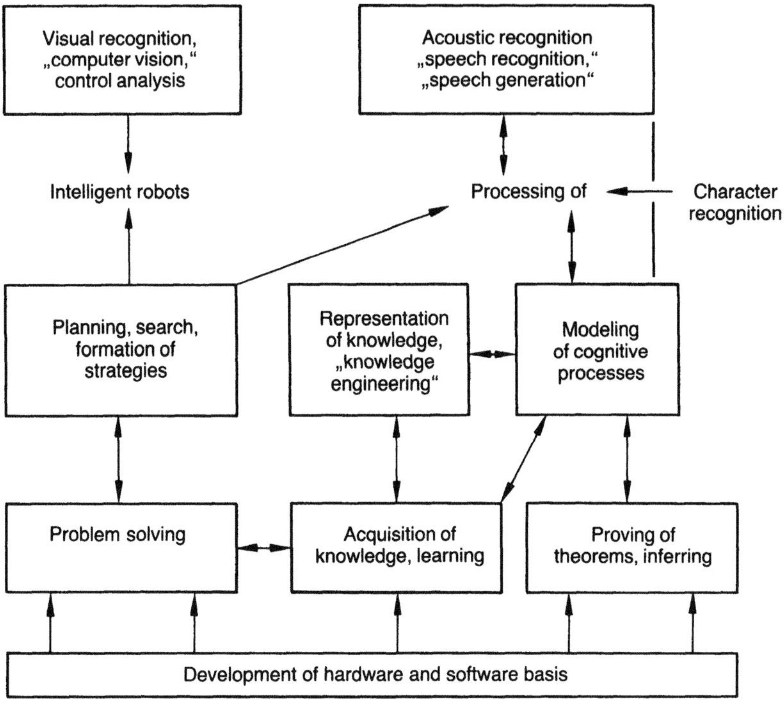

Fig. 136. Main fields of application for artificial intelligence [219]

related) independently, or it can make available to the operator well-founded proposals for decisions within a very short time.

In addition to the efficient processing of knowledge, the aspect of communication is of central importance. Users must be able to communicate their requirements to the system in the way that is simplest for them and must also get the required information in such a way. It is clear that human language, graphic representations, manual sketches, and the direct input of characters constitute the most effective forms of input, since human beings are most familiar with them. However, this makes demands on the hardware and software which by far exceed the usual functions of computers.

Computer systems with such a potential which are, moreover, capable of deductive and inductive information processing are today generally designated as computers of the fifth generation. Thus, computers of the fifth generation have a high degree of artificial intelligence. However, intelligent functions not connected with a computer are also conceivable.

In the following section essential targets of development and possibilities for applying the principles of artificial intelligence are discussed.

9.2 Fields of Application

9.2.1 Effective Forms of Communication

Proceeding from the technical languages which have generally been used so far, the logical further development leads to natural language (word and sentence identification), to character identification, input of sketches, and to image recognition and evaluation systems [233]. Computer-aided design is considerably facilitated by the input of manual sketches. The reconstruction technique described in Sect. 7.7 is capable of generating exact computer-internal 3-D models from simple sketches. These models can be processed further.

9.2.2 Extended Possibilities for Problem Solving

There are extended possibilities for problem solving as a result of increased storage capacities and computing speeds. Completely new methods which have so far not been economical or executable by computer engineering may soon become feasible.

9.2.3 Efficient Storage of Information and Knowledge

This category includes expert systems, knowledge stores, and inquiry systems. By means of highly developed programming languages (e.g., PROLOG [36], PROSCODE-2 [232]) it has become possible to process predicative logic and to generate user-owned knowledge cells.

An example of this is given by Gero [80]. It is based on a purely geometric knowledge which can be developed from the definition of elementary concepts such as area, edge, and corner, through so-called meta-knowledge (relations between objects, e.g., definition of "above" or "below"), up to the description of total components.

A concrete example of the application of these principles to technical conditions is given by Maeda et al. [153]. By means of a programming language (frame operating language), which is again special, individual machine elements (e.g., bolts) as well as relations between individual elements (e.g., creation of a bolted connection) can be defined. Based on this, it should be possible to describe the structure and mode of action of total components and machines.

9.2.4 Deductive and Inductive Information Processing on the Basis of Stored Knowledge and Semantic Transformations

This complex is also designated as "inference ability", and it constitutes the precondition for the generation and use of knowledge cells. For the user there are

highly developed programming languages by which predicative logic is dissolved into a "machine-intelligible" form. The automated software generation is an example.
Starting from an external problem description by human beings (possibly in a natural language), the system is capable of developing an internal model and of converting it physically into a program. This makes it possible to handle, by computer engineering, problems for which algorithmic elaboration and programming have so far been regarded as too expensive. In the automobile industry, for example, there are many more or less unique problems of detail in connection with very specific designs to which these conditions fully apply.

9.2.5 Deductive and Inductive Acquisition of Knowledge (Self-Learning Systems)

Without a doubt, this is one of the most interesting new possibilities. If deductive and inductive acquisition of knowledge really became a utilizable function it would be machine intelligence far exceeding any of the forms known to date: storage, comparison, analysis, evaluation, and offering of results.
Such performance of intelligence — without any preliminary models on a purely machine basis — is indeed regarded as possible, without reducing the need for *man's* creativity.

9.2.6 Process-Related Intelligence with Improved Capabilities of Self-Diagnosis

Storage-programmed CNC controls doubtless provide the best preconditions for realizing higher intelligence performance in connection with processes (cf. Sect. 3.4.1.1). This form of machine intelligence can be utilized directly in the prepara-

Theory/methods/techniques

Specialized AI hardware and software
(e.g., data bank machines, languages as LISP, PROLOG)

- - -

Representation of knowledge
Deductive solving of problems
Heuristic searching
Learning

- - -

Image recognition	Semantic interpretation of natural language (in acoustic or written form)

Applications

Expert systems
Image-processing systems
Intelligent industrial robots
Automatic systems

Fig. 137. Survey of the field of artificial intelligence

tion of manufacture and the manufacturing process [134] and for self-diagnosis in connection with suitable sensor systems and specific knowledge bases. Since such functions are used in a more and more decentralized way with processes, system stability increases at the same time.

In general, it can be stated that artificial intelligence requires the integration of purely formal logic with heuristic principles of decision finding. Fig. 137 is a survey of the general structure of the new discipline "artificial intelligence" and its most important applications at present (according to Schubert [203]). The following are expected to be the essential fields of application in the coming years [166]:

— Expert systems for solving manufacturing problems
— Situation recognition, supervision, diagnostics, and the determination of remedial strategies
— Vision and manipulation
— Optimization of the man-machine interface

10 Development from CAD/CAM to CIM

10.1 The State of the Art of CAM

The state of the art of automated manufacturing described in the preceding sections and the possibilities of versatile computer-aided engineering are the preconditions for even more complex solutions of automation. With examples from several industrial countries it has been demonstrated that even for small and medium-sized batch production a relatively broad basis of flexible automation and computer-aided engineering has been established.

Computer-aided manufacturing — including the automatic production of numerically controlled machining, forming, welding, assembly, and other manufacturing tools; machining centers; flexible manufacturing systems and machining cells as the core of flexible factory automation — is a reality today in many companies.

Computer-aided design — including design, simulation, and optimization of product performance; analysis of material and other manufacturing requirements; production of tools, dies, and special features — is also reality, with commercially available hardware and software products able to support the creative engineer in all these functions effectively.

Computerized manufacturing planning and control systems which can use their internally stored knowledge of a product's subassembly and component structure to generate appropriate production and supplier purchase orders, and to apply NC programming systems and methods of group technology, are also very helpful tools in industry today.

The efficiency of computer-aided engineering is increasing as methods of artificial intelligence in the form of expert systems, knowledge bases, and other kinds of knowledge assistance are being applied. There are also new, efficient principles in software development which increase modularity and portability.

On the whole, the technical level of development presents the preconditions for integrating all the devices of automation employed into closed (interconnected) systems.

10.2 Computer-Integrated Manufacturing

The term "computer-integrated manufacturing" (CIM) refers to the integrated use of computers in all sections of enterprise, from the planning of production, through the design and manufacture of a product, up to the ensuring of good quality.
The interconnected functions of all processes are illustrated in Fig. 138. On the process level symbols represent the stations through which materials, individual parts, subassemblies, etc. pass. The terminals indicate that an exchange of information is necessary or possible at the individual stations.
The data connections between the most important sections and fields of tasks can be characterized in the following way.

10.2.1 Part Geometry, Product Specification

Within the more modern, solid modeling-based CAD/CAM systems a data base of complete data on geometry and attributes can be created, maintained, and used by all necessary engineering functions (including manufacturing engineering). For actual fabrication of the parts (using NC tools) and assembly of the complete product or logical subassemblies, the information on part geometries and the relationships of parts to each other must also be passed to the machines on the shop floor. Today, this is typically done in the form of an NC language description of the machining operations required to create the part. In the future, machine controllers are expected to link directly with the product data base to interpret the geometry and material information and create their own instructions. Information on such things as surface finish may be used to select appropriate tools. Other operations, such as welding or fastening to other parts for more complex product structures, may also be interpreted directly from the hierarchy of components in the data base and their relationship to each other.

10.2.2 Manufacturing Process Data

To the extent that parts and products cannot be "automatically" manufactured from the CAD/CAM data as discussed above, the design process can be used to facilitate semi-automated (or even manual) production. The design engineer, perhaps in a team relationship with a manufacturing engineer, or conceivably the CAD system itself, can assign a group technology code describing the parts' basic geometric and manufacturing characteristics. From this, a computer-aided process planning (CAPP) system can use its data base and logic rules to generate manufacturing process plans (sometimes called routings or manufacturing data sheets). These might include data on the tools required, including any that are unique to the particular part, such as dies, molds, or templates conforming to the part geometry.

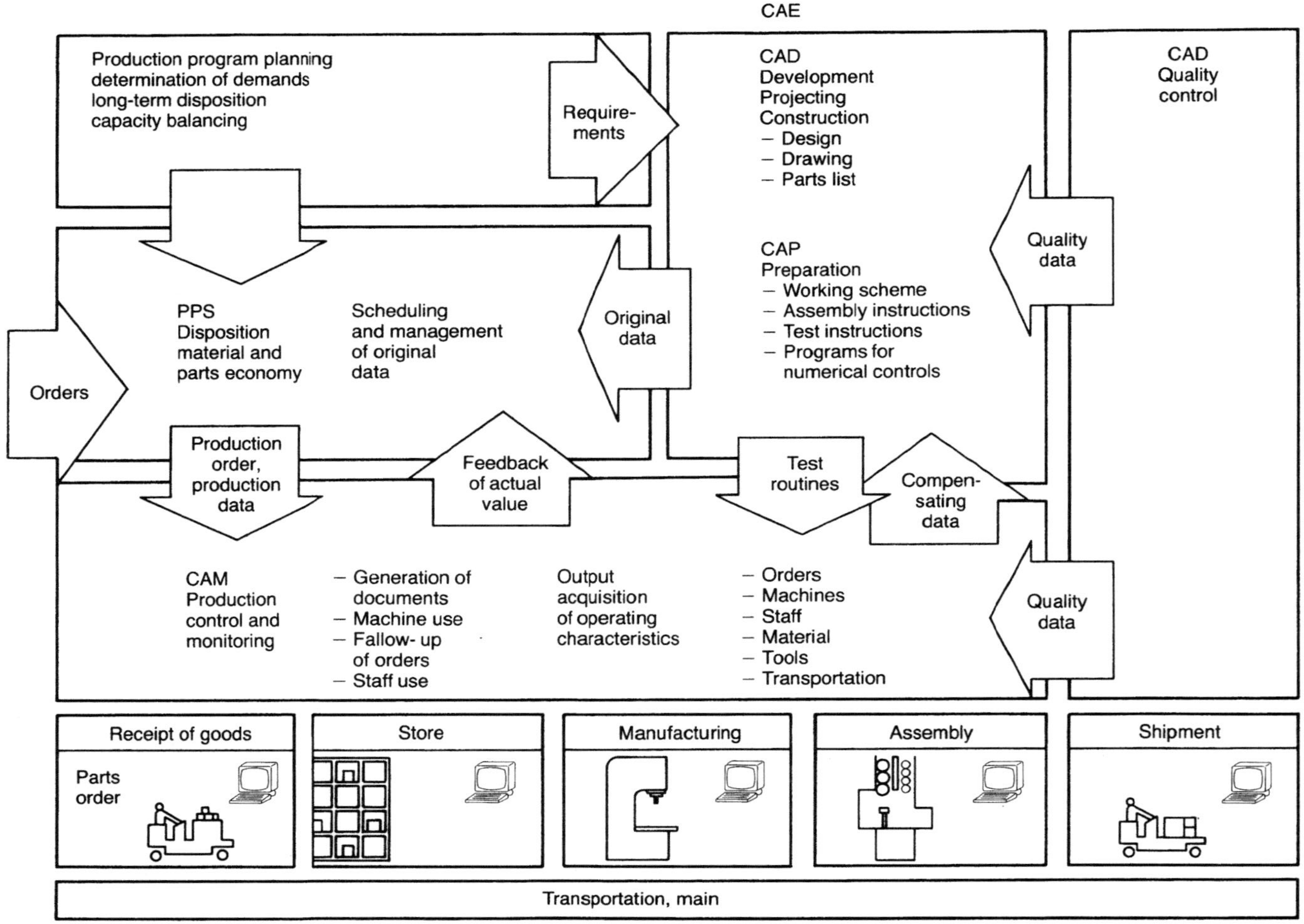

Fig. 138. Main functions of computer-integrated manufacturing (from Waller [241])

10.2.3 Product Structure (Bill of Materials)

The CAD/CAM system should also develop the complete product structure, or bill of materials, for the benefit of the material planning and control systems (MPCS). This information is needed for the material requirements planning (MRP) portion of the MPCS, which must explode the Master Production Schedule generated from orders received and/or finished goods inventory decisions into detailed production schedules and purchase orders for all of the component parts of the product.

10.2.4 Configuration Control

Most complex products today go through many variations in their life cycle as a result of special customer requirements or engineering changes to improve performance, reduce cost, correct error, etc. (In some cases, such as commercial aircraft production, it can easily happen that no two members of a product family are exactly alike.) Information on the effectivity of engineering changes, by model/serial number, date of engineering release, date of manufacturing release, etc., must be maintained both in the engineering records of the company and in the production scheduling records. If these data bases are separate (as is assumed for most present-day systems), this necessitates a flow of data back and forth concerning the specific configuration of all subassemblies and components of each model or range of serial numbers. Often, changes in the configuration of a product will be made by the manufacturing department, which may substitute allowable alternative components or vary the actual product design from what was specified by the engineering department (for various reasons of expediency), so it is important that this be a two-way data flow.

10.2.5 Production Schedules, Results

An obvious two-way linkage must exist between the MPCS systems and the manufacturing automation systems regarding the *planned* schedule of production to meet order requirements and the *actual* production results, which may differ because of machine down time, misestimation of operation time, unavailability of material, or many other factors. In fully automated systems at both ends of this linkage, such discrepancies should theoretically not exist (i.e., machines are properly maintained or alternate routings found, inventory is prescheduled and therefore always available, estimates are corrected on-line) but in practice *today,* variations and consequent interactions of production scheduling are inevitable; hence, the two-way flow.

10.2.6 Inventory Availability

This term implies both the planned inventory of components and semi-finished products generated by the MRP portion of the MPCS system, and information on the actual state of inventory and locations of items as determined by the manufacturing automation system (which actually controls the handling and transformation of material). This linkage is somewhat analogous to the scheduling information described above, but in this case it deals with physical counts, locations, movements, and transformations of inventory from one state to another. With full feedback from the production automation equipment, data on rejects, scrap, and consumables (lubricants, coolants, etc.) can be monitored, as well as usable parts and products, so complete control of all material can be maintained.

The different computer systems for the preparation and execution of production have to be interconnected in a way that will establish optimal flexibility in the entire range of operations. This can be done with the aid of so-called local networks — local area network (LAN). In a spatially delimitated range, a LAN includes workstation-related systems, which can communicate, for example, through coaxial cables or fiber-optic light guides. By means of such networks it is possible to interconnect devices for the processing, storage, or representation of information as a system in such a way that specific tasks can be assigned to individual co-operating system components [241]. In this way free access to various data or program ranges can be provided to users (see Fig. 139). It should be pointed out that the model conceptions represented in Figs. 138 and 139 should be regarded as a target for the years to come.

In the course of this process of development toward computer-integrated manufacturing a number of the theoretical and practical problems must still be solved. There is a connection between system size, necessary volume of information, and system stability. Proceeding from the purpose, there will be an optimal range or a compromise solution for every technical system with respect to these three components. If the system size is regarded as given, maximum stability will be achieved when the coupling relations within the system and outwards are minimized and the extent of the necessary exchange of information is reduced to a minimum. The determination of these minimum values constitutes one of the essential problems. Since no algorithms or general principles for the determination of such limits exist, the experience gained from working with such systems has to be used as the basis. On principle, as seen from the viewpoint of system theory, a so-called multistable system must be generated, consisting of ultrastable subsystems (the system is independently capable of finding a stable state of equilibrium). This is accomplished by the continuous rearrangement of the system structure in response to each disturbing influence (deviation from the desired state), which initiates appropriate counterreactions.

For computer-guided production systems this means that a shift of tasks has to be controlled in an operation-connected way. Control systems which are efficient in this respect aim at informational restructuring with the concept of "priority-controlled shifting intelligence".

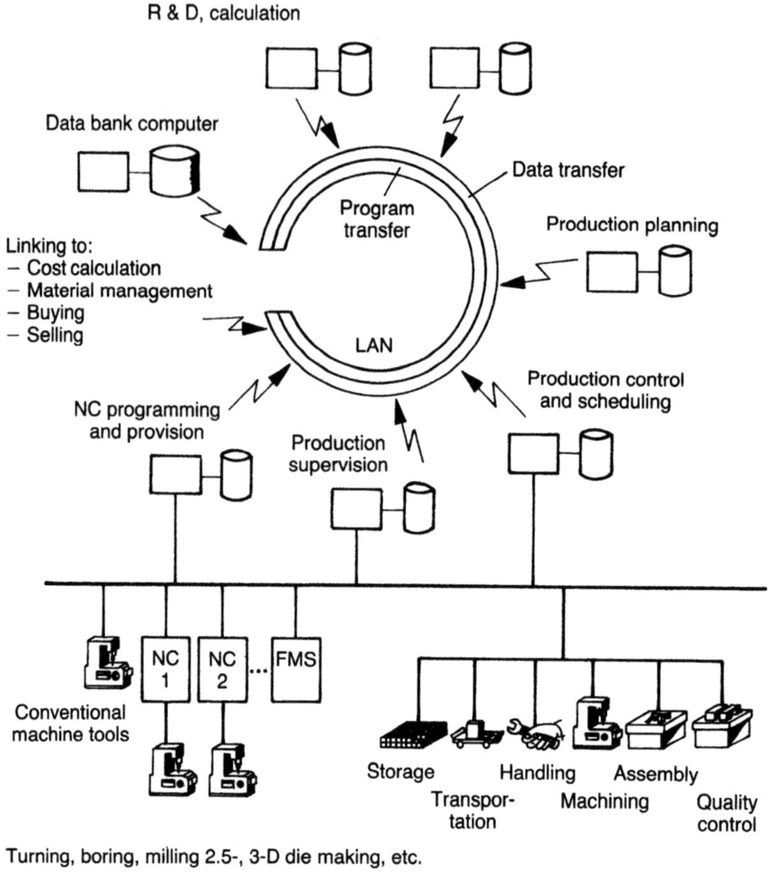

Fig. 139. Example of an integrated production system

10.3 First Attempts at Computer-Integrated Manufacturing

As stated in the preceding section, there is no enterprise in the world at present in which perfect information-technical control of all function complexes shown in Fig. 138 has been achieved. In the United States these aims are being pursued with the ICAM project (integrated computer-aided manufacturing). One of the first attempts in the Federal Republic of Germany uses the CIAM system (computerized integrated and automated manufacturing) for the requirements of the aircraft industry [53].

The aim of the concept presented in Fig. 140 is the logical consequence of individual processes of automation in development, planning, and manufacture, which are based on the versatile use of computer engineering for the control, coordination and monitoring of manufacture [53].

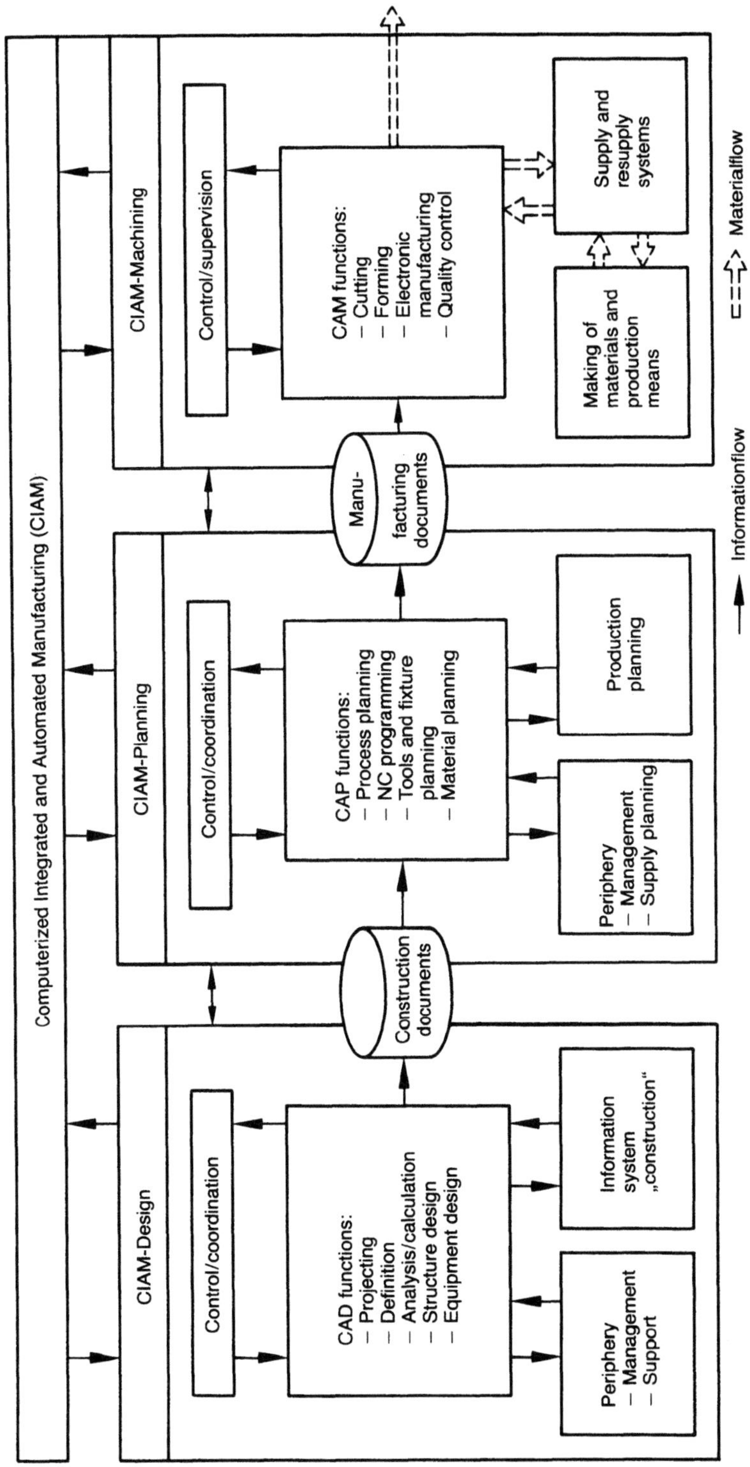

Fig. 140. Components of an integrated and automated production system

The subsystem CIAM-Design comprises several functions at its core which are executed at present with computer assistance. In aircraft construction, for instance, extensive CAD systems, such as CADAM for producing drawings, GEOLAN for surface design, and NASTRAN for finite element computation, are used at the different stages of development, from project drawing to detailed design. All of these programs remained single solutions as long as no integrating geometry model for the manipulation of geometric data was available.

The subsystem CIAM-Planning is influenced mainly by the conceptions of design and manufacture. EDP is generally used for the administration of large data stocks at present, such as work-plan data or parts lists, as well as for NC programming. The computerization of planning functions to a higher degree leads to higher stages of integration with the advantages of (a) faster execution of planning, (b) higher quality of planning, and (c) improved product quality.

The highest level of automation so far has been realized with the third system component — CIAM-Machining (cf. Sect. 3.4.2.3). This system includes automated machining, the making available of material and tools and all supply and removal operations connected with manufacture, control of manufacture, continuous stock recording, and recording of operating data.

Appendix 1

Activities of the IFIP WG 5.3

1. PROLAMAT Conference 1969 — Rome, Italy
   Programming Languages for Machine Tools
2. PROLAMAT Conference 1973 — Budapest, Hungary
3. PROLAMAT Conference 1976 — Stirling, Scotland
   Change of titel to
   Programming Research and Operations Logistic in Advanced Manufacturing Technology
4. PROLAMAT Conference 1979 — Ann Arbor, Michigan, USA
5. PROLAMAT Conference 1982 — Leningrad, USSR
6. PROLAMAT Conference 1985 — Paris, France
   Working Conferences
   1. Man-Machine Communication in CAD/CAM — 1980 Tokyo, Japan
      IFIP WG 5.2; 5.3
   2. Integration of CAD/CAM — 1983 — Gaussig,
      GDR IFIP WG 5.3; 5.2
   3. Industrial robotics in discrete manufacturing — 1984 Como, Italy
      Planned events:
7. PROLAMAT Conference 1988 — Dresden, GDR
8. PROLAMAT Conference 1991 — Tokyo, Japan

Appendix 2

Engineering Research Association

This national research project, called "Flexible Manufacturing System Complex Provided with Laser", was launched by the Agency of Industrial Science and Technology, MITI in December, 1977. The Engineering Research Association (ERA) was set up in March 1978 for the commissioning of this project to the private sector. At present, 20 companies are participating in the research association, sharing in the 7-year R & D funding to develop areas of high-technology potential.

The total budget for this project is estimated at 13 billion yen, consisting of allocations by three government organizations including MEL, ETL and GIRI Kyushu.

The fundamental purpose of this project is to take the lead in the development of an automated manufacturing system capable of performing all manufacturing processes, from metal forming to flexible and rapid product inspection, for the production of machinery components in small batches.

The R & D has been divided into four subgroups: (a) machining and assembly, (b) metal forming (c) laser application, and (d) diagnosis and management, with each subgroup consisting of four to six companies. With the aid of these subgroups, a total system is being developed. Coordination of the research in each subgroup and responsibility for the total system is determined by the ERA committees.

The development of the elemental technologies of each subgroup was nearly completed in 1979/80, and the design and construction of an FMC test plant has been in progress since 1981.

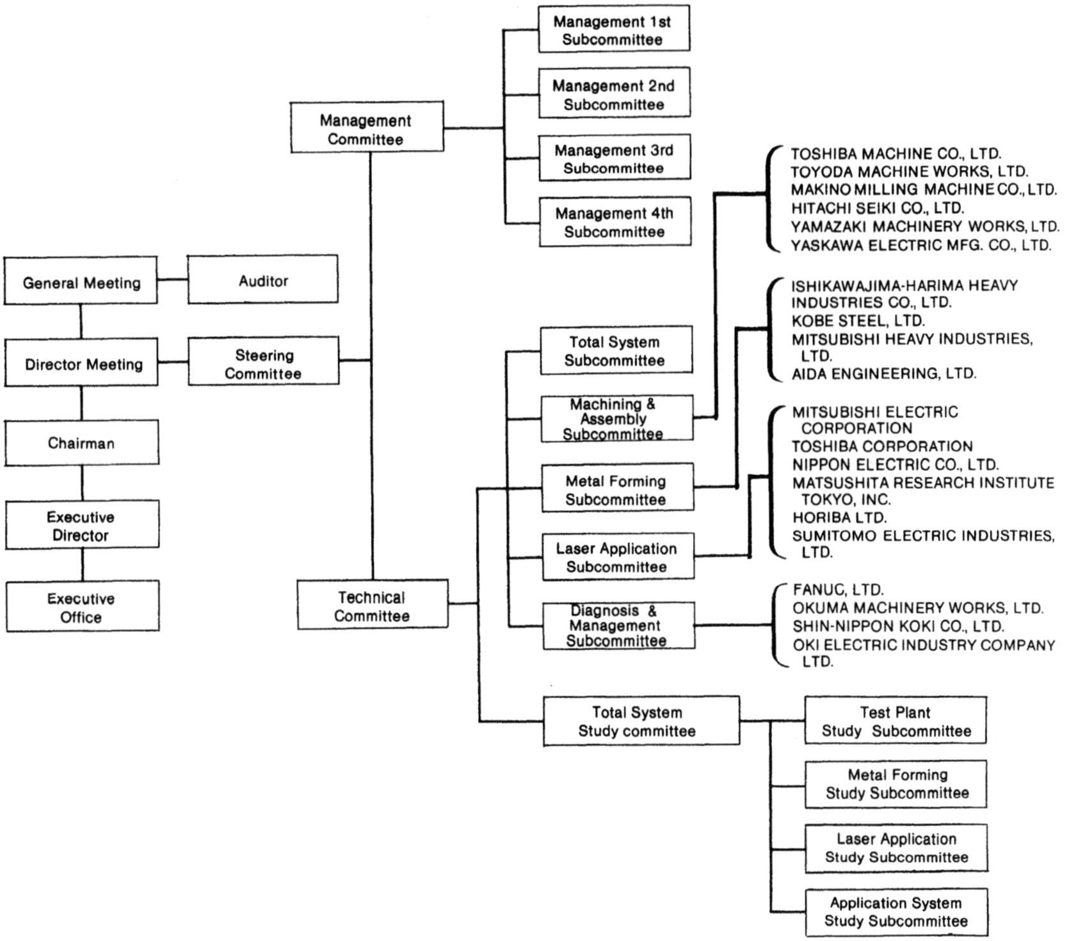
Management 1st Subcommittee
Management 2nd Subcommittee
Management Committee
Management 3rd Subcommittee
Management 4th Subcommittee
TOSHIBA MACHINE CO., LTD.
TOYODA MACHINE WORKS, LTD.
MAKINO MILLING MACHINE CO., LTD.
HITACHI SEIKI CO., LTD.
YAMAZAKI MACHINERY WORKS, LTD.
YASKAWA ELECTRIC MFG. CO., LTD.
General Meeting
Auditor
ISHIKAWAJIMA-HARIMA HEAVY INDUSTRIES CO., LTD.
KOBE STEEL, LTD.
MITSUBISHI HEAVY INDUSTRIES, LTD.
AIDA ENGINEERING, LTD.
Director Meeting
Steering Committee
Total System Subcommittee
Machining & Assembly Subcommittee
Chairman
MITSUBISHI ELECTRIC CORPORATION
TOSHIBA CORPORATION
NIPPON ELECTRIC CO., LTD.
MATSUSHITA RESEARCH INSTITUTE TOKYO, INC.
HORIBA LTD.
SUMITOMO ELECTRIC INDUSTRIES, LTD.
Metal Forming Subcommittee
Executive Director
Laser Application Subcommittee
Executive Office
Technical Committee
Diagnosis & Management Subcommittee
FANUC, LTD.
OKUMA MACHINERY WORKS, LTD.
SHIN-NIPPON KOKI CO., LTD.
OKI ELECTRIC INDUSTRY COMPANY LTD.
Total System Study committee
Test Plant Study Subcommittee
Metal Forming Study Subcommittee
Laser Application Study Subcommittee
Application System Study Subcommittee

Research and Development Program

- R & D Diagram for Flexible Manufacturing System Complex Provided with Laser
- Long-Term R & D Program for Flexible Manufacturing System Complex Provided with Laser
- Targets of R & D

Items	Targets
Products to be manufactured	Machine components with maximum of 300 parts, maximum dimensions of 1000 mm and maximum weight of 500 kg
Time required for production	Less than 50% that of current production systems
Number of stages required for production	Less than 60% that of current production systems
High-power laser oscillator	CO_2 gas laser oscillator with output of 20 kW, electrode life of over 2000 h and excitation energy efficiency of more than 15%
Medium-power laser oscillators	Rare gas laser oscillator with output of 200 W, solid-state laser oscillator with output of 300 W

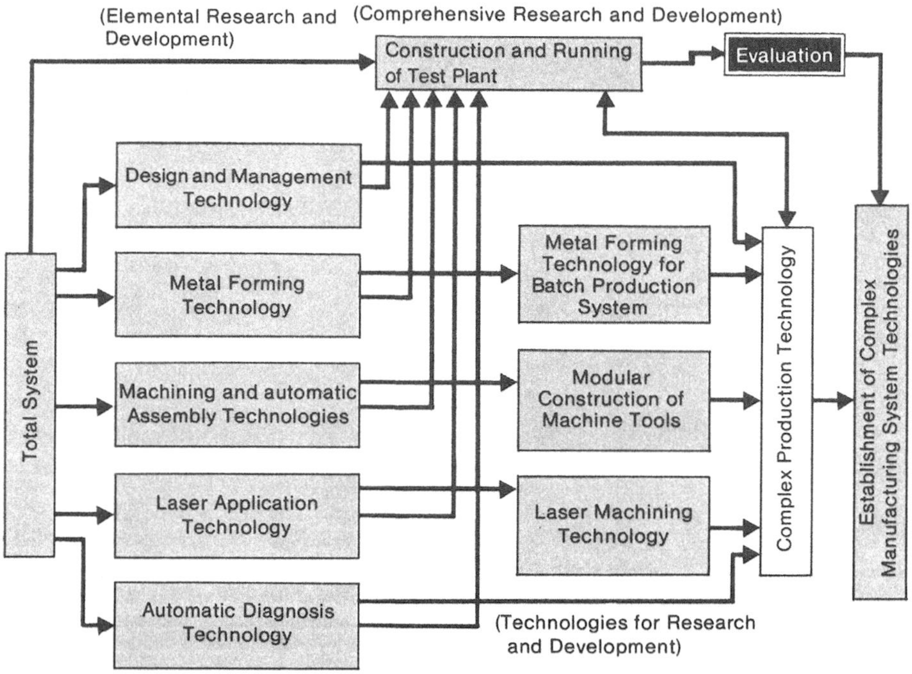

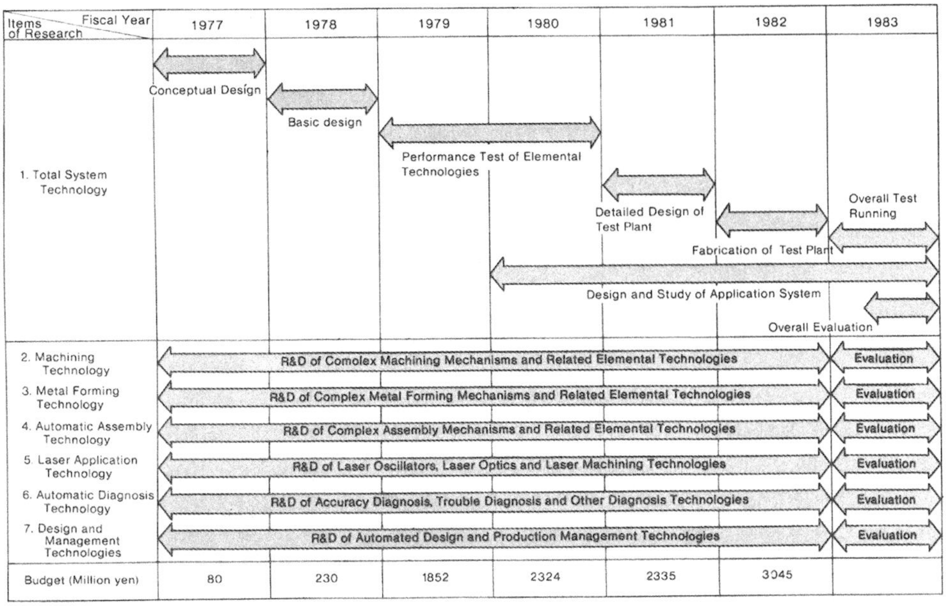
Items of Research
Fiscal Year
1977
1978
1979
1980
1981
1982
1983
1. Total System Technology
Conceptual Design
Basic design
Performance Test of Elemental Technologies
Detailed Design of Test Plant
Fabrication of Test Plant
Overall Test Running
Design and Study of Application System
Overall Evaluation
2. Machining Technology
R&D of Comolex Machining Mechanisms and Related Elemental Technologies
Evaluation
3. Metal Forming Technology
R&D of Complex Metal Forming Mechanisms and Related Elemental Technologies
Evaluation
4. Automatic Assembly Technology
R&D of Complex Assembly Mechanisms and Related Elemental Technologies
Evaluation
5. Laser Application Technology
R&D of Laser Oscillators, Laser Optics and Laser Machining Technologies
Evaluation
6. Automatic Diagnosis Technology
R&D of Accuracy Diagnosis, Trouble Diagnosis and Other Diagnosis Technologies
Evaluation
7. Design and Management Technologies
R&D of Automated Design and Production Management Technologies
Evaluation
Budget (Million yen)
80
230
1852
2324
2335
3045

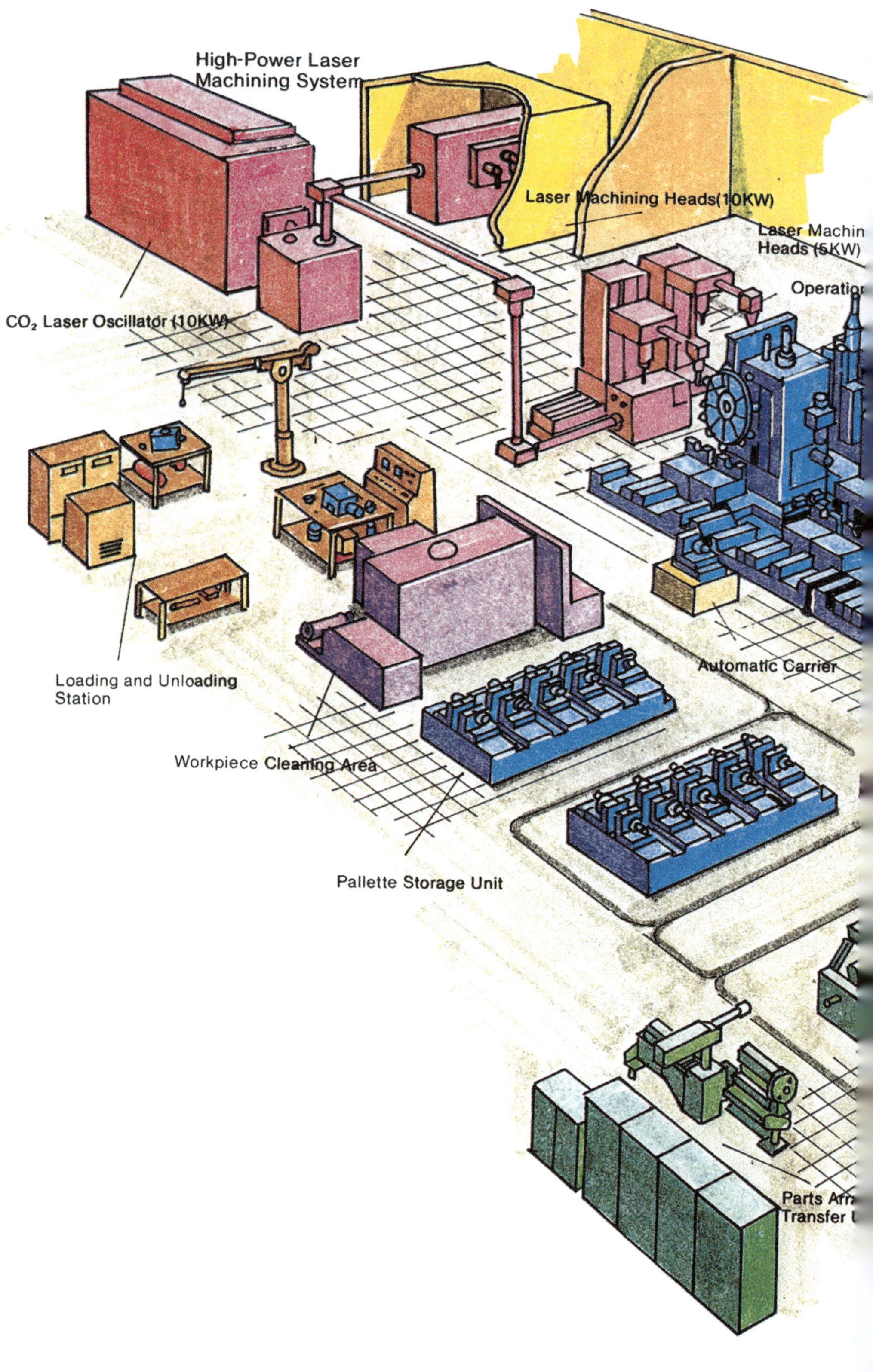
High-Power Laser Machining System
Laser Machining Heads(10KW)
Laser Machin
Heads (5KW)
Operatio
CO$_2$ Laser Oscillator (10KW)
Automatic Carrier
Loading and Unloading Station
Workpiece Cleaning Area
Pallette Storage Unit
Parts Arra
Transfer U

Tsukuba Test Plant

A test plant integrating the necessary elemental technologies required by the FMC with systematization technology is currently under construction at MEL in Tsukuba Science City. The plant will be operated with the aim of both evaluating the system's productivity, flexibility, reliability, and safety, etc., and collecting design data for practical systems.

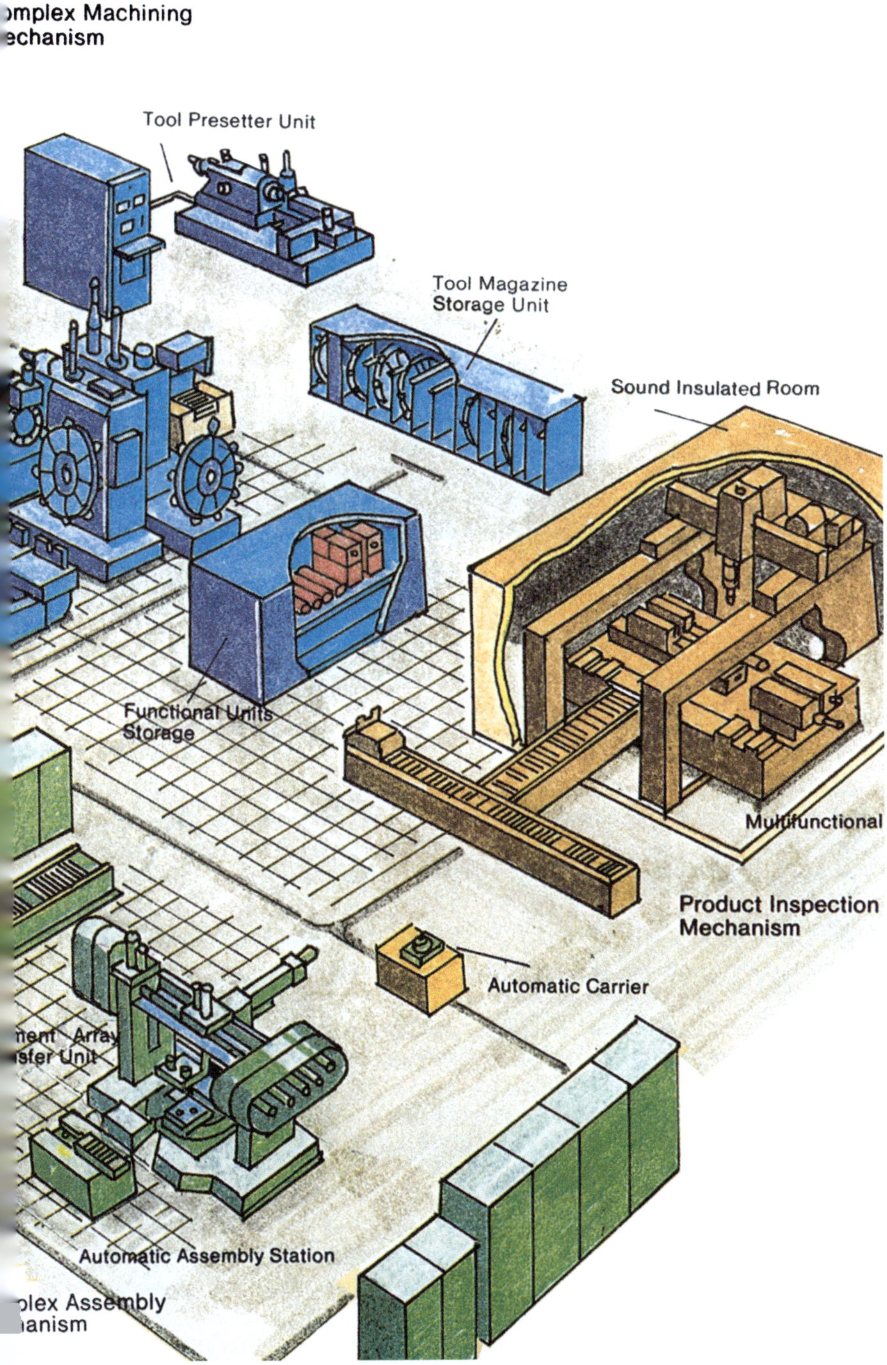

The Tsukuba test plant will be composed of mechanical units capable of complex machining, complex assembly, product inspection, and laser processing, along with associated equipment and control systems. The main elements are as follows:

- Complex Machining Mechanism
  - Operational Cells
  - Automatic Tool Changers
  - Measurement Cell
  - Operational Units
  - Associated Equipment
- Complex Assembly Mechanism
  - Automatic Assembly Mechanism (insertion, press fitting, thread fastening)
  - Parts Array and Transfer Unit
  - Part Procurement Array and Transfer Unit
  - Parts Cassette Storage Unit
- Product Inspection Mechanism
  - Multifunctional Inspection Machine (dimension, shape, shaft vibration, lost motion, torque, temperature, vibration, etc.)
  - Sound-Insulated Room
  - Automatic Conveyer Unit

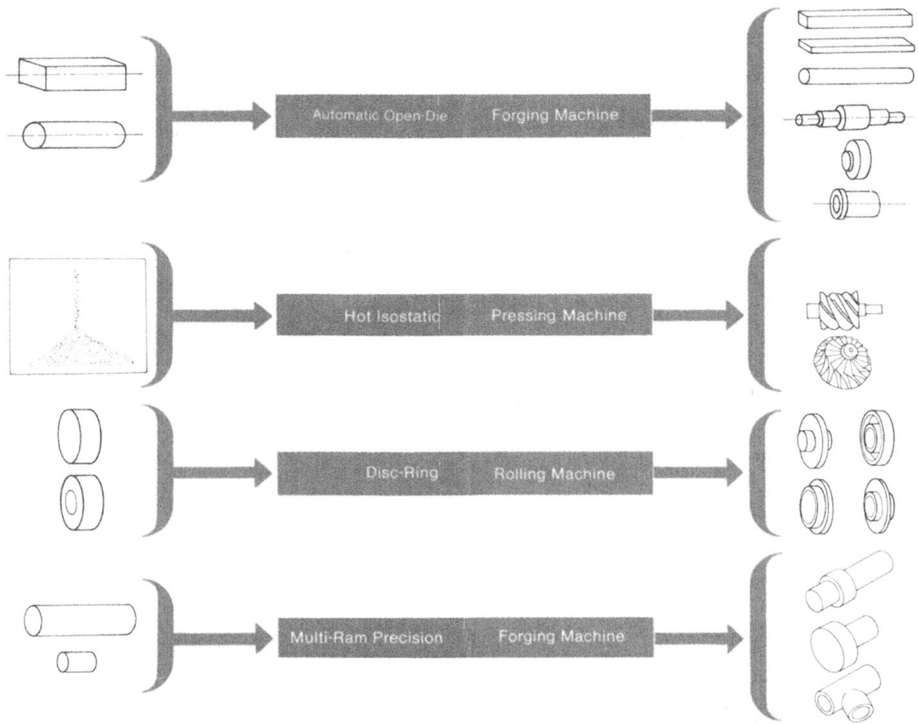

— Laser Machining Mechanism
  1. High-Power Laser Machining System
  — 10 kW CO_2 Laser Oscillator
  — 5 kW Machining Heads (for on-line welding and surface heat treatment)
  — 10 kW Machining Heads (for off-line welding and cutting)
  2. Medium-Power Laser Machining System
  — 300 kW YAG Laser Oscillator
  — Laser Machining Heads (for chip breaking and deburring)
— Control System
  — Hierarchic Control System
— Peripheral Equipment
  — Automatic Carrier
  — Workpiece Cleaning Area
  — Loading and Unloading Station
  — Tool Magazine Storage Unit

Metal Forming Technology

A machine is an assemblage of various parts including machined parts (casings, gears, shafts, etc.), pressed parts (covers, bolts, etc.), rubber products (oil seals, O-rings, etc), and others, which are produced through a variety of processes. The study of metal forming is aimed at discovering techniques for fabricating blanks for machining both freely and with a degree of precision that will reduce both machining time and metal removal rates. Four forming machines, described in detail below, are being developed. Each of them will be capable of flexible forming of materials into certain specific shapes using a small number of generalized forming tools.

Machining Technology

The following elemental mechanisms and techniques are being developed in small-batch production to improve machining efficiency, reduce requirements for skilled workers and inventory parts, and develop the capacity to perform multiple machining operations.

- A complex machining mechanism capable of performing various kinds of processing (cutting, grinding, gauging, etc.) setting up flexible arrangements of interchangeable machining units.
- A compact spindle mechanism with advanced spindle functions that decrease the total number of spindle units required, and a feed mechanism capable of workpiece installation and separation functions.

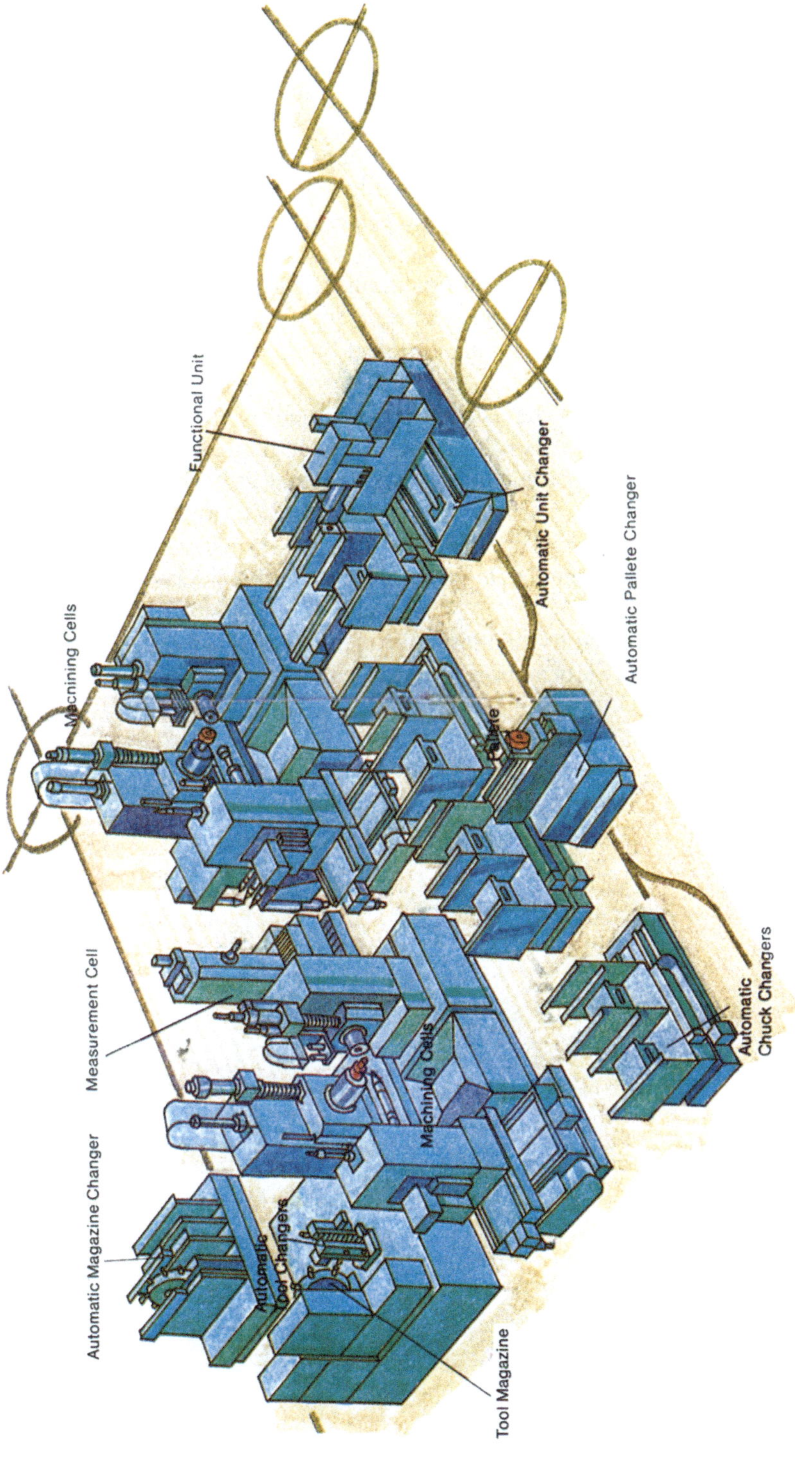
Functional Unit
Machining Cells
Automatic Unit Changer
Automatic Pallete Changer
Pallete
Measurement Cell
Machining Cells
Automatic Chuck Changers
Automatic Magazine Changer
Automatic Tool Changers
Tool Magazine

System Control Technology

The FMC will be capable of constructing a variety of machining mechanisms from modularized units to meet different manufacturing needs, so the control system for these mechanisms must be a highly flexible system satisfying the following conditions:

— Highly independent functions should be modularized.
— Modularized functions should be individually usable.
— The total control system should take a hierarchical form made up of abovementioned modules.
— The method of on-line and real-time control should be employed.

The control system under development has a four-rank hierarchy. NC machining data, scheduling commands, etc. are supplied by an off-line information processing system.

The complex machining mechanism is then operated by a controller for a complex machining system that oversees the operation of individual machining space controllers, each of which in turn supervises the several unit controllers operating in each machining space. Each unit controller is composed of a one-axis NC together with a programmable controller. Based on instructions received from the one-axis NC, a servo-controller then directs the operation of an AC servo-motor to perform the desired machining.

Electric power is supplied to the unit controllers by a high-frequency power transformer, while control information is sent by a wireless optical transmitter.

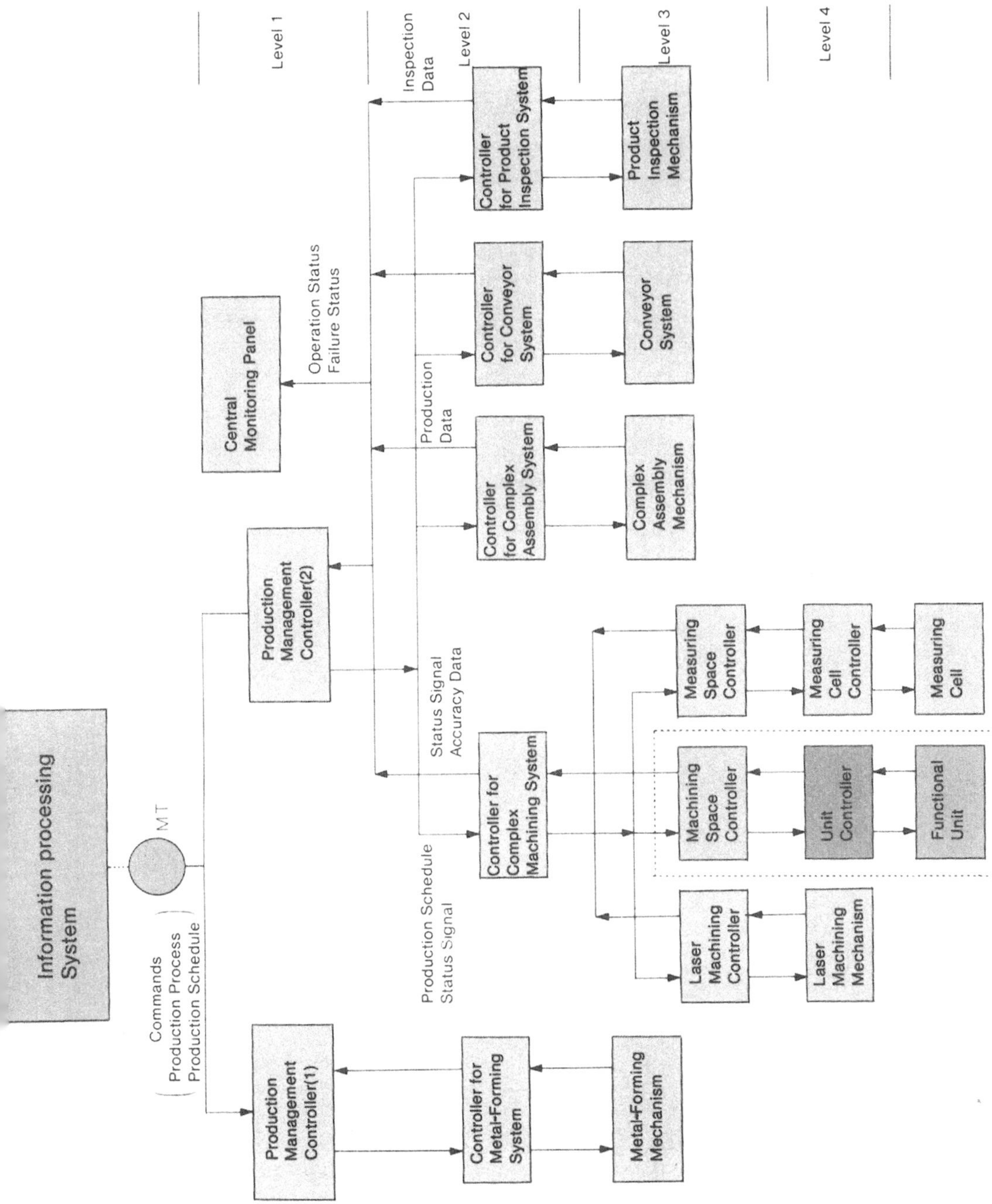
Level 1
Level 2
Level 3
Level 4
Information processing System
MT
Commands
Production Process
Production Schedule
Production Management Controller(1)
Production Management Controller(2)
Central Monitoring Panel
Operation Status
Failure Status
Inspection Data
Production Data
Status Signal
Accuracy Data
Production Schedule
Status Signal
Controller for Metal-Forming System
Metal-Forming Mechanism
Controller for Complex Machining System
Controller for Complex Assembly System
Complex Assembly Mechanism
Controller for Conveyor System
Conveyor System
Controller for Product Inspection System
Product Inspection Mechanism
Laser Machining Controller
Laser Machining Mechanism
Machining Space Controller
Unit Controller
Functional Unit
Measuring Space Controller
Measuring Cell Controller
Measuring Cell

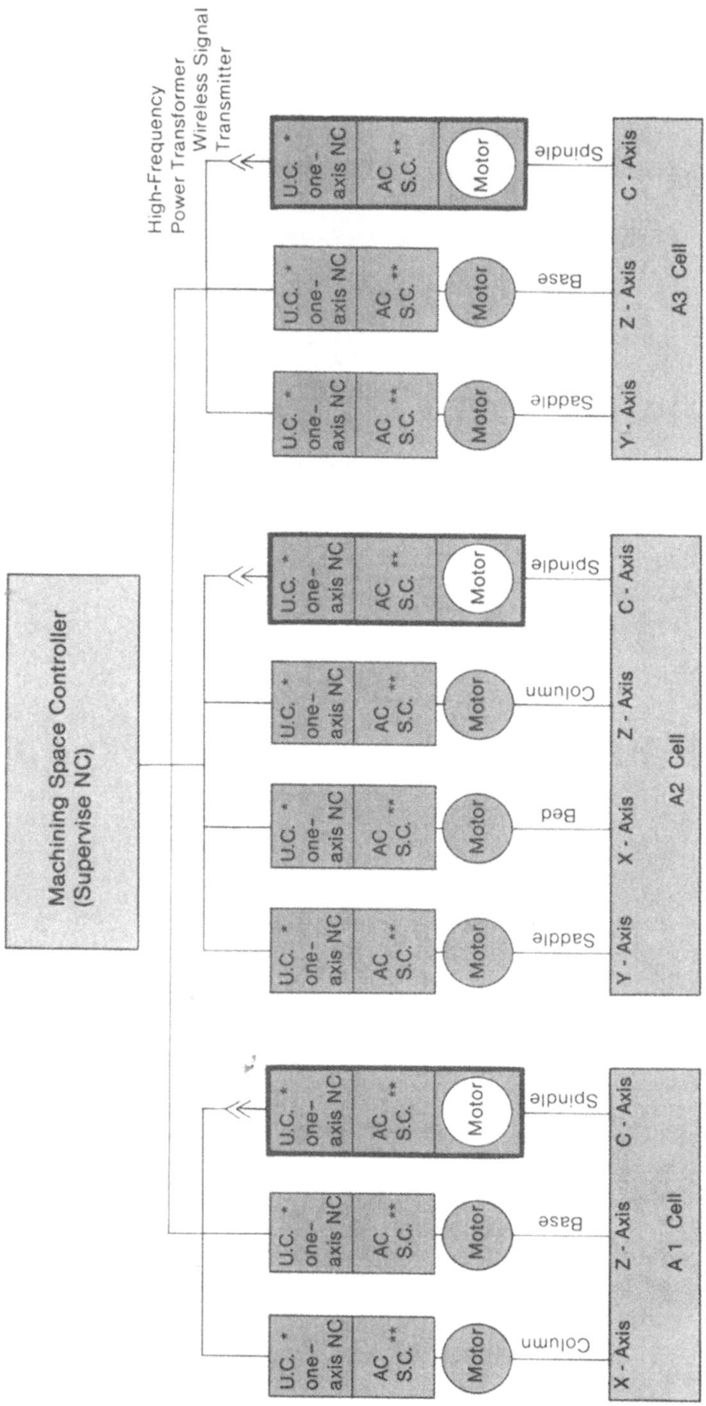
Machining Space Controller
(Supervise NC)
High-Frequency
Power Transformer
Wireless Signal
Transmitter
U.C. *
one-
axis NC
AC **
S.C.
Motor
Column
Base
Spindle
Saddle
Bed
X - Axis
Y - Axis
Z - Axis
C - Axis
A 1 Cell
A2 Cell
A3 Cell

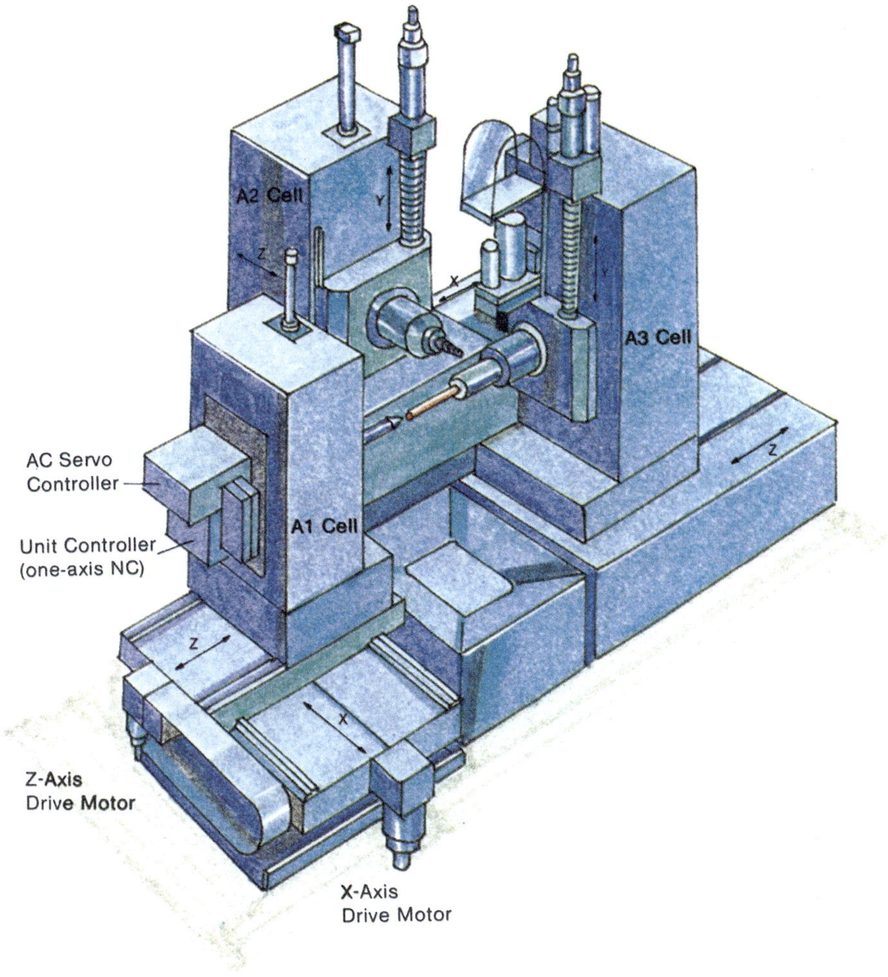
A2 Cell
A3 Cell
A1 Cell
AC Servo Controller
Unit Controller (one-axis NC)
Z-Axis Drive Motor
X-Axis Drive Motor
X
Y
Z

References

1. Abeln O, Bauhuber F, Eitel H, Klug HG (1977) Verknüpfungen von Programmsystemen zur integrierten Informationsverarbeitung im Betrieb. wt-Z Ind Fertig 67 (3):139-143
2. ACM SIGGRAPH Graphics Standards Planning Committee Status Report (1979) Comp Graphics 13 (3)
3. ADA (1980) Reference manual of the programming language ADA. Department of Defense, USA
4. Alban G (1984) Praxiserfahrungen eines CAD/CAM-Anwenders. Industrie-Anzeiger Essen, No. 22, March 16, pp 42-46
5. Albano A (1977) A method to improve two-dimensional layout. Computer-Aided Design 9 (1):48-52 IPC London
6. Ambacher G, Hammer H (1983) Rüstzeitfreier Auftragswechsel und selbstüberwachter Automatikbetrieb beim Bohren und Fräsen. Informationsschrift Werner und Kolb, W. Berlin
7. American National Standards Institute (1981) Engineering drawing and related documentation practices — digital representation for communication of product definition data (IGES). ANS
8. Anderl R, Tröndle M (1983) Modellaustausch, Notwendigkeit für die Integration von CAD/CAM-Anwendungen. VDI Z 125 (4):91-95
9. APS progress report 2. Kernforschungszentrum, Karlsruhe, February 1983
10. APT part programming manual. IIT Research Institute, Chicago, 1964
11. Arai YY, Takeda KK (1981) NC-Daten-Teil- und Aufbereitungseinrichtung für NC-Werkzeugmaschinensysteme. Hitachi, Ltd. Offenlegungsschrift 3146342 A1, November 23
12. Backus J (1978) The history of FORTRAN I, II, and III. ACM SIGPLAN Notices, 13 (8):165-180
13. Baer A, Eastman C, Henrion M (1979) Geometric modeling: a survey. CAD 11 (5):253-272
14. Bathor M, Siegler A (1983) Graphic simulation for robotic programming. Computer and Automation Institute, Hungarian Academy of Sciences, Budapest
15. Beeby W (1982) The future of integrated CAD/CAM systems: the Boeing perspective. IEEE CG&A, January, pp 51-56
16. Bernard F (1981) CATIA: système tridemensionnel de conception et de fabrication assistées par ordinateur. In: 3è journées scientifiques et techniques de la production automatisée. ADEPA, Montrouge
17. Berner A (1977) Geometrieverarbeitung und Beschreibung geometrischer Elemente in NCMES. Ind Anz 99 (33):593-594
18. Bezier PE (1983) UNISURF, from styling to tool-shop. In: Warman EA (ed) Computer applications in production and engineering. North-Holland, Amsterdam, pp 593-594
19. Bobrow DG, Raphael B (1974) New programming languages for artificial intelligence. Computing Surveys 6 (3):153-174
20. Böhm W, Kahman J (1983) Grundlagen von Kurven und flächenorientierten Modellierung. In: Nowacki H, Gnatz R (eds) Geometrische Modellierung. Informatik Fachberichte, vol. 65. Springer-Verlag, Berlin Heidelberg New York Tokyo, pp 173-218
21. Boeing experiments with integrated design and manufacturing. American Machinist Special Report 746, June 1982, pp 170-171
22. Böttger H (1984) Expertensysteme — Zielstellung, Prinzipien, Tendenzen. Conference INFO 84, Dresden University of Technology

23. Bourret P, Demolombe R (1983) Estimation et mesure de performances des SGBD relationnels en vue de la réalisation d'un prototype d'un tel système. ONERY-CERT, Toulouse, rapport No. 2/3177/DERT
24. Boyse JW, Gilchrist JE (1982) GM Solid: interactive modeling for design and analysis of solids. IEEE Comp. Graphics and Application 2 (2):27-40
25. Brauner K, et al. (1981) Initial Graphics Exchange Specification. American National Standards Institute Y14 26M, New York
26. Brown CM (1982) PADL-2: a technical summary. In: IEEE Computer Graphics and Applications, Special Issue: Solid Modeling March, pp 69-84
27. Budde W, Adamczyk P (1980) Anwendungstechniken des neuen EXAPT-Modulsystems. ZwF 75 (6):280-284
28. Buzacott JA, Shanthikumar JG (1980) Models for understanding flexible manufacturing systems. AIIE Trans 12 (4):339-350
29. CAPEX (1982) Firmenschrift der EXAPT-NC-Systemtechnik GmbH, Aachen
30. Castle DH (1982) Allis-Chalmers Corporation's experience with FMS. Flexible Manufacturing Systems seminar, SME, April
31. CERT (1974) Proceedings of workshop on general-purpose computer-aided design systems. CERT, Toulouse
32. Chan BTF (1982) ROMAPT — a new link between CAD and CAM. CAD 14 (5):261-266
33. Church J (1982) Simulation aspects or flexible manufacturing systems. Proc. Fall Ind. Eng. Conf., Cincinnati, November
34. Church JJ (1982) Simulation is a necessary tool for design and evaluation of flexible manufacturing systems. Illinois Institute of Technology Research Institute
35. Clementson AT (1978) ECSL CAPS, detailed reference manual. University of Birmingham, England
36. Clochsin W, Mellish C (1981) Programming in PROLOG. Springer-Verlag, Berlin Heidelberg New York
37. Colding B, Colwell LV, Smith DN (1977) Delphi forecast of manufacturing technology — manufacturing systems, material removal. Society of Manufacturing Engineers, Dearborn, MI
38. Colmerauer A, Kanoui H, Pasero R, Roussel P (1973) Un système de communication homme-machine en français. Rapport de recherche Groupe d'Intelligence Artificielle, Université d'Aix-Marseille II, June
39. Borrell J (1982) The solid modeling marketplace. Computer Graphics World, November, pp 50-51
40. Crestin JP, Paillard JP (1974) A small graphical system for programming NC machines. In: Hatvany J (ed) Computer languages for numerical control. North-Holland, Amsterdam
41. Crestin JP, Queinnec C (1980) Towards semantics for graphical interactions. In: Guedj RA, et al. (eds) Methodology of interaction. North-Holland, Amsterdam pp 171-181
42. Crestin JP (1983) Syntax and semantics of programming languages versus CAD/CAM requirements: the role of artificial intelligence languages. Proc. of 1st ISAI conference, Leningrad, IFAC
43. Dallery Y, Descotes-Genon B, Bonette R (1982) Simulation of a flexible manufacturing system: Case study of the Citroen factory, Meudon. Proc. of APMS meeting, Bordeaux, August
44. Der Einsatz flexibler Fertigungssysteme. KfK-PFT 41, Kernforschungszentrum, Karlsruhe, August 1982
45. Descottes Y, Latombe JC (1981) GARI: a problem solver that plans how to machine mechanical parts. IJCAI 7:766-772
46. DIN 66027 (1975) Informationsverarbeitung. Programmiersprache FORTRAN. Beuth, Berlin
47. DIN 66215 (1974) Programmierung numerisch gesteuerter Arbeitsmaschinen. CLDATA. Beuth, Berlin
48. DIN 66246 (1983) Programmierung numerisch gesteuerter Arbeitsmaschinen. Prozessor-Eingabesprache. Beuth, Berlin
49. DIN 66252 (1983) Informationsverarbeitung. Graphisches Kernsystem (GKS). Entwurf. Beuth, Berlin
50. Dincbas M (1980) A knowledge-based expert system for automatic analysis and synthesis in CAD. Proc. IFIP Congress 80, IFIP, pp 705-710
51. Dooner M, Taylor NK, Bonney MC (1982) Manning robot installations by CAD. Proc. of CAD 82

52. Draft International Standard ISO/DIS 7642 (1982) Information processing. Graphical Kernel System (GKS), functional description. Version 7.2, November 14
53. Dronsek J (1979) Technische und wirtschaftliche Probleme der Fertigung im Flugzeugbau. Proc. Produktionstechnisches Kolloquium Berlin, Hanser, Munich, pp 107-115
54. Dupont-Gatelmand C (1982) A survey of flexible manufacturing systems. J Manufact Systems 1:1-16
55. Eastman CM (1980) The representation of design problems and maintenance of their structure. In: Guedj RA et al. (eds) Methodology of interaction. North-Holland, Amsterdam
56. Eigner M (1982) Einführung und Anwendung von CAD-Systemen. Hanser, Munich
57. El Maraghy HA (1982) Simulation and graphical animation of advanced manufacturing systems. J Manufact Systems 1:8-11
58. El Maraghy HA (1983) FMSSIM. Presentation at SME Seminar, Detroit
59. Encarnacao J, Hornung C, Kuhlman H (1983) Geometrische Modellierungssysteme, Stand und Trends. In: Nowacki H, Gnatz R (eds) Geometrische Modellierung. Informatik-Fachberichte, vol. 65. Springer-Verlag, Berlin Heidelberg New York Tokyo, pp 3-24
60. EUCLID. Matra Datavision, 1983
61. Eversheim W, Fuchs H, Zons HH (1980) Anwendung des Systems AUTAP zur Arbeitsplanerstellung. Ind Anz 102 (55):29-33
62. Eversheim W, Loersch U, Esch H (1981) Arbeitsplanerstellung im Dialog mit dem System DISAP. Ind Anz 97:29-32
63. Ewen W (1981) Einführung von CAD in einen mittleren Maschinenbaubetrieb. BW 4744. VDI-Bildungswerk, Düsseldorf
64. EXAPT I (1968) Geometrical processor. Verein zur Förderung des DXAPT Programmiersystems e.v., Aachen
65. Fair DG, Beaverstad H, Erikson CF (1969) Einrichtung zum direkten Steuern von Werkzeugmaschinen durch einen Computer. Offenlegungsschrift 1908757, September 18
66. Falkoff AD (1978) The evolution of APL. ACM SIGPLAN Notices, 13 (8):47-58
67. Farney B (1983) Geometrisches Modellieren technischer Objekte durch Rekonstruktion der dreidimensionalen Werkstückgeometrie aus zweidimensionalen Zeichnungsrissen. In: Nowacki H, Gnatz R (eds) Geometrische Modellierung. Informatik-Fachberichte, vol. 65. Springer-Verlag, Berlin Heidelberg New York Tokyo, pp 25-35
68. Farny B, Müller E (1980) Grafische Datenverarbeitung im Konstruktionsprozeß. VDI-Z 122 (1/2):21-26
69. Fitzgibbon T (1983) FLEXPLAN — a decision support system for FMS's. Charles Stark Draper Laboratory
70. Foisseau J, Jacquart R, Vignat JC (1981) PRAIRIAL. Manuel de référence de LDSS. ONERA-CERT, Toulouse, Rapport no. 5/3121/DERI
71. Foley JD van dam (1982) Fundamentals of interactive computer graphics. Addison-Wesley, Reading
72. Fotland HC, Wärvägen T, Lillehagen F, Losnedal J (1981) ICAN ICUE-Reference manual. April
73. Frachet JP (1981) Equipements de production automatisée: nécessité d'outils méthodologiques. In: 3è Journées scientifiques et techniques de la production automatisée. ADEPA, Montrouge
74. Freist c, Granow R (1983) Erweiterung eines DNC-Systems zum Werkstattkommunikationssystem. Industrie-Anzeiger Essen No. 7, HFG-Bericht 624
75. Freist C, Granow R, Inclen W (1984) Leitstand für ein flexibles Fertigungssystem. ZWF 79 (5):219-222
76. Fuchs H (1981) Automatische Arbeitsplanerstellung — Ein Baustein im Rahmen der integrierten Fertigungsunterlagenerstellung. Dissertation, RWTH Aachen
77. Furlani CM (1979) Summary of the simulation effort for the AMRF. Automated Manufacturing Research Facility at the National Bureau of Standards Washington, D.C.
78. Gausemeier BJ (1977) Eine Methode zur rechnerorientierten Darstellung technischer Objekte im Maschinenbau. Dissertation, Technische Universität, Berlin
79. Geometric product models. (1980) An internordic joint project. KTH, Stockholm
80. Gero JS (1984) Object modeling through knowledge engineering. CIRP seminar proc., Tokyo
81. Gettelman KM, Albert MD (1983) Fundamentals of NC/CAM modern machine shop. NC/CAM guidebook. Cincinnati, 54244

82. GKS (1982) Graphical kernel system. Functional description. Draft international standard. ISO/DIS, ISO TC 97/SC5/WG 2N
83. Glushkov OI (1983) A linguistic approach to problems in connection with descriptions in geometry and programming. In: Automated process planning. ITK AN BSSR, Minsk, vol 1, pp 65-76 (in Russian)
84. Glushkov OI, Konoplyanikov HA (1984) Modeling processes of workpiece synthesis using the SPLINK system. In: Automated process planning. ITK AN BSSR, Minsk, vol 2, pp 80-86 (in Russian)
85. Goebl R, Pacha F (eds) (1982) CAD/CAM — Rechnergestütztes Konstruieren und Fertigen. Oldenburg, Vienna
86. Goldman R (1977) Recent work with the AL system. Proc. 5th IJCAJ, Boston
87. Goldstein RA, Malin L (1979) 3-D Modeling with the Synthvision system. Proc. 1st Annual Conference on Computer Graphics in CAD/CAM Systems. Cambridge, MA, April 9-11, pp 244-247
88. Goldstein Ra, Nagel R (1971) 3-D Visual simulation. Simulation 16 (1):25-31
89. Gorelik AG (1980) Automated engineering-graphical work with the help of IBM. Minsk University, p 206 (in Russian)
90. Grabowski H, Anderl R, Seiler W (1983) DICAD — ein CAD-System für 3D-Modellierung technischer und technologischer Information. In: Nowacki H., Gnatz R (eds) Geometrische Modellierung Informatik-Fachbericht, vol. 65. Springer-Verlag, Berlin Heidelberg New York Tokyo, pp 219-236
91. Granow R, Hesselmann U, Weller H (1983) Möglichkeiten zur Erstellung und Verteilung von NC-Daten. RWT Co. GmbH, Krailling
92. Hammer H (1983) Verbesserung der Wirtschaftlichkeit durch flexible Automatisierung beim Bohren und Fräsen. ZWF 78 (2):77-86
93. Hatvany J (1983) Dreams, nightmares and reality. In: Warman EA (ed) Computer applications in production and engineering. North-Holland, Amsterdam, pp 3-10
94. Hatvany J (ed) (1983) World survey of CAM. Butterworths, Sevenoaks, Kent
95. Hernandez JT (1983) Moniteur de dialogue; thèse de 3è cycle. ENSTA, Paris VI
96. Hewitt C, Smith B (1975) A PLASMA primer. MIT, Boston
97. Hillgard RC, Braid IC (1978) The analysis of dimensions and tolerances in computer-aided mechanical design. Computer Graphics 10 (3):161-166
98. Hillgard RC, Braid IC (1978) Characterizing on ideal shapes in terms of dimensions and tolerances. Computer Graphics 12 (3):234-238
99. Hübner H (1979) Integration und Informationstechnologie. Minerva, Munich
100. Hughes JJ (1979) Application of flexible manufacturing systems. Presentation at 17th IEEE Conference on Decision and Control, San Diego, January
101. Hutchinson GK (1978) Simulation: an aid to RMS design. Presentation at IKM 78, March
102. Hutchinson GK (1979) Activity cycles: a basis for manufacturing systems modeling and control. Proc. of ASME annual meeting, New York, December
103. Hutchinson GK (1979) Flexible manufacturing systems in the United States. University of Wisconsin, Milwaukee
104. Hutchinson GK (1981) System components. University of Wisconsin, Milwaukee
105. Hutchinson GK (1982) Flexible manufacturing systems and simulation. Lecture no. 25, IKB 82, Leipzig, March 10 — April 12
106. Hutchinson GK (1982) Manufacturing cell model. University of Wisconsin, Milwaukee
107. Hutchinson GK (1982) A generalized model for robotic work cell design. University of Wisconsin, Milwaukee
108. Hutchinson GK, Hughes JJ (1977) A generalized model of flexible manufacturing systems. Presentation at workshop on multistation, digitally controlled manufacturing systems, University of Wisconsin, Milwaukee
109. ICAM Program Prospectus (1978) Wright-Patterson Air Force Base, Ohio, June
110. ICEM (1982) Integrated computer-aided engineering and manufacturing solution. Control Data Corporation, Minneapolis
111. IEEE Computer Graphics and Applications (1982) Special issue: solid modeling. March
112. IKM (1982) International Congress on Metalworking. Leipzig, March
113. Ingenierie assistée par ordinateur progiciel SDRC I-DEAS. CAE International, Nanterre, 1982

114. IPAD report (1984) Industry software for company-wide access to design and computer programs.
115. Ito Y (1984) Recent and future trends of FMS in Japan. Bull Jpn Soc Prec Eng 16 (4)
116. Iwata K (1984) FMS in Japan. Bull Jpn Soc Prec Eng 18 (2):186-192
117. Jacquart T, Regnier P, Valette FR (1974) GERMINAL. Towards a general and integrated system for computer-aided design. Proc. 11th design automation workshop, Denver, pp 352-358
118. Jansen H, Krause FL (1982) Dialog zwischen Mensch und Maschine. Wissenschaftsmagazin, Technische Universität Berlin 1 (2):69-73
119. Jansen H, Meyer B (1983) Rekonstruktion von volumenorientierten 3D-Modellen aus handskizzierten 2D-Ansichten. Informatik-Fachberichte, vol 65. Springer-Verlag, Berlin Heidelberg New York
120. Johnson ISC (1983) FMS operations management decision aids. Proc. of FMS-West Society of Manufacturing Engineers, San Francisco
121. Johnson TL, Milligan SD, Fortmann TE, Bloom HM, McLean CR, Furlani C (1982) Emulation/simulation of a modular hierarchical feedback system. 21st IEEE Conference on Decision and Control, December 8-10, San Francisco
122. Kay A (1980) SMALLTALK. In: Guedj RA et al. (eds) Methodology of interaction. North-Holland, Amsterdam
123. Kegg RL (1984) Flexible manufacturing trends and data base requirements. Proc. of 5th international conference on production engineering, Tokyo, ISPE
124. Kelly JC et al. (1982) A technical briefing on the initial graphics exchange specification (IGES). Sandia Labs Report
125. Kessner H (1983) Beitrag zur Systematik und Gestaltung diagnostizierbarer DNC-Systeme. Doctoral thesis A, Ingenieurhochschule, Dresden
126. Kief H (1982) NC-Handbuch. NC-Handbuch Verlag, Michelstadt Stockheim, Federal Republic of Germany
127. Kimura M, Osaki S, Soda C, Yoshida Y, Tatsue Y, Kanai M (1984) Flexible manufacturing system complex provided with laser. 1. Concept of the system. Proc. of 5th international conference on production engineering, Tokyo
128. Kimura K (1984) Flexible manufacturing systems complex. 2. Tsukuba FMSC test plant. Proc. of 5th international conference on production engineering, Tokyo
129. Klaus G (1973) Kybernetik und Gesellschaft. VEB Deutscher Verlag der Wissenschaften, Berlin
130. Klimov W, Klishin WW (1982) Rechnergestützter Entwurf der Geometrie dreidimensionaler Objekte im System SIMAK. Studientexte Computergeometrie und -graphik. Technische Universität, Dresden, June:44-49
131. Klimov W, Klishin WW (1983) Axiomatic problems of synthesis in the geometry of three-dimensional objects. Bull AN SSSR Technical Cybernetics 4:57-62 (in Russian)
132. Klimov V, Vagin V, Lebedeva TF (1983) A formal model of assembly arrangement in CAD of machine structures. Proc. of ICED-83, Copenhagen
133. Kochan D (1977) Fertigungsprozeßgestaltung und Informationsverarbeitung. Verlag Technik, Berlin
134. Kochan D, Zehe KH (1984) Improved intelligence of CNC controllers for sculpture surface milling. 16th CIRP seminar on Utilization of artificial intelligence and pattern recognition techniques in manufacturing engineering, July 13-14, Tokyo
135. Kochan D, Enderlein H, Fichtner D (1980) Mehrmaschinenbedienung im Maschinenbau. Verlag Technik, Berlin
136. Koller R, Farwick H, Spiegels G (1972) Konstruieren von Kurvenscheibengetrieben mit Unterstützung elektronischer Datenverarbeitungsanlagen und einer Bildschirmeinheit. Ind Anz 94 (44):1011-1017
137. Krämer F, Webbarth T (1983) Flächenorientiertes Modellieren mit den Systemen CD/2000 und Duct und ihre Anwendung. In: Informatik-Fachberichte, vol 65. Springer-Verlag, Berlin Heidelberg New York Tokyo, pp 285-295
138. Krause FL (1980) Systeme der CAD-Technologie für Konstruktion und Arbeitsplanung. Reihe Produktionstechnik Berlin, vol. 14. Hanser, Munich
139. Krause FL, Schliep W (1982) Methoden der Geometrieverarbeitung für CAD-Systeme. Symposium CAD/CAM. Rechnergestütztes Konstruieren und Fertigen, Österreichische Computer Gesellschaft, Vienna, May 25-27

140. Krause I, Wilsing N (1979) COMPAC. In: Geometric modeling seminar, Bournemouth, CAM-I 179-228
141. Kupka I, Wilsing N (1974) Functions describing interactive programming. In: International computing symposium. North-Holland, Amsterdam
142. Kupper H (1984) Integration of CAD/CAM. Proc. IFIP working conference. North-Holland, Amsterdam ROTA/NCM
143. Kurth J (1971) Rechnerorientierte Werkstückbeschreibung. Ein Beitrag zur Rationalisierung produktionsbezogener Planungsprozesse. Dissertation, Technische Universität, Berlin
144. La Duke NL, Proctor T (1967) Vorrichtung zum Steuern von Werkzeugmaschinen von einer Zentrale aus. Offenlegungsschrift 1588083, of October 26
145. Laffrey D, Bezacier P, d'Oliveira JL, May P (1977) ELAN: a conversational processor for generation, checking and correction of NC tapes. In: McPherson D (ed) Advances in computer-aided manufacture. North-Holland, Amsterdam, pp 43-51
146. Latombe JC (1978) Artificial intelligence and pattern recognition in computer-aided design. North-Holland, Amsterdam
147. Latombe JC, Mazer E (1981) A high-level programming language for controlling assembly robots. Proc. of the 11th JSIR, Tokyo, pp 683-690
148. Lemoine Y, Mutel B (1982) Automatic recognition of production cells and part families. Proc. PROLAMAT 82. IFIP-IFAC. North-Holland, Amsterdam
149. Lemon JR, Tolani SK, Klostermann AL (1980) Integration and implementation of computer-aided engineering and related manufacturing capabilities into the mechanical product development process. Informatik-Fachberichte, vol. 34. Springer-Verlag, Berlin Heidelberg New York Tokyo
150. Liebermann LJ, Weseley MA (1977) AUTOPASS: an automatic programming system for computer-controlled mechanical assembly. IBM J Devel Res 21:321-333
151. Lindsey CH, van der Meulen SG (1971) Informal introduction to ALGOL 68. North-Holland, Amsterdam
152. Lozano-Perez T, Winston J (1977) LAMA: a language for automatic mechanical assembly. Proc. 5th international joint conference on artificial intelligence. Boston, pp 710-716
153. Maeda G, Takeshige A, Koguchi T, Tomiyama T, Yoshikava H (1984) Frame operating system for CAD. International symposium on design and synthesis, Tokyo
154. Mäntyla M (1983) Compiler-compiler as a tool for CAD systems development. In: Warman EA (ed) Computer applications in production and engineering. North-Holland, Amsterdam
155. Mäntyla M, Sulonen (1982) GWB: a solid modeler with Euler operators. IEEE Comp Graphics Appl 2 (7):17-31
156. Mayer RJ, Talavage JJ (1976) Simulation of a computerized manufacturing system. Report No. 4, NSF grant no. APR74-15256
157. McCarthy (1978) History of LISP. ACM SIGPLAN Notices 13 (8):217-224
158. McElory J (1981) CAD/CAM comes of age. Automotive Industries, July
159. McGoldrick PF, Gibson RK (1980) PICTORAL representation of the cutting action of 3-axes NC milling machines. Proc. of CAD 80
160. Medusa — a system overview. Applied Graphics Systems, Rotterdam, 1981
161. Metcut Research Institute (1980) Machinability handbook, chap. 24. Computer-aided manufacturing technologies. Cincinnati
162. Michel FJ (1979) Flexible machining systems. Presentation at FMS conference, USA TARADCOM, Warren, MI, July
163. Mitrofanov SP (1960) Wissenschaftliche Grundlagen der Gruppenbearbeitung. VEB Verlag Technik, Berlin
164. Modulares Programmsystem für die Steuerung eines FMS. (1984) Forschungszentrum des Werkzeugmaschinenbaus, Karl-Marx-Stadt
165. Mollon A (1981) PACMECA: un système cohérent à la portée des PMI. 3è Journées Scientifiques et Techniques de la Production Automatisée. ADEPA, Montrouge
166. Moto-oka T (1984) New-generation computing. OHMSA, Ltd. Tokyo, Springer-Verlag, Tokyo
167. Myers W (1982) An industrial perspective on solid modeling. In: IEEE Computer Graphics and Application. Special Issue: Solid Modeling March, pp 86-97
168. Naur P (1978) The European side of the last phase of the development of ALGOL 60. ACM SIGPLAN Notices 18 (8):15-44

169. Newman WM, Sproull RF (1973) Principles of interactive computer graphics. McGraw Hill, New York
170. Newman WM, Sproull RF (1979) Principles of interactive computer graphics, 2nd edn. McGraw Hill, New York
171. Nishimura HK, Ishigaki T, Kariya A (1972) Programmsteuerung zum numerischen Rechnerdirektsteuern mehrerer Werkzeugmaschinen einer Fertigungsanlage. Toyoda Koki, Auslegeschrift 2240170
172. Nowacki H, Gnatz R (eds) (1983) Geometrische Modellierung. Informatik-Fachberichte, vol. 65. Springer-Verlag, Berlin Heidelberg New York Tokyo
173. Nygaard K, Dahl OJ (1978) The development of the SIMULA languages. ACM SIGPLAN Notices 13 (8):245-272
174. Okino N et al. (1978) TIPS-1. Institute of Precision Engineering, Hokkaido University, Sapporo
175. Pässler E, Hutchinson GK, Rudolph K, Stanek W (1983) Production systems design: a directed graph approach. Research Center of the Machine Tool Industry, Karl-Marx-Stadt, and University of Wisconsin, Milwaukee
176. Paying a visit to FMS plant (1982) Metalworking Engineering and Marketing 4 (1):72-76
177. Percebois C, Sansonnet JP (1982) A LISP machine to implement PROLOG. Proc. of 1st international logic programming conference. Fac Sc Luminy
178. Petri M (1962) Kommunikation mit Automaten. Thesis, Frankfurt/Main
179. Phillips DT, Heisterbey RJ (1977) Definition, development, and implementation of a generalized manufacturing simulator. Report no. GEMS-3-77, Texas A&M University, College Station
180. Pistorius E (1983) Grafische Simulation von Drehprozessen. Tagungsunterlagen CAMP' 83. VDE, Berlin
181. Polozov WS et al. (1983) Automated projection. Geometrical and graphic problems. Mechanical Engineering, Moscow, p 280 (in Russian)
182. Poppelstone RJ, Ambler AP, Bellos J (1978) RAPT: a language for describing assemblies. Industrial Robot 5 (3):131-137
183. Prager KP (1983) Kopplung externer und interner Programmiersysteme für Industrieroboter. Doctoral thesis, Technische Universität, Berlin. Hanser, Munich
184. Prince D (1971) Interactive graphics for computer-aided design. Addison-Wesley, Reading
185. Prior H, Fuchs H (1980) Integrierte Erstellung von Fertigungsunterlagen. Ind Anz 102 (82):120-127
186. Prior H (1980) Rechnerunterstützte Erstellung von Einzelteilzeichnungen — Ein Baustein zur integrierten Fertigungsunterlagenerstellung. Dissertation, RWTH, Aachen
187. Radin G (1978) The early history of PL/I. ACM SIGPLAN Notices 13 (8):227-242
188. Reboh R, Sacerdoti E (1973) A preliminary Q LISP manual. SRI AI Center technical note
189. Reinhardt A (1983) Modeling material handling systems with SIMELEX/2. University-GhK, Kassel, June 10 (draft)
190. Requicha AAG (1980) Representation for rigid solids: theory, methods and systems. ACM Computing Surveys 12 (4):437-464
191. Requicha AAG, Voelcker HB (1977) Constructive solid geometry technical memo 25. Production Automation Project
192. Requicha AAG, Voelcker HB (1982) Solid modeling: a historical summary and contemporary assessment. In: IEEE Computer Graphics and Application. Special Issue: Solid Modeling March, pp 9-24
193. Romulus (1983) Shape Data Ltd., Cambridge, England
194. Roos D (ed) (1967) ICES system — General description. MIT, Boston, pp 67-149
195. Ross DT (1978) Origins of the APT language for automatically programmed tools. ACM SIGPLAN Notices 13 (8):61-100
196. Roth M (1982) Der kühne Weg zur intelligenten Maschine. Wissenschaft und Fortschritt 32:12
197. Salmon M (1978) SIGLA: a programming language for the Olivetti robot. Controlli numerici machine a CN robot industriali 10 (3):56-61
198. Salzmann RM, Little AD (1984) The evolution from CAD/CAM to CIM: possibilities, problems and strategies for the future. Proc. CAMP' 84, West Berlin, September 25-28
199. Sarin SC, Wilhelm WE (1983) Flexible manufacturing systems: a review of modeling approaches for design, justification, and operation. Ohio State University, January

200. Schaller J (1982) Funktionserweitertes DNC für die Nutzung in flexiblen Fertigungs- und durchgängig automatisierten Systemen. Doctoral thesis B, Technische Universität, Dresden
201. Schaller J (1984) Zuverlässigkeit und Risiko beim Einsatz eines DNC-Systems. Fertigungstechnik und Betrieb 11:648-651
202. Schaller J, Rudolph J, Sieg CD (1984) Schaltungsanordnung zur situationsabhängigen Driftkompensation Prozeßrelevanter Parameter in automatisierte Bearbeitungsstationen. Patentschrift 2632810 at AfEP of the German Democratic Republic
203. Schubert D (1984) Intelligente Technik als Arbeitsmittel. Thesis for Colloquium Dresdense, Technische Universität, Dresden, October
204. Seifert DJ, Low JA (1982) Simulation: a dynamic tool for evaluating advanced manufacturing concepts. Presentation at IEEE Systems Man Cybernetics Conference, Seattle, October
205. Seifert H, Bergele N, Fritsche B (1978) Different ways to design three-dimensional representations of engineering parts with PROREN-2. Proc. conference interactive techniques in computer-aided design, Bologna, September, pp 335-343
206. Shaw AC (1980) On the specification of graphics command languages and their processors. In: Guedj RA et al. (eds) Methodology of interaction. North-Holland, Amsterdam
207. Simpson JA, Hocken RJ, Albus JS (1982) The automated manufacturing research facility of the National Bureau of Standards. J Manufact Systems 1 (1)
208. Smith JM, Chang PYT (1975) Optimizing the performance of a relational algebra database interface. CACM. 18 (10)
209. Sokolovski AP (1938) Problems connected with the unification of production-technological processes (in Russian). LONITOMASH
210. Special Projects 1979. CAM-1, Arlington
211. Spur G (1979) Produktionstechnik im Wandel. Produktionstechnisches Kolloquium, Berlin. Hanser, Munich
212. Spur G (1983) Aufschwung, Krisis und Zukunft der Fabrik. Produktionstechnisches Kolloquium, Berlin. Hanser, Munich
213. Spur G, Debler H (1974) COMPAC: a system for computer-aided design of mechanical workpieces. Mecanique 300
214. Spur G, Krause FL (1981) Aufbau und Einordnung von CAD Systemen. VDI-Bericht 413, Düsseldorf
215. Spur G, Kurz O (1982) Weiterentwicklung des CAD-Systems COMVAR. ZwF 77 (3):130-135
216. Spur G, Auer BH, Sinning H (1979) Industrieroboter-Steuerung, Programmierung und Daten von flexiblen Handhabungseinrichtungen. Hanser, Munich
217. Spur G, Stute G, Weck M (1977) Rechnergeführte Fertigung. Hanser, Munich
218. Stanek W, Rudolph K (1983) MAOSIM: an operation research model system for automated manufacturing simulation. Research Center of Machine Tool Building Industry, Karl-Marx-Stadt
219. Steinbuch K (1961) Automat und Mensch. Springer-Verlag, Berlin Göttingen Heidelberg
220. Steinke P, Heinrichs H (1980) Praxisgerechter Einsatz der Finite-Elemente-Methode zur Bauteilanalyse. APS-Seminar, Aachen, December
221. Straayer DH (1983) Graphics standards: the pace quickens. Computer Graphics Forum 2 (1)
222. Sussman GJ, McDermott DV (1972) From PLANNER to CONNIVER. A genetic approach. Proc. Fall joint computer conference, AFIPS, vol. 41, part 2, pp 1171-1179
223. System 360 (1968) APT numerical control processor (360 A-CN-10X), Version 2 part programming manual. IBM Technical Publishing Department, New York
224. Szabo ZF (1977) Systematische Planung von Programmsystemen zur Erstellung von Fertigungsunterlagen. VDI, Düsseldorf
225. Takeyama H, Sawada K, Ito M, Oboshi S, Asano K (1981) Development of programmable precision manufacturing systems (PPMS) for small-lot production. CIRP Annals 30 (1):343-347
226. Talavage J (1983) Simulating manufacturing systems. IEEE Spectrum, May
227. Temming R, Seeliger C (1979) Numerische Programmsteuerung für Werkzeugmaschinen. Siemens, Offenlegungsschrift 2932790 of August 13
228. Thornton J (1980) Use of computer simulation is expanding. also, How some companies are using simulation. American Market/Metalworking News, June
229. Tilove RB (1980) Set membership classification: a unified approach to geometric intersection problems. IEEE Trans Computers C-29 (10):874-883

230. Tilove RB (1983) Extending solid modeling systems for mechanism design and kinematic simulations. IEE Comp Graphics&Applications 3 (3)
231. Tilove RB, Requicha AAG (1980) Closure of Boolean operations on geometric entities. CAD 12 (5):219-220
232. Tomiyama T, Yoshikava H (1984) Requirements and principles of intelligent CAD systems. Department of Precision Machinery Engineering, University of Tokyo
233. Tsuji S, Nakono H (1982) Knowledge-based identification of artery branches in cine-audiograms — an image-understanding system which utilizes production-type knowledge. Computer Science and Technology 1982. North-Holland, Amsterdam.
234. Users' Guide to VAL. (1978) A robot programming and control system 744-1800. Unimation, Inc. Danbury, CN 06810 USA
235. Usui N (1982) Untended machines build machines. Am Machinist 125 (6):142-145
236. Valette R (1978) Etude comparative entre deux outils: le GRAFCET et le réseau de Petri. Automatisme 23 (12):377-382
237. van Wijngaarden A, Mailloux EJ, Peck JEL, Koster CHA (1968) Report on the algorithmic language ALGOL 68. Springer-Verlag, Berlin Heidelberg New York Tokyo
238. VDI-Richtlinien 2213 (1983) Integrierte Herstellung von Konstruktions- und Fertigungsunterlagen. Draft, 2nd edn. VDI, Düsseldorf
239. Voelcker HB, Hunt WA (1981) The role of solid modeling in machining — process modeling and NC verification. Proc. SAE 1981, International congress and exposition, Detroit, February 23-27, pp 1-8
240. Vought to purchase a Milacron FMS (1983) Iron Age 226 (19):18
241. Waller S (1983) Die automatisierte Fabrik. VDI Z 125 (20)
242. Warren DH, Pereira LM (1977) PROLOG. The language and its implementation compared with LISP. ACM SIGPLAN Notices 12 (8)
243. Weck M, Zühlke D (1980) Programmiersprache für NC-Handhabungsgeräte. Industrie Anzeiger 102 (37):27-30
244. Wesley MA (1980) Construction and use of geometric models. In: Encarnacao J (ed) Computer-aided design. Springer-Verlag, Berlin Heidelberg New York Tokyo, pp 79-136
245. Will P, Grossmann D (1974) An experimental system for computer-controlled mechanical assembly. IBM, T.J. Watson Research Center, New York
246. Wirth N (1973) Systematic programming, an introduction. Prentice-Hall, Englewood Cliffs
247. Wiseman NE, Cheney CG, Etherton M, Miles JD, Lemke HU (1972) RAINBOW: a multipurpose CAD system. Software Practice and Exper 2:359-375
248. Woodsford P (1970) GINO, graphical input/output, 3rd edn. University of Cambridge, CAD Group, June
249. Zadeh JA (1983) Commonsense knowledge representation based on fuzzy logic. Computer October: 61-65
250. Zemanek (1979) Die Zukunft der Informationsverarbeitung. Industriemagazin, April

GPSR Compliance

The European Union's (EU) General Product Safety Regulation (GPSR) is a set of rules that requires consumer products to be safe and our obligations to ensure this.

If you have any concerns about our products, you can contact us on ProductSafety@springernature.com

In case Publisher is established outside the EU, the EU authorized representative is:

Springer Nature Customer Service Center GmbH
Europaplatz 3
69115 Heidelberg, Germany

Batch number: 10620854

Printed by Printforce, the Netherlands